# Study Guide for
# CHEMISTRY
## ━ THIRD EDITION ━
## Steven S. Zumdahl

## Paul B. Kelter
### University of Wisconsin, Oshkosh

**D. C. Heath and Company**
Lexington, Massachusetts          Toronto

Published simultaneously in Canada.

Printed in the United States of America.

International Standard Book Number: 0-669-32868-5

10 9 8 7 6 5 4

# PREFACE

Thank you for choosing this study guide. The key to using the guide effectively is to recognize that it was specifically written for Chemistry, Third Edition, by Zumdahl. The study guide is intended to be used in conjunction with your textbook. It is not a substitute. You will often be asked in this guide to refer to specific tables or figures in the textbook. It is therefore most useful to work with the guide and textbook together.

# ACKNOWLEDGMENTS

Thanks go to the following friends and loved ones for putting up with me during the preparation of this guide.

To Joe and Maureen, who, like me, never forget.

To Mary Le Quesne, who was always there.

To Anne, every coworker's dream.

To Tisha, Miss Kathy, Paulette, and Kim, who prove that public education is thriving.

To Ken, who shows the progress that can be made in science education when two basket cases work together for mutual gain.

To Mom, for teaching me the positive aspects of guilt.

To Barb, who is brilliant and cute.

To Seth and Aaron, who are just like their mommy.

# TABLE OF CONTENTS

| Chapter | Title | Page # |
|---|---|---|
| 1 | Chemical Foundations | 1 |
| 2 | Atoms, Molecules, and Ions | 21 |
| 3 | Stoichiometry | 46 |
| 4 | Types of Chemical Reactions and Solution Stoichiometry | 75 |
| 5 | Gases | 110 |
| 6 | Thermochemistry | 135 |
| 7 | Atomic Structure and Periodicity | 151 |
| 8 | Bonding - General Concepts | 174 |
| 9 | Covalent Bonds - Oribitals | 199 |
| 10 | Liquids and Solids | 210 |
| 11 | Properties of Solutions | 226 |
| 12 | Chemical Kinetics | 244 |
| 13 | Chemical Equilibrium | 267 |
| 14 | Acids and Bases | 289 |
| 15 | Applications of Aqueous Equilibria | 320 |
| 16 | Spontaneity, Entropy and Free Energy | 352 |
| 17 | Electrochemistry | 371 |
| 18 | The Representative Elements - Groups 1A Through 4A | 391 |
| 19 | The Representative Elements - Groups 5A Through 8A | 401 |
| 20 | Transition Metals and Coordination Chemistry | 413 |
| 21 | The Nucleus - A Chemist's View | 427 |
| 22 | Organic Chemistry | 440 |
| 23 | Biochemistry | 460 |

# CHAPTER 1

# Chemical Foundations

| Section | Contents | Page in this Study Guide |
|---------|----------|--------------------------|
| 1.2* | The Scientific Method | 1 |
| 1.3 | Units of Measure | 2 |
| 1.4 | Uncertainty in Measurement | 4 |
| 1.5 | Significant Figures and Calculations | 5 |
| 1.6 | Dimensional Analysis | 8 |
| 1.7 | Temperature | 11 |
| 1.8 | Density | 13 |
| 1.9** | Classification of Matter | 14 |

This chapter will help guide your study of **what it means to think like a scientist in terms of strategy and measurement.**

## 1.2 - The Scientific Method

When you finish this section you will be able to:

- List and define each of the major steps in the **scientific method.**
- Identify the limitations of the **scientific method.**

The **scientific method** is a general procedure by which scientists gain understanding of our universe. Structured thinking is as applicable to biology, physics or geology as it is to chemistry. Thoughtful work is important to any scientific experiment.

As explained in your textbook, the **scientific method** is characterized by the following steps:

1. Making **observations** (gathering data).
2. Looking for **trends** in your data.
3. Formulating a **hypothesis**. Your hypothesis gives a possible explanation of **how** a process occurs. It is not the same as a **fact,** which is some observation or process that is so correct that it is unreasonable to dispute. If your hypothesis survives scrutiny by many experiments, it becomes a **theory**.
4. Performing experiments. This is done to determine if your hypothesis has **predictive value.**

*Section 1.1 is called "Thinking Like a Chemist." In this section the author wants you to know that although chemistry has a reputation for being super hard, it is not if you take the time to learn a rational approach to problem solving. He implies that one of the goals of his text is to teach you this rational thinking.*
**Section 1.10 discusses the types of jobs available to people with a degree in chemistry.*

## Example 1.2 - Scientific Method

One of the most important laws in chemistry, called Boyle's Law, states that **as the pressure (P) is increased on a gas, the volume (V) that the gas occupies will shrink proportionately, so that at a constant temperature, PV = a constant.** A chemistry student, unaware that this pressure-volume relationship exists, sought to "discover" it. His procedure was:

    a.   Hypothesize that, at constant temperature, pressure increases as volume increases.
    b.   Run one experiment where he observes the pressure on a gas as he changes the volume.
    c.   His result was inconclusive, so he proposed a theory that pressure and volume of a gas are unrelated.
    d.   He put away his equipment and went on to his next project.

List and explain those things that were wrong with this student's "scientific method."

## Solution

1.   He made a hypothesis based on no previous observations. A hypothesis can only be reasonable if it is based on some real-life experience.

2.   As we will see in **Section 1.4**, one experiment gives very little information. The student should have performed many experiments so that sources of random error could have been eliminated.

3.   He called his conclusion a theory. Hypotheses become theories only after long periods of testing to see if the hypothesis is correct. In addition, by putting away his equipment and going on to the next project, he did not test his own conclusions. It is always better if you can find your own mistakes before someone else does.

Your textbook points out that scientists, just as everyone else, have biases and prejudices. Science is self-correcting. If our hypotheses are invalid, we will eventually run enough tests to find out. The best that we chemists and chemistry students can do is to keep asking questions about theories. Questioning represents the best "scientific method".

Problems 1-2 at the end of this study chapter cover material from this section.

## 1.3 - Units of Measurement

When you finish this section, you will be able to:

- List the fundamental SI units of measurement.
- List the important SI prefixes.
- Differentiate between fundamental and derived units.

The United States is the last country in the industrial world that uses the English system. To be consistent with every other country, we are slowly adopting the **International System** (le Systeme International in French), or **(SI System)**.

The SI system is based on the metric system. We see examples of the SI system when we go to the grocery store to buy soda. The volume is listed as 12 fluid ounces (English system) and 354 milliliters (SI system).

You should know the 7 fundamental SI units. These are given in <u>Table 1.1 of your textbook.</u> It is important to develop "a feel" for units in the SI system. For example, in Earth's gravitational field, if you weigh **220 pounds,** your mass is **100 kilograms (kg)**. If you are **5.00 feet tall**, your height is **1.52 meters (m)**.

Measurement in meters is fine if you are measuring how tall you are. On the other hand, if you are measuring the distance from the Earth to Jupiter, the value of 1,000,000,000,000 m gets rather cumbersome. For large distances, we often use **kilometers, or km.** Kilo is the prefix meaning 1000, or $1 \times 10^3$. Jupiter is about 1 billion km from Earth.

We use prefixes with SI units to denote small values as well as large ones. The mass of a styrofoam peanut, which is used as a packing material, is about 0.040 g. It is easier to describe it as 40 milligrams (mg).

<u>Table 1.2 in your textbook</u> lists SI prefixes. You should know the ones that are listed in color.

---

## Example 1.3 A - Practice with Prefixes

Put the following prefixes in order, from smallest to largest:

centi    mega    milli    deka    deci    nano    kilo    micro

<u>Nano</u>    <u>micro</u>    <u>milli</u>    <u>centi</u>    <u>deci</u>    <u>deka</u>    <u>kilo</u>    <u>mega</u>
smallest                                                                                largest

## Solution

These prefixes along with their values, need to be committed to memory. You will get used to working with them soon enough.

<u>nano</u>        <u>micro</u>    <u>milli</u>    <u>centi</u>    <u>deci</u>    <u>deka</u>    <u>kilo</u>    <u>mega</u>
smallest                                                                    largest

---

Distance (in meters) is a **fundamental SI unit**. Speed can be measured as distance per unit time, or **m/s**. Because speed is derived from fundamental units, it is called a **derived unit**. Volume, in $m^3$, is another example of a derived unit, because it is derived from a fundamental unit, length.

---

## Example 1.3 B - Fundamental vs. Derived Units

State whether the following units are **fundamental** or **derived**:

density (g/mL)          _____

area ($m^2$)              _____

distance (cm)          _____

newton (N = $\dfrac{kg/m}{s^2}$)    _____

volume (mL)            _____

weight (pounds)        _____

time (ms)              _____

## Strategy

- Any unit that combines several fundamental units is derived.
- Any unit that has any unit to a power (such as area = m$^2$) is derived.
- Fundamental units may have prefixes attached to them. They are still viewed as being fundamental.

## Solution

| | |
|---|---|
| density | derived |
| area | derived |
| distance | fundamental |
| newton | derived (This is a unit of force, as is weight below) |
| volume | derived |
| weight | derived (This is a unit of force, equalling mass × acceleration) |
| time | fundamental |

---

Problems 3-5 at the end of this study chapter cover material from this section.

---

## 1.4 - Uncertainty in Measurement

When you finish this section, you will be able to:

- Define accuracy and precision.
- Determine if a set of measurements is accurate and/or precise.
- State the difference between implied and estimated error.

Among the most important skills that chemists have is to measure quantities. We can measure some things better than others. For example, taking into account the table of contents, the index, appendices and so forth, you can "measure" the number of pages in your textbook. If you can count separate units (such as pages of a book or eggs in dozen) you can get an **exact number.**

Now pick up your text. How much does it weigh? On a bathroom scale the book weighs 5 pounds, (or a little over 2 kg). That is **not an exact number**, because the bathroom scale is hard to read and is not very consistent in its values. Using a bathroom scale to measure weight gives rise to **uncertainty in your measurement.** The better your scale, the better your measurement. If you have an excellent scale, or have access to a high-quality balance, your answer will be very accurate.

- **Accuracy: How close you are to the true value.** If you take repeated measurements, your average value is likely to become accurate.
- **Precision: How close your values are to one another (internal consistency).**

Figure 1.7 in your textbook illustrates the possible combinations of accuracy and precision. The accompanying text discusses **random and systematic errors.** Know how these terms relate to accuracy and precision.

---

## Example 1.4 - Accuracy and Precision

Each of 4 general chemistry students measured the mass of a chemistry textbook. They each weighed the book 4 times. Knowing that the true mass is 2.31 kg, which student weighed the book:

a. accurately and precisely
b. inaccurately but precisely
c. accurately but imprecisely
d. inaccurately and imprecisely

| weighing | student #1 | student #2 | student #3 | student #4 |
|---|---|---|---|---|
| 1 | 2.38 kg | 2.06 kg | 2.32 kg | 2.71 kg |
| 2 | 2.23 | 1.94 | 2.30 | 2.63 |
| 3 | 2.07 | 2.09 | 2.31 | 2.66 |
| 4 | 2.55 | 2.40 | 2.32 | 2.68 |
| Average - | 2.31 ± 0.16 kg | 2.12 ± 0.14 kg | 2.31 ± 0.01 kg | 2.67 ± 0.03 kg |

## Strategy

You have to ask yourself two questions about each data set:

1.  Is the average close to the accepted (true) value? If it is, then the result is accurate.
2.  Is the average deviation small relative to the actual value? If it is, then the result is precise.

Keep in mind that accuracy and precision are relative terms. What is considered accurate and precise in one procedure may not be good enough under other experimental conditions.

## Solution

Data set #1 - **accurate but imprecise**
Data set #2 - **inaccurate and imprecise**
Data set #3 - **accurate and precise**
Data set #4 - **inaccurate but precise**

In the previous example, each of the calculations had a calculated average deviation with it. When we can calculate the deviation, it is called the **stated deviation**, or **stated error**. When we don't have enough information for such a calculation, we estimate that the error is probably ±1 in the last digit. This is called the **estimated error**. If someone gives you a value without giving a stated error, you must assume an estimated error. For example:

| value | estimated error |
|---|---|
| 5 grams | ±1 gram |
| 3.006 mL | ±0.001 mL |
| $3.8 \times 10^2$ K | $±0.1 \times 10^2$ K |

---

Problems 6-8 at the end of this study chapter cover material from this section.

---

## 1.5 - Significant Figures and Calculations

When you finish this section you will be able to:

- Determine the number of significant figures in a single value.
- Calculate results of mathematical expressions to the proper number of significant figures.

When you report a measured value, it is assumed that all the figures are correct except for the last one, where there is an uncertainty of ±1. If your value is expressed in proper **exponential notation**, all of the figures in the pre-exponential value are significant, with the **last digit** being the **least significant figure (LSF)**.

$$\text{"}7.143 \times 10^{-3}\text{ grams" contains 4 significant figures (SF)}$$
$$\uparrow$$
$$\text{(LSF)}$$

If that value is expressed as 0.007143, it still has 4 significant figures. Zeros, **in this case**, are place holders. **If you are ever in doubt about the number of significant figures in a value, write it in exponential notation.**

The rules for counting significant figures are given in your textbook. Let's use those rules in the following exercise.

---

## Example 1.5 A - Counting Significant Figures

Give the number of significant figures in the following values:

    a.   38.4703 mL
    b.   0.00052 g
    c.   0.05700 s
    d.   $6.19 \times 10^8$ years

### Helpful Hint

Convert to exponential form if you are not certain as to the proper number of significant figures.

### Solution

    a.   38.4703 mL = **6 SF**.
    b.   0.00052 g = $5.2 \times 10^{-4}$ g = **2 SF**. The leading zeros are not significant.
    c.   0.05700 s = $5.700 \times 10^{-2}$ s = **4 SF**. Again, the leading zeros are not significant. The trailing zeros are.
    d.   $6.19 \times 10^8$ years = **3 SF**.

There is uncertainty in all measured values (we assume that exact numbers are not "measured", but rather "counted"). Performing mathematical operations with measured values NEVER decreases uncertainty. Such operations only exacerbate error (make it more severe).

Your textbook states the rules for retaining the proper number of significant figures in your calculations.

A very important idea is that you **DO NOT ROUND OFF YOUR ANSWER UNTIL THE VERY END OF THE PROBLEM.** Rounding off in the middle just leads to greater error.

---

## Example 1.5 B - Significant Figures in Calculations

Perform the indicated calculations on the following measured values, giving the final answer with the correct number of significant figures.

    a.   12.734 - 3.0

    b.   $61 \times 0.00745$

c.   $\dfrac{5 \times 10^{16}}{(4.78 - 2.314)}$

d.   $(6.022 \times 10^{23} + 4.14 \times 10^{17}) \times (8.31 \times 10^{-11} - 9.2 \times 10^{-9})$

## Helpful Hint

Recall that you never round off intermediate answers.  Only round off at the very end of the problem.

## Solution

a.   12.734 has **3** figures past the decimal point.  3.0 has only **1** figure past the decimal point.  Therefore, your final result, **where only addition or subtraction is involved**, should round off to **one figure past the decimal point.**

$$\begin{array}{r} 12.734 \\ -3.0 \\ \hline 9.734 \end{array} \xrightarrow{\text{round off to}} \mathbf{9.7}$$

b.   In multiplication and division, your answer should have the **same number of significant figures as the least precise measurement.**  (the one with the fewest significant figures will be the least precise).

$$61 \times 0.00745 = 0.45445 = \mathbf{0.45}$$
$$\uparrow \qquad\quad \uparrow \qquad\qquad\qquad\qquad \uparrow$$

(2 SF)     (3 SF)                    (rounded
                                        to 2 SF)
least precise

c.   You will often have to solve problems where there is a combination of mathematical operations.  To get reasonable answers, you need to recall your **order of operations**, given in the table below:

Order of Operations
1. parentheses
2. exponents and logs
3. multiplication and division
4. addition and subtraction

Therefore, in our problem, we might think that we do the subtraction last.  However, the numbers are **inside parentheses**.  Therefore, the subtraction becomes our **highest** priority.  Our problem really is:

$$5 \times 10^{16} / 2.46_6 = 2.028 \times 10^{16}$$
$$\uparrow$$
not really a SF

In this division, $5 \times 10^{16}$, having only 1 significant figure, is the least precise (compared to 3 figures from our subtraction).  Therefore, our answer must round to $\mathbf{2 \times 10^{16}}$.

d.   Here again, the parentheses dictate our order of operations, working left to right,

$$(6.022 \times 10^{23} + 4.14 \times 10^{17}) = 6.022 \times 10^{23}$$

Your calculator did not make a mistake!  The value of $4.14 \times 10^{17}$ is so small relative to $6.022 \times 10^{23}$ that it is insignificant and can be ignored.  Therefore, in this special case, you are entitled to retain 4 significant figures from this operation.

In the next operation,

$$(8.31 \times 10^{-11} - 9.2 \times 10^{-9}) = -9.1169 \times 10^{-9}$$

Although this rounds off to **-9.1 × 10⁻⁹** (2 SF) we will retain the entire value **until the very end.**

Finally,

$$6.022 \times 10^{23} \times -9.1169 \times 10^{-9} = -5.4902 \times 10^{15},$$

which rounds to 2 significant figures to equal **-5.5 × 10¹⁵**.

You must not rush with significant figure calculations. Make sound, careful decisions and your answers will be reasonable.

---

# Problems 9-13 at the end of this study chapter cover material from this section.

---

## 1.6 - Dimensional Analysis

When you finish this section, you will be able to:

- Convert between English and metric units.
- Convert values from one prefix to another.

Dimensional analysis is the single most valuable mathematical technique that you will use in general chemistry. The method involves **using conversion factors to cancel units until you have the proper unit in the proper place.**

When you are setting up problems using dimensional analysis, you are more concerned with **units** than with numbers. Let's illustrate this by finding out the mass of a 125 pound box (1 kg = 2.2046 pounds).

**PROBLEM SOLVING STEPS**

1. List the relevant **conversion factors.**
2. Set up the problem as follows*

$$\text{kg} \quad \left\| \quad \frac{\overset{\text{exact number*}}{\downarrow}{1 \text{ kg}}}{2.2046 \text{ pounds}} \quad \times \quad \frac{125 \text{ pounds}}{1} \quad = \quad 56.7 \text{ kg} \right.$$

| The unit(s) you want on the left hand side | Double lines mean "solution to follow" | Conversion factor with desired unit in proper place (kg in numerator) | Your original value, in pounds, needs to be in the numerator to cancel pounds from the conversion factor |
|---|---|---|---|

3. Multiply all the values in the numerator and divide by all those in the denominator.

4. Double check that your units cancel properly. If they do, your numerical answer is probably correct. If they don't, your answer is certainly wrong.

### UNITS ARE THE KEY TO PROBLEM SOLVING

Let's extend the method to the following example.

---

*This problem solving scheme was taught to me by Prof. David Brooks of the University of Nebraska. (Thanks Dave!)

*"Exact number means perfect integer, that is, it has infinite significance and doesn't affect significant figures in the calculation."

## Example 1.6 A - Practice with Dimensional Analysis

It takes, including other ingredients, exactly 1 egg to make 8 pancakes.  A pancake eating contest was held at which the winner ate 74 pancakes in 6.0 minutes.  At this rate how many **eggs** (in the pancakes) would be eaten by the winner **in one hour?**

## Strategy

Following our problem solving procedure,

Step 1.  Conversion Factors:

    a.  1 egg = 8 pancakes.  Keep in mind that this is **exactly the same** as 8 pancakes = 1 egg.  You can therefore either use

$$\frac{1 \text{ egg}}{8 \text{ pancakes}} \quad \text{or} \quad \frac{8 \text{ pancakes}}{1 \text{ egg}}$$

       However, it is **NOT CORRECT** to use $\dfrac{8 \text{ eggs}}{1 \text{ pancake}}$ or $\dfrac{1 \text{ pancake}}{8 \text{ eggs}}$. When you flip units, the numbers must flip with them.

    b.  Although it is not stated in the problem, you need a conversion factor from minutes to hours.

$$\frac{60 \text{ minutes}}{1 \text{ hour}} \quad \text{or} \quad \frac{1 \text{ hour}}{60 \text{ minutes}} \quad \text{(exactly)}$$

    c.  74 pancakes per 6 minutes can be expressed as

$$\frac{74 \text{ pancakes}}{6 \text{ minutes}} \quad \text{or} \quad \frac{6 \text{ minutes}}{74 \text{ pancakes}}$$

Step 2.  Setting up the Problem:

The units that you want go on the left hand side.

$$\frac{\text{eggs}}{\text{hour}} \quad \Bigg\| \quad \frac{1 \text{ egg}}{8 \text{ pancakes}} \quad \times \quad \frac{74 \text{ pancakes}}{6 \text{ minutes}} \quad \times \quad \frac{60 \text{ minutes}}{1 \text{ hour}}$$

| ↑ | ↑ | ↑ | ↑ |
|---|---|---|---|
| read: "eggs per hour" | has proper unit (eggs) in numerator | cancels pancakes, puts time in denominator | cancels minutes, puts hours in denominator |

## Solution and Double Check

Step 3.  Multiply numerator and divide by denominator

$$\frac{74 \times 60}{8 \times 6} = \frac{4440}{48} = \frac{92.5 \text{ eggs}}{\text{hour}}, \quad \text{rounds to} \quad \frac{\textbf{92 eggs}}{\textbf{hour}}$$

Step 4.  Double check your units:

$$\frac{\text{egg}}{\text{pancake}} \times \frac{\text{pancakes}}{\text{minutes}} \times \frac{\text{minutes}}{\text{hour}} = \frac{\text{eggs}}{\text{hour}}$$

Dimensional analysis will often involve interconversion between prefixes of the same unit.  You must be very careful to **think** about the validity of your final value.

## Example 1.6 B - Practice with Prefixes

How many µm are there in 1 km?

### Critical Concept

Keep in mind that the prefix **"micro"** means $10^{-6}$. It is a very small number. There are many µm in a meter. You can write the conversion between µm and m as follows:

$$\text{(A)} \ 1 \ \mu m = 10^{-6} \ m$$
$$\text{or} \quad \text{(B)} \ 10^6 \ \mu m = 1 \ m.$$

It is generally easier to use "(A)" (because $\mu = 10^{-6}$) and your conversion becomes a simple substitution. Similarly, 1 km = $10^3$ because **kilo = $1 \times 10^3$**. Since both µm and km relate to <u>m</u>, we should use m as our **bridge** between µm and km.

### Solution

Conversion factors:    $\dfrac{1 \ \mu m}{1 \times 10^{-6} \ m}$    and    $\dfrac{1 \ km}{1 \times 10^3 \ m}$

You want µm.

$$km \text{----------}> m \text{----------}> \mu m.$$

The bridge between
prefixes.

$$\mu m \ = \ 1 \ km \ \times \ \frac{1 \times 10^3 \ m}{1 \ km} \ \times \ \frac{1 \ \mu m}{1 \times 10^{-6} \ m} \ = \ \mathbf{1 \times 10^9 \ \mu m \ in \ 1 \ km}$$

### Does the Answer Make Sense?

A µm is very small. A km is very large. Therefore, we would expect that there would be a lot of µm in a km. Carelessly inverting prefix conversions (such as incorrectly stating that 1 m = $10^3$ km) is among the major sources of incorrect answers in general chemistry.

## Example 1.6C - Summing it All Up

In the vacuum of space, light travels at a speed of 186,000 miles per hour. How many centimeters can light travel in a year?

### Strategy

We have $\dfrac{miles}{hour}$. We want $\dfrac{cm}{year}$.

Step 1.  Conversion factors: (You know some of these. You may have to check the inside back cover of your textbook for others)

186,000 miles per 1 hour
24 hours per 1 day
365 days per 1 year
1 mile per 1.6093 kilometers
$1 \times 10^3$ m per kilometer
$1 \times 10^{-2}$ m per 1 cm

Step 2.  Setting up the problem:

$$\frac{mi}{hr} \text{ ----> } \frac{km}{hr} \text{ ----> } \frac{m}{hr} \text{ ----> } \frac{cm}{hr} \text{ ----> } \frac{cm}{day} \text{ ----> } \frac{cm}{year}$$

start                              top is                       finish
                                   converted
                                   to cm

$$\frac{cm}{year} = \frac{186{,}000 \text{ mi}}{1 \text{ hour}} \times \frac{1.6093 \text{ km}}{1 \text{ mi}} \times \frac{1 \times 10^3 \text{ m}}{1 \text{ km}} \times \frac{1 \text{ cm}}{1 \times 10^{-2} \text{ m}} \times \frac{\overset{(*)}{24 \text{ hours}}}{1 \text{ day}} \times \frac{365 \text{ days}}{1 \text{ year}}$$

(*) =Note that identical terms don't have to be <u>next</u> to each other to cancel.  As long as a term appears in both the numerator and denominator, it cancels.

Step 3.  Multiply numerator and divide by denominator:

$$\frac{(186{,}000)(1.6093)(1 \times 10^3)(1)(24)(365)}{(1)(1)(1)(1 \times 10^{-2})(1)(1)} = \frac{2.62 \times 10^{12}}{1 \times 10^{-2}}$$

$$= \mathbf{2.62 \times 10^{14} \text{ cm/year}}$$

Step 4.  Double check your units:

$$\frac{mi}{hour} \times \frac{km}{mi} \times \frac{m}{km} \times \frac{cm}{m} \times \frac{hours}{day} \times \frac{days}{year} = \frac{cm}{year}$$

---

We did not deal here with conversions involving units to a power, such as $1 \text{ L} = 1000 \text{ cm}^3$.  Pay special attention to <u>Problems 19c and 20d in your textbook</u>.  This important problem deals with such conversions.

---

Problems 14-23 at the end of this study chapter cover material from this section.

---

## 1.7 - Temperature

When you finish this section you will be able to interconvert among Fahrenheit, Celsius and Kelvin temperature scales.

Your textbook gives information regarding the Fahrenheit, Celsius and Kelvin temperature scales.  Among the things you should be able to state regarding these three temperature scales are:

- how to convert among them,
- the freezing and boiling points of water on each scale, and
- the units of each one:  °F, °C and K (no <u>degree</u> sign in Kelvin).

A helpful hint - you can **derive** the conversion factor between °C and °F (and vice-versa) just by remembering that the boiling point of water can be expressed as 100°C or 212°F.  The best way to get from 100°C to 212°F, and involve a factor of 9/5 is:

$$100°C \times \frac{9°F}{5°C} = 180 + 32 = 212°F$$

$$\text{or} \quad \text{°C} \times \frac{9}{5} + 32 = \text{°F}$$

Let's try some practice problems.

---

## Example 1.7 A - Conversions Among Temperature Scales

The boiling point of water on top of Long's Peak in Colorado (14,255 feet above sea level) is about 86°C.  What is the boiling point in:

  a.  Kelvins,
  b.  degrees Fahrenheit?

### Solution

  a.  $T_K = T_C + 273.15$

   substituting,

   $T_K = 86 + 273.15 = 359.15$ K

   which rounds to **359 K**.

  b.  $T_F = T_C \times \dfrac{9°F}{5°C} + 32°F$

   substituting,

   $T_F = 86 \times 9/5 + 32 = 154.8 + 32 = 186.8°F$

   which rounds to **187°F**.

In the next problem, we convert °C to °F using an alternate method.

---

## Example 1.7 B - Conversions Among Temperature Scales

New materials can act as superconductors at temperatures above 250 K.  Convert 250 K to degrees Fahrenheit.

### Solutions

There are two ways of solving this problem.  Both involve converting

$$K \text{ ----> } °C \text{ ----> } °F$$

Method 1:

   Convert Kelvin to degrees Celsius:

$$T_C = T_K - 273.15 = 250 - 273.15 = \textbf{-23.15°C}$$

We don't round off, because this is an **intermediate** calculation.  Now convert degrees Celsius to degrees Fahrenheit as in the last problem.

$$T_F = T_C \times \frac{9°F}{5°C} + 32°F = -23.15 \times \frac{9}{5} + 32 = -9.67°F$$

Which rounds to **-10°F**. Note that if we would have rounded to -23°F, our answer would have been -9°F. That's not far off, but it's not correct either.

Method 2:

Convert kelvin to degrees Celsius as above. Convert degrees Celsius to degrees Fahrenheit as given in your textbook.

$$\frac{T_F + 40}{T_C + 40} = \frac{9°F}{5°C}$$

Substituting in degrees Celsius from above, solve for $T_F$.

$$\frac{T_F + 40}{-23.15 + 40} = \frac{T_F + 40}{16.85} = \frac{9}{5} \quad \Rightarrow \quad T_F + 40 = 30.33$$

$$T_F = -9.67°F, \text{ or } \mathbf{T_F = -10°F}$$

Either method is correct for converting degrees Fahrenheit to degrees Celsius.

---

## Problems 24-25 at the end of this study chapter cover material from this section.

---

## 1.8 - Density

When you finish this section, you will be able to perform density calculations.

As discussed in your textbook,

$$\text{density} = \frac{\text{mass}}{\text{volume}} .$$

It is a derived unit, most often expressed as grams/cm³ or grams/mL. When solving a density problem, the key is to **keep it simple**. Find out **how many grams** you have, determine the **total volume** your mass occupies, calculate the **ratio** of the two, and report your answer using the proper units.

---

## Example 1.8 A - Density Calculations

The density of mercury is 13.6 g/cm³. How many pounds (1 pound = 453.6 g) would 1 liter of mercury weigh?

## Solution

There are 13.6 g per cm³. There are 1000 cm³ in 1 liter. Therefore, there are

$$\frac{13.6 \text{ g}}{\text{cm}^3} \times \frac{1000 \text{ cm}}{1 \text{ L}} = \frac{13,600 \text{ g}}{1 \text{ L}}$$

Now convert g/L to pounds/L:

$$\frac{13,600 \text{ g}}{1 \text{ L}} \times \frac{1 \text{ pound}}{453.6 \text{ g}} = \frac{29.982 \text{ pounds}}{1 \text{ L}} = \frac{\mathbf{30.0 \text{ pounds of mercury}}}{1 \text{ L}}$$

An alternative strategy is to set the entire problem up using dimensional analysis:

$$\text{pounds} \quad \left\| \quad \frac{1 \text{ pound}}{453.6 \text{ g}} \quad \times \quad \frac{13.6 \text{ g}}{1 \text{ cm}^3} \quad \times \quad \frac{1000 \text{ cm}}{1 \text{ L}} \quad = \quad \frac{30.0 \text{ pounds of mercury}}{1 \text{ L}} \right.$$

## Example 1.8 B - Density Calculations

An unknown metal having a mass of 287.8 g was added to a graduated cylinder that contained 31.47 mL of water. After the addition of the metal, the water level rose to 56.85 mL. Using this information and Table 1.5 in your textbook, identify the unknown metal.

### Strategy

As with any density problem, your goal is to determine the ratio of mass/volume. You know the mass of the metal (287.8 g). The volume is equal to the amount that the water has been displaced.

### Solution

Mass of the metal = 287.8 g

Displacement of water = 56.85 mL - 31.47 mL = **25.38 mL**

$$\underset{\substack{\text{water} \\ + \\ \text{metal}}}{} \quad \underset{\substack{\text{original} \\ \text{volume} \\ \text{of water}}}{}$$

Density of the metal = $\dfrac{287.8 \text{ g}}{25.38 \text{ mL}}$ = **11.34g/mL**

According to Table 1.5 in your textbook this metal is **lead**.

---

Problems 26-30 at the end of this study chapter cover material from this section.

---

## 1.9 - Classification of Matter

This study section will help guide you through some of the terms and techniques that are discussed in these sections of Chapter One of your textbook.

Terms you should be able to define:

1. Matter
   a. solid
   b. liquid
   c. gas

2. Mixtures
   a. homogeneous (solution)
   b. heterogeneous

3. Physical and Chemical Changes

4. Compounds and Elements

<u>You should be able to describe the following separation techniques</u>

1.  Distillation

2.  Filtration

3.  Chromatography

---

Problems 31-32 at the end of this study chapter cover material from this section.

---

**END OF CHAPTER ONE**

# SELF-TEST QUESTIONS

1. In a simulated boxing match done on a computer several years ago, Joe Louis (champion in the 1940s) defeated Muhammad Ali (champion in the 1960s) in 15 rounds. The computer simulation took into account the fighters' height, weight, styles, punching power, and speed. Was the simulation realistic? What other variables would have to be included to make the simulation more valid?

2. If your chemistry teacher said she **knew** that the Earth is flat, and challenged you to prove, using scientific tests, that she was wrong, what tests might you perform?

3. List the fundamental units that you would combine to get the following derived units (you may need to look up the meaning of some of the terms):

    a. velocity
    b. acceleration
    c. volume
    d. heat capacity
    e. density
    f. pressure

4. Put the following prefixes in order, from smallest to largest:  femto, tera, mega, deci, kilo, atto, exa.

5. Put the following prefixes in order, from smallest to largest:  giga, hecto, micro, peta, milli, nano, pico.

6. Three students weighed the **same** sample of copper shot five times. Their results were as follows:

    | weighing | Student #1 | Student #2 | Student #3 |
    |----------|-----------|-----------|-----------|
    | 1 | 17.516 g | 15.414 g | 13.893 g |
    | 2 | 17.888 g | 16.413 g | 13.726 g |
    | 3 | 19.107 g | 14.408 g | 13.994 g |
    | 4 | 21.456 g | 15.637 g | 13.810 g |
    | 5 | 19.983 g | 15.210 g | 13.476 g |

    a. Calculate the mean (average) mass of the sample, as determined by each student.
    b. The **average deviation (a.d.)** is a measure of **precision**. It is calculated by the formula:

$$a.d. = \frac{\sum_{i=1}^{n} |x_i - \bar{x}|}{n}$$

    That means that you take each of the values in a data set (5 total values), subtract the mean, and take the absolute value. Add up all 5 absolute values and divide the total by 5. That number is your a.d. Calculate the average deviation for each of the students' data.

    c. If the true mass of copper shot is **15.384 g**, which of the students was most accurate? Which was most precise? What could be the possible sources of error in the determinations?

7. A student weighed 15 pennies on a balance and recorded the following masses:

    | | | | | |
    |---|---|---|---|---|
    | 3.078 | 3.055 | 3.060 | 3.066 | 3.102 |
    | 3.107 | 3.121 | 2.518 | 3.052 | 2.476 |
    | 2.546 | 3.050 | 3.073 | 3.080 | 3.128 |

    a. Calculate the mean mass.
    b. What might cause the difference in weights?

8. Define the following terms:

    a. accuracy
    b. precision
    c. random error
    d. systematic error

9.  How many significant figures are there in each of the following values?

    a.  $6.07 \times 10^{-15}$  3
    b.  $0.003840 \times 10^{-3}$  4
    c.  17.00   4

    d.  $8 \times 10^8$   1
    e.  463.8052   7

    $3.6 \times 10^{-5}$

10. How many significant figures are there in each of the following values?

    a.  1406.20  6
    b.  0.0007   1

    c.  1600.0   5
    d.  0.0261140  6

    e.  $1.250 \times 10^{-3}$   4

11. Use exponential notation to express the number 37,100,000 with:

    a.  one sig fig
    b.  two sig figs

    c.  three sig figs
    d.  six sig figs

12. Perform the indicated calculations on the following measured values, giving the final answer with the correct number of significant figures.

    a.  $16.81 + 3.2257 = 20.03$
    b.  $324.6 \times 815.991$  264.9
    c.  $(3.8 \times 10^{-12} + 4 \times 10^{-13}) / (4 \times 10^{12} + 6.3 \times 10^{13})$  $1.05 \times 10^{-24}$
    d.  $3.14159 \times 68 / (5.18 \times 10^{-11} - 6 \times 10^{-4})$

13. Perform the indicated calculations on the following measured values, giving the final answer with the correct number of significant figures.

    a.  $2.85 + 3.4621 + 1.3$
    b.  $7.442 - 7.429$
    c.  $1.65 \times 14$

    d.  $27/4.148$
    e.  $[(3.901 - 3.887)/3.901] \times 1.00$
    f.  $6.404 \times 2.91 \times (18.7 - 17.1)$

14. Using the conversion factors in the inside back cover of your textbook, convert 3.5 quarts to:

    a.  liters (L)
    b.  milliliters (mL)

    c.  microliters (μL)
    d.  cubic centimeters ($cm^3$)

15. Using the conversion factors in the inside back cover of your textbook, convert 4.2 yards to:

    a.  meters
    b.  centimeters

    c.  micrometers
    d.  kilometers

16. If you put 8 gallons of gas in your car and it cost you a total of $9.20, what is the cost of gas per liter?

17. A runner can run a 5.0 kilometer race in a time of 21 minutes and 22 seconds.

    a.  What is the runner's speed in miles per hour?
    b.  How long, on the average, did it take for the runner to run one mile?

18. A student made the 27.0 kilometer drive to school in 16 minutes.

    a.  How many miles did the student drive?
    b.  If the speed limit is 55 mph, was the student speeding?  How fast was the student driving?

19. A 5 lb. bag of flour costs $.89.  What is the cost of flour per kilogram?

20. A radio station broadcasts at a frequency of 107.9 megahertz (MHz), (cycles per second).

    a.  How many seconds per cycle are there in 107.9 MHz?
    b.  What is the broadcast frequency in gigahertz (GHz)?

21. During a recent baseball game, a pitcher threw a fastball that had a velocity of 93.7 mph.

    a.   Calculate the velocity in meters/second.
    b.   Calculate how long it took this pitch to travel from the mound to home plate (60 ft 6 in.)

22. Which of the following is greater:

    a.   35 kg or 3500 g?
    b.   $6 \times 10^4$ mL or $6 \times 10^3$ L?

23. If a student weighs 185 lb, what is his mass in µg?

24. Perform the following temperature conversions:

    a.   300 K to °F          d.   -100°C to K
    b.   300°F to K           e.   1555 K to °C
    c.   -40°F to °C          f.   0 K to °F

25. Perform the following temperature conversions:

    a.   16°C to °F              c.   0.0°F to °C              e.   -45°C to K
    b.   305 K to °F             d.   150°F to K               f.   920 K to °C

26. A sample of motor oil with a mass of 440 g occupies 500 mL.  What is the density of the motor oil?

27. The density of an object is 16.3 g/mL.  Its volume is 0.27 L.  What is the mass of the object?

28. An object weighing 4.0 lbs occupies 1700 mL.  What is the density of the object in g/mL?

29. The density of the Earth is about 3.5 g/cm³.  If the earth has a radius of 7000 miles, what is its mass? (volume = $[4\pi r^3/3]$)

30. Which of the following is less:

    a.   8.7 g/mL or 6.1 µg/µL?
    b.   $4 \times 10^5$ kg/cm³ or $4 \times 10^{-1}$ mg/cm³?

31. Define the following:

    a.   mixture              c.   homogeneous
    b.   pure substance       d.   heterogeneous

32. List five physical methods of separating mixtures.

# ANSWERS TO SELF-TEST QUESTIONS

1. The simulation was not especially realistic. The simulation could not take into account such variables as the mental condition of the fighters, whether they overtrained, whether they had to struggle to make the weight or the differences in styles between 1940s and 1960s fighters and equipment.

2. You might travel around the Earth and end up at the same point (if you had the money), or use a pendulum, among other possibilities.

3. a.  m/s
   b.  $m/s^2$

   c.  $m^3$
   d.  J/g K (or J/g °C)

   e.  g/mL (or $g/cm^3$)
   f.  $kg/m^{-1} s^{-2}$

4. atto, femto, deci, kilo, mega, tera, exa.

5. pico, nano, micro, milli, hecto, giga, peta

6. a.  means:   student #1 = 19.190 g
                 student #2 = 15.416 g
                 student #3 = 13.780 g

   b.  average deviations:   student #1 = 1.224 g
                             student #2 = 0.487 g
                             student #3 = 0.143 g

   c.  Student #2 is the most accurate. Student #3 is the most precise. Among the sources of error could be some spillage of the sample, an inaccurate balance or a wet beaker.

7. a.  2.967
   b.  Beginning in 1982, "copper" pennies were made from zinc with a copper coating outside (97% Zn, 3% Cu).

8. a.  the agreement of a particular value with the true value
   b.  the degree of agreement among several measurements of the same quantity
   c.  an error that has an equal probability of being high or low
   d.  an error that always occurs in the same direction and is likely to have a procedural cause

9. a.  3
   b.  4

   c.  4
   d.  1

   e.  7

10. a.  6
    b.  1

    c.  5
    d.  6

    e.  4

11. a.  $4 \times 10^7$
    b.  $3.7 \times 10^7$

    c.  $3.71 \times 10^7$
    d.  $3.71000 \times 10^7$

12. a.  20.04
    b.  $2.649 \times 10^5$

    c.  $6.3 \times 10^{-26}$
    d.  $-4 \times 10^5$

13. a.  7.6
    b.  0.013

    c.  23
    d.  6.5

    e.  0.359
    f.  30

14. a.  3.3 L
    b.  $3.3 \times 10^3$ mL

    c.  $3.3 \times 10^6$ µL
    d.  $3.3 \times 10^3$ $cm^3$

15. a.  3.8 m
    b.  $3.8 \times 10^2$ cm

    c.  $3.84 \times 10^6$ µm
    d.  0.0038 km

16. $.30/liter

17. a.   8.7 mi/hr          b.   6 minutes and 53 seconds

18. a.   16.8 miles          b. yes, roughly 63 mph

19. $.39 kg

20. a.   $9.268 \times 10^{-3}$          b.   0.1079 GHz

21. a.   41.9 m/s          b.   0.440 seconds

22. a.   35 kg          b.   $6 \times 10^3$ L

23. $8.39 \times 10^{10}$ µg

24. a.   80.3°F          c.   -40°C          e.   1282°C
    b.   422 K          d.   173 K          f.   -460°F

25. a.   61°F          c.   -18°C          e.   228 K
    b.   89.3°F          d.   339 K          f.   647°C

26. 0.88 g/mL

27. 4400 g

28. 5.2 g/mL

29. $2.1 \times 10^{28}$ g

30. a.   6.1 µg/µL          b.   $4 \times 10^{-1}$ mg/cm$^3$

31. a.   substance that has a variable composition
    b.   substance that has a constant composition
    c.   mixtures that are the same throughout
    d.   mixtures containing regions with different properties

32. distillation, filtration, liquid-column chromatography, gas chromatography, paper chromatography

# CHAPTER 2

# Atoms, Molecules, and Ions

| Section | Contents | Page in this Study Guide |
|---|---|---|
| 2.2* | Fundamental Chemical Laws | 21 |
| 2.3 | Dalton's Atomic Theory | 23 |
| 2.5# | The Modern View of Atomic Structure | 24 |
| 2.6 | Molecules and Ions | 25 |
| 2.7 | An Introduction to the Periodic Table | 27 |
| 2.8 | Naming Simple Compounds | 28 |

This chapter will help guide your study of **the nature of atoms and molecules, and ions, how they combine and how we name them.**

## 2.2 - Fundamental Chemical Laws

When you finish this section you will be able to use:

- The Law of Conservation of Mass
- The Law of Definite Proportion
- The Law of Multiple Proportions

Your textbook defines and discusses **the law of conservation of mass.** Read that material, then try the following exercise.

---

*Section 2.1, dealing with "the Early History of Chemistry" is fascinating and important reading. Do not neglect this section.

#Section 2.4 deals with "Early Experiments to Characterize the Atom." You need to be familiar with experiments using the cathode ray tube, radioactivity, and oil drops, all of which helped elucidate the structure of the atom.

## Example 2.2 A - The Law of Conservation of Mass

In a combustion reaction, 46.0 g of ethanol reacts with 96.0 g of oxygen to produce water and carbon dioxide. If 54.0 g of water is produced, how much carbon dioxide is produced?

### Solution

If the law of conservation of mass holds, the total of the mass of reactants must equal the product masses. Therefore,

**Total mass reacted = Total mass produced**

46.0 g + 96.0 g = 54.0 g + g carbon dioxide

142.0 g - 54.0 g = **88.0 g carbon dioxide produced**

**The law of definite proportion** is described in your textbook. After reading that paragraph, try the following problem.

## Example 2.2 B - The Law of Definite Proportion

A sample of chloroform is found to contain 12.0 g of carbon, 106.4 g of chlorine and 1.01 g of hydrogen. If a second sample of chloroform is found to contain 30.0 g of carbon, how many grams of chlorine and grams of hydrogen does it contain?

### Strategy

Assuming that the law of definite proportion is true, the **ratios of the weights of the elements in chloroform** are constant. Therefore, if the amount of carbon is increased by a factor of 2.5 (30.0 g/12.0 g, as above), then the same must hold true for chlorine and hydrogen.

### Solution

Carbon is increased by a factor of 2.5 (30.0 g/12.0 g). Therefore,

g chlorine = 106.4 g × 2.5 = **266.0 g chlorine**
g hydrogen = 1.01 g × 2.5 = **2.53 g hydrogen**

**The law of multiple proportions** is discussed in your textbook. The key idea with this law is that given a compound such as $H_2O$, if $H_2O_2$ forms, the **ratio of the masses of oxygens** that combine with one gram of hydrogen will be a small whole number.

## Example 2.2 C - The Law of Multiple Proportions

Water, $H_2O$, contains 2.02 g of hydrogen and 16.0 g of oxygen. Hydrogen peroxide, $H_2O_2$, contains 2.02 g of hydrogen and 32.0 g of oxygen. Show how these data illustrate the law of multiple proportions.

## Solution

In $H_2O$, 7.92 g (16.0 / 2.02) of oxygen combines with each 1.0 g of hydrogen.  In $H_2O_2$, 15.84 g (32.0 / 2.02) of oxygen combines with each 1.0 g of hydrogen.  The **ratio of the masses** of oxygen in the two compounds is:

$$\frac{15.84 \text{ g}}{7.92 \text{ g}} = 2.00$$

This is a "small whole number" that illustrates the law of multiple proportions.

---

## Problems 1-2 at the end of this study chapter cover material from this section.

---

## 2.3 - Dalton's Atomic Theory

When you finish this section you will be able to select those ideas in Dalton's atomic theory that apply to specific chemical problems.

Read the four statements relating to Dalton's atomic theory that are given in your textbook.  Use the following exercise to test your understanding of the theory.

---

## Example 2.3 - Dalton's Atomic Theory

Match the chemical statement on the left side with the atomic theory statement on the right side.

### Chemistry

a.  Although graphite and diamond have different properties (due to the nature of interatomic bonding), they are both composed solely of carbon. The carbon atoms are identical.

b.  $2H_2 + O_2 \longrightarrow 2H_2O$, not $CS_2$ or NaCl

c.  There are $6.02 \times 10^{23}$ atoms in 55.85 g of iron.

d.  $C + O_2 \longrightarrow CO_2$.  $CO_2$ is not CO, $CO_3$ or $Fe_2O_3$.

### Atomic Theory

1.  Each element is made up of tiny particles called atoms.

2.  The atoms of a given element are identical; the atoms of different elements are different in some fundamental way or ways.

3.  Chemical compounds are formed when atoms combine with each other.  A given compound always has the same relative numbers and types of atoms.

4.  Chemical reactions involve reorganization of the atoms - changes in the way they are bound together.  The atoms themselves are not changed in a chemical reaction.

## Solution

a.  2
b.  4

c.  1
d.  3

The statements in Dalton's atomic theory are **not necessarily mutually exclusive**.  As a result, statement "b" can fit #3, and statement "d" can fit #4.

---

In reading this section in your text, don't neglect the discussion of **Avogadro's hypothesis**! That stated relationship between volumes and numbers of particles of gases set the stage for realizing the *diatomic* nature of gases such as $O_2$, $H_2$ and $Cl_2$.

---

Problem 3 at the end of this study chapter  covers material from this section.

---

## 2.5 - The Modern View of Atomic Structure

When you finish this section you will be able to:

- Assign the correct number of protons, neutrons, and electrons to an atom,
- Write the symbol for an atom, and
- Do the above operations for any isotopes of an atom.

The names, sizes and masses of the parts of the atom are discussed in your textbook.  Pay special attention to Table 2.1 in your textbook.

The key point is that although the mass of an atom is concentrated in its nucleus, most of the volume of the atom is taken up by electron movement.  Since electrons are involved in bonding, we need a way to symbolize the number of protons, neutrons and electrons in an atom.  Given

$$^A_Z X$$

**X** = The element symbol, as read from the periodic table.
**Z** = the number of protons.
**A** = The number of protons plus neutrons.  (The number of neutrons = A - Z.)

The number of **protons** determines the element.  **Potassium** has **19 protons**.  If an element has **20 protons**, it is **calcium**.  The number of neutrons and electrons do not determine what element you have.  Only the number of protons is important.  For example,

$$^{31}_{15} P$$

contains **15 protons, 15 electrons** (in an <u>atom</u>, which is electrically neutral, the number of protons equals the number of electrons), and **16 neutrons** (31 - 15).  The atomic mass is 31.  If instead, we had

$$^{30}_{15} P$$

everything would be the same as in the previous example, except for the <u>number of neutrons</u>, which would be **15 neutrons** (30 - 15).  This phosphorous atom is therefore an **isotope** of the previous atom.

## Example 2.5 - Protons, Neutrons, Electrons and Symbols

Fill in the missing information in the following table:

| Symbol | Element | Protons | Neutrons | Electrons |
|---|---|---|---|---|
| $^{14}_{6}C$ | _____ | _____ | _____ | _____ |
| $^{235}_{92}$___ | _____ | _____ | _____ | _____ |
| ___ | U | _____ | 146 | _____ |
| $^{55}_{25}$___ | _____ | _____ | _____ | 25 |

## Helpful Hints

1.  # of protons = # of electrons **in an atom**.
2.  The element is determined by the number of protons.
3.  In the periodic table, the elements are placed in order of increasing number of protons (although the groupings are really much more involved and meaningful).

## Solution

| Symbol | Element | Protons | Neutrons | Electrons |
|---|---|---|---|---|
| $^{14}_{6}C$ | C | 6 | 8 | 6 |
| $^{235}_{92}U$ | U | 92 | 143 | 92 |
| $^{238}_{92}U$ | U | 92 | 146 | 92 |
| $^{55}_{25}Mn$ | Mn | 25 | 30 | 25 |

Determining the number of each kind of particle becomes more challenging when **ions** are introduced. Both your textbook and this study guide deal with these ideas in our next sections.

## Problems 4-7 at the end of this study chapter cover material from this section.

## 2.6 - Molecules and Ions

When you finish this section you will be able to:

- Determine if a species is an atom, molecule, or ion.
- Determine the charge, and the number of protons, neutrons, and electrons, in monatomic ion.

Some very fundamental ideas and terms are discussed in your textbook.  Read this section and be able to define:

1.  covalent bonds
2.  molecule
3.  chemical formula
4.  structural formula
5.  ion
6.  cation
7.  anion
8.  ionic bond
9.  polyatomic ion

A key point of the section is that **molecules contain atoms**.  The molecule $SF_6$ contains 7 atoms (6 fluorine and one sulfur).  These atoms are not ionized.

Certain atoms tend to ionize.  For example, $CaF_2$ contains 3 ions; a $Ca^{2+}$ and 2 F⁻.  We call $CaF_2$ an **ionic solid** or a **salt**.  Ionic solids contain ions, whose charges exactly balance.  In $CaF_2$, there are 2+ and 2⁻ that balance.

You will learn, in future chapters, how to predict whether molecules or salts will form.  At this point it is only expected that you know them when you see them.

---

## Example 2.6 A - Molecule, Atom or Ion?

Which of the following is an **atom**, an **ion**, or a **molecule**?

| | |
|---|---|
| $C_6H_{12}O_6$ | *Molecule* |
| $N_2$ | *Molecule* |
| $CO_3^{2-}$ | *Ion* |
| $Ag$ | *Atom* |
| $Fe^{3+}$ | *Ion* |
| $NH_4^+$ | *Ion* |

## Solution

| | |
|---|---|
| $C_6H_{12}O_6$ | molecule |
| $N_2$ | molecule (contains more than one atom) |
| $CO_3^{2-}$ | ion (polyatomic anion) |
| $Ag$ | atom |
| $Fe^{3+}$ | ion (cation) |
| $NH_4^+$ | ion |

---

For the next exercise, you need to keep in mind that the **total charge** of an ion is determined by whether you have **more protons or electrons**.

$Fe^{3+}$     means that iron has **3 more protons than electrons**
(26 protons and 23 electrons)

$O^{2-}$  means that oxygen has **2 more electrons than protons**
(8 protons and 10 electrons)

## Example 2.6 B - Protons, Neutrons and Electrons in Ions

Fill in the missing numbers in the following table:

| Symbol | Protons | Neutrons | Electrons | Charge |
|---|---|---|---|---|
| $^{32}_{16} S^{2-}$ | 16 | 16 | 18 | 2- |
| $^{137}_{56} —^{2+}$ | 56 | 81 | 54 | 2+ |
| $—Cl^-$ | 17 | 20 | 18 | -1 |
| $^{55}_{24}—^{7+}$ | 24 | 31 | 17 | 7+ |

## Solution

| Symbol | Protons | Neutrons | Electrons | Charge |
|---|---|---|---|---|
| $^{32}_{16} S^{2-}$ | 16 | 16 | 18 | -2 |
| $^{137}_{56} Ba^{2+}$ | 56 | 81 | 54 | +2 |
| $^{37}_{17} Cl^-$ | 17 | 20 | 18 | -1 |
| $^{52}_{24} Cr^{7+}$ | 24 | 28 | 17 | +7 |

Problems 8-11 at the end of this study chapter cover material from this section.

## 2.7 - An Introduction to the Periodic Table

When you finish this section you will be able to state whether an element is a metal or non-metal, and list what group it belongs to.

Your textbook points out that the periodic table is organized the way it is because certain elements have **properties in common** with one another. From the textbook, **and your own experience** you should be able to identify the properties of metals vs. nonmetals.

Look at the **periodic table** in this section of your textbook. Note the following things:

1. The bold line that runs underneath boron and all the way to the right side of polonium. That line separates metals and nonmetals.
2. Group names. You will be using these designations all year.
3. The various properties of each group, as listed in you textbook.
4. **Groups** go **down**, **periods** go **across**.

Note also that the symbols of several elements are based on original Latin or Greek names. These elements, and their original names, are given in **Table 2.2** in your textbook.

---

## Example 2.7 - Identify the Element

Given the following information, identify each element:

 a.   This is the only metal in group 6A.
 b.   This alkali metal is in the same period as iodine.
 c.   Two atoms of this element, which is in the same period as magnesium, combine with a calcium ion to form a salt.

### Solution

 a.   The only metal in group 6A is **polonium**. It is to the left of the bold line on the periodic table. Elements that are next to that line are sometimes called **metalloids** because they have some properties of both metals and nonmetals.

 b.   The only alkali metal that is in the same period as iodine is **rubidium**.

 c.   Calcium forms $Ca^{2+}$. Therefore, if two atoms combine with it to make an electrically neutral salt, they each must have a -1 charge. **Halogens** readily form ions with a -1 charge. The halogen that is in the same period as magnesium is **chlorine**.

---

Problems 12-14 at the end of this study chapter cover material from this section.

---

## 2.8 - Naming Simple Compounds

When you finish this section you will be able to name, or give formulas for, the following classes of compounds:

 • binary salts
 • salts with polyatomic ions
 • binary covalent compounds
 • acids

There are really two keys to naming, or assigning formulas, to compounds. The first key is to **know the names of the elements** (typically, knowing #1 - #54 is sufficient). The second key is to **correctly assign charges to ions**.

IUPAC (International Union of Pure and Applied Chemistry) nomenclature standards are used in your textbook and in this study guide. That way, any chemist (or chemistry student) should know exactly what you are talking about when you name, or give the formula of, a compound.

### A. Naming Compounds

  1.  Binary Salts

      **A binary salt contains only one kind of cation (positive ion) and one kind of anion (negative ion).**

a.   **Cations:**  Ions with positive charges fall into two general classes.  **The first class** contains elements that **ionize to only one oxidation state**, as given in the following list:

| Elements that Ionize to a Single Oxidation State (Cation) | Oxidation State | Examples |
|---|---|---|
| All group 1A (alkaline) | +1 | $Na^+$, $Rb^+$ |
| All group 2A (alkaline earths) | +2 | $Mg^{2+}$, $Ba^{2+}$ |
| All group 3A | +3 | $Al^{3+}$, $Ga^{3+}$ |

Your textbook calls these **"Type I Binary Ionic Compounds."**  When a cation forms from an element in groups 1A, 2A, or 3A, we give **the ion the name of the element**, and add "**ion**."  For example, $K^+$ is called **potassium ion**, and $Be^{2+}$ is called **beryllium ion**.  The reverse process is also true.  **Calcium ion** is labelled $Ca^{2+}$.

The **second class** of cations contains elements that can ionize to one of several oxidation states.  Your textbook calls these **"Type II Binary Ionic Compounds."**  For example, iron, Fe, can form stable oxidation states of $Fe^{6+}$, $Fe^{3+}$, and $Fe^{2+}$.  (At this point, you are not expected to predict which one will form.  You should just be able to know one when you see one, and tell what you see.)  **Most transition elements form several positive oxidation states.  Some halogens and noble gases also form positive oxidation states.**

We use **roman numerals in parentheses after the element** to name cations of elements that can form several positive oxidation states.  Some examples are given in the following list:

| Element | Oxidation State | Symbol | Name |
|---|---|---|---|
| iron | +3 | $Fe^{3+}$ | iron (III) ion |
| iron | +2 | $Fe^{2+}$ | iron (II) ion |
| chromium | +7 | $Cr^{7+}$ | chromium (VII) ion |
| chromium | +3 | $Cr^{3+}$ | chromium (III) ion |
| nitrogen | +5 | $N^{5+}$ | nitrogen (V) ion |

(Your textbook points out that although silver is a transition element, virtually all silver compounds contain the cation as $Ag^+$.  Therefore, we normally omit the roman numeral, and use the "Type I" designation, ie; AgBr = silver bromide, <u>NOT</u> silver (I) bromide.)

Try the following example regarding cation nomenclature.

---

# Example 2.8 A - Naming Cations

Fill in the following table:

| Element | Oxidation State | Symbol | Name |
|---|---|---|---|
| phosphorous | +5 | $P^{5+}$ | _____ |
| _____ | _____ | $Mn^{7+}$ | _____ |
| _____ | _____ | _____ | rubidium ion |

|  |  |  | bromine (V) ion |
|---|---|---|---|
| vanadium | +2 | _____ | _____ |
| strontium | +2 | _____ | _____ |

## Strategy

Keep in mind that certain elements form cations of **only one oxidation state**. These **do not** require roman numerals. All others do.

## Solution

| Element | Oxidation State | Symbol | Name |
|---|---|---|---|
| phosphorous | +5 | $P^{5+}$ | phosphorous (V) ion |
| manganese | +7 | $Mn^{7+}$ | manganese (VII) ion |
| rubidium | +1 | $Rb^+$ | rubidium ion |
| bromine | +5 | $Br^{5+}$ | bromine (V) ion |
| vanadium | +2 | $V^{2+}$ | vanadium (II) ion |
| strontium | +2 | $Sr^{2+}$ | strontium ion |

---

b. **Anions**: negative ions (anions) that are made up of a single element are all named in the same fashion. You simply **strip off the ending and add "ide."** For example,

| Symbol | Neutral Element | Anion Name |
|---|---|---|
| $Cl^-$ | chlorine | chloride |
| $H^-$ | hydrogen | hydride |
| $S^{2-}$ | sulfur | sulfide |
| $N^{3-}$ | nitrogen | nitride |

Some anions can have different oxidation states. For purposes of naming the anion part of compounds, the oxidation state is unimportant. The "ide" rule always applies. Similarly, when you see an "ide" in a name, it indicates an anion.

In order to determine the oxidation state of anions, you can use the following general rule (which will be justified in chapter 7 of your textbook):

**THE OXIDATION STATE OF AN ANION WILL BE
EQUAL TO THE GROUP NUMBER MINUS EIGHT.**

**Oxidation state = group # - 8**

For example, if nitrogen (in group 5A) acts as an anion (it can also be a cation in many cases),

$$\text{oxidation state} = 5 - 8 = -3 \text{ or } N^{3-}.$$
$$\underset{\text{group \#}}{\uparrow}$$

Looking at bromine (if it acts as an anion),

$$\text{oxidation state} = 7 - 8 = -1 \text{ or } Br^-.$$

Try your hand at the following exercise, which tests your ability to go from anion to name and name to anion.

## Example 2.8 B - Naming Anions

Fill in the following table:

| Element | Oxidation State | Symbol | Anion Name |
|---------|-----------------|--------|------------|
| oxygen | -2 | _____ | _____ |
| _____ | _____ | $Se^{2-}$ | _____ |
| iodine | _____ | _____ | _____ |
| _____ | _____ | _____ | phosphide |

## Solution

| Element | Oxidation State | Symbol | Anion Name |
|---------|-----------------|--------|------------|
| oxygen | -2 | $O^{2-}$ | oxide |
| selenium | -2 | $Se^{2-}$ | selenide |
| iodine | -1 | $I^-$ | iodide |
| phosphorous | -3 | $P^{3-}$ | phosphide |

c.  **Naming binary salts:** The processes of naming binary salts involves combining names which you separately assign to the cation and anion halves. The only complication can be with determining whether the <u>cation</u> requires a roman numeral. Keep in mind:

1.  The cation **always comes first** in a name (it is written on the left hand side).
2.  The total **charge** on a compound **must equal zero.**

Many students find the second point to be troublesome, but you don't have to. Just remember that the total positive count must equal the total negative count.

Let's work together on naming the compound whose formula is $CaF_2$. Let's work first on the anion (right) side. The fact that there are two fluorines rather than one or three is important so that we can establish a total negative charge due to fluorine (-1 for each fluorine × 2 fluorines = -2 total). However, the **name** of the anion is still **fluoride.**

Let's now go to the cation (left) side. Because the anion side contributes a total charge of "-2," the cation side must be "+2" to make a neutral compound. Therefore, we have $Ca^{2+}$ as the cation. (We really knew this already, because calcium is an alkaline earth, and can only ionize to a +2 oxidation state.) The name for the cation is therefore <u>calcium</u> ion, <u>not</u> calcium (II) ion, because calcium ion can have only one oxidation state. Combining the anion and cation gives us

calcium + fluoride = **calcium fluoride**

↑       ↑

cation   anion

Let's try another one. Name the compound whose formula is $PbS_2$. Working on the anion side as before, we have two <u>sulfides</u>. Because each sulfur is in group <u>6A</u>, its oxidation state is <u>-2</u> (6 - 8). Therefore, the two sulfides contribute **-4** to the compound.

On the cation side, the <u>lead</u> can have more than one oxidation state. The oxidation state on the lead must be **+4**, because the positive and negative charges must balance, and there is only one lead atom in the compound. Therefore, the cation name is <u>lead (IV)</u> ion.

Combining the cation and anion parts, the compound is named **lead (IV) sulfide**.

---

# Example 2.8 C - Naming Binary Salts

Name the following binary compounds:

    a. KCl          b. $TiCl_4$          c. $IF_5$          d. $Fe_2O_3$

## <u>Strategy</u>

1. Name the anion.
2. Establish the total charge due to the anion.
3. Name the cation.
4. Determine the charge on each cation. (Remember that the compound must be neutral.)
5. If the cation is not in groups 1A, 2A, or 3A, its oxidation should be specified with roman numerals.
6. Combine the cation and anion to get the name of the compound.

## <u>Solution</u>

a. potassium chloride   (potassium is in group 1A)
b. titanium (IV) chloride
c. iodine (V) fluoride (Iodine can act as a cation!  We will see later that we can also call this iodine pentafluoride.)
d. iron (III) oxide

---

2.   Salts With Polyatomic Ions

    a.   Conventional **polyatomic ions**: These contain **more than one atom**. Table 2.5 contains formulas for, and names of, the polyatomic ions that you should know. With the exception of ammonium ion, $NH_4^+$, all of the listed ions are anions.

        The same rules are followed when naming compounds involving polyatomic ions, as when naming binary salts. The only thing to keep in mind is that you don't change the name of the polyatomic ion. Let's work together to name the compound whose formula is $NaC_2H_3O_2$. The <u>Na$^+$</u> is <u>sodium</u> ion. The $C_2H_3O_2^-$ is the <u>acetate</u> ion. The compound name is therefore **sodium acetate**.

    b.   Oxyanions: Certain polyatomic anions contain **different numbers of oxygen atoms** combined with an atom of a different element. These are called **oxyanions**. An oxyanion series normally contains either 2 or 4 members. In a two member series, the anion with **more oxygens** gets the suffix "**ate**," and the one with **fewer oxygens** gets the suffinx "**ite**."

            $NO_2^-$ = nitrite      $SO_3^{2-}$ = sulfite
            $NO_3^-$ = nitrate      $SO_4^{2-}$ = sulfate

When there are 4 atoms in the series (as in many halogen containing oxyanions), the sequence is:

$IO^-$ = **hypoiodite**
$IO_2^-$ = **iodite**
$IO_3^-$ = **iodate**
$IO_4^-$ = **periodate**

Just remember to approach the naming in a systematic fashion (as in example 2.8D) and relax - you are not being timed.

---

## Example 2.8 D - Naming Compounds with Polyatomic Ions

Name the following compounds:

a. $KMnO_4$
b. $Ba(OH)_2$
c. $Fe(OH)_3$

d. $NaH_2PO_4$
e. $(NH_4)_2Cr_2O_7$
f. $NaBrO_4$

g. $KIO_3$
h. $Ca(OCl)_2$
i. $Cr(NO_3)_3$

## Solution

| Formula | Name |
|---|---|
| a. $KMnO_4$ | potassium permanganate |
| b. $Ba(OH)_2$ | barium hydroxide (Barium is in group 2A - it is not a transition element, and doesn't require a roman numeral.) |
| c. $Fe(OH)_3$ | iron (III) hydroxide (Iron is a transition metal.  A roman numeral is required.) |
| d. $NaH_2PO_4$ | sodium dihydrogen phosphate |
| e. $(NH_4)_2Cr_2O_7$ | ammonium dichromate |
| f. $NaBrO_4$ | sodium perbromate ($BrO_4^-$ is part of an oxyanion series.) |
| g. $KIO_3$ | potassium iodate ($IO_3^-$ is part of an oxyanion series.) |
| h. $Ca(OCl)_2$ | calcium hypochlorite ($OCl^-$ is also part of an oxyanion series.) |
| i. $Cr(NO_3)_3$ | chromium (III) nitrate (Chromium is a transition element.) |

---

3. Binary Covalent Compounds

Such compounds are formed between **two nonmetals**.  Your textbook calls these **"Type III Covalent Compounds."**  We have already considered IUPAC naming of such compounds (see example 2.8 C in this study guide, where we named $IF_5$).  However, Zumdahl proposes a more common, and equally acceptable, nomenclature system.  Use the prefixes in <u>Table 2.6</u>, for both the cation and anion part of your binary covalent compound, remember **not** to put **mono** in front of a **single cation**.

The binary covalent compound $IF_5$, for example, could be named **iodine pentafluoride**.  Because Zumdahl uses common systematic names for binary covalent compounds, we will do that in **this study guide** as well.

(REMEMBER, COMMON SYSTEMATIC NAMES ARE TO
BE USED <u>ONLY</u> WITH BINARY COVALENT COMPOUNDS!!)

## Example 2.8 E - Naming Binary Covalent Compounds

Name the following compounds:

a.  $CO_2$
b.  $P_2O_5$

c.  $N_2O_3$
d.  $Cl_2O_7$

e.  $SO_3$
f.  $BrF_3$

## Solution

a.  carbon dioxide
b.  diphosphorous pentoxide (drop the "a" in penta - the o or a in the prefix is normally dropped when the element begins with a vowel.)
c.  dinitrogen trioxide
d.  dichlorine heptoxide
e.  sulfur trioxide
f.  bromine trifluoride

---

4.  Acids

You will learn about several different types of acids in chapters 14 and 15. At this early point, **for purposes of nomenclature only,** you may consider an **acid** to be any compound in which hydrogen is the only cation. This is a very liberal definition. We will modify and tighten this definition in later chapters. HCl is an acid. $HNO_3$ is an acid. $NH_4OH$ is not, because $NH_4^+$ is the cation, not just $H^+$. The names of some important acids are given in Tables 2.8 and 2.9 of your textbook.

To name an acid, determine if the anion part is polyatomic or not.

IF THE ANION PART IS MONATOMIC (such as $Cl^-$ or $S^{2-}$)

a.  change the "**ide**" ending to "**ic acid,**" and
b.  add the prefix "**hydro**" to the beginning of the name.

For example, given $H_2S$

a.  $S^{2-}$ is sulfide. This part gets changed to **sulfuric acid**.
b.  Add "hydro" to the front.

$H_2S$ is **hydrosulfuric acid.**

The one exception to the rule is $CN^-$. Even though the anion is polyatomic, the acid is known as hydro-cyanic acid.

IF THE ANION PART IS POLYATOMIC (such as $NO_3^-$ or $C_2H_3O_2^-$)

a.  If the anion suffix is "**ite**," change it to "**ous acid.**"
b.  If the anion suffix is "**ate**," change it to "**ic acid.**"

For example, given $HIO_4$, $IO_4^-$ is the periodate ion. We change the "ate" to "ic acid," giving **periodic acid**.

## Example 2.8 F - Naming Acids

Name the following acids:

a.  HF
b.  $HC_2H_3O_2$

c.  $HBrO_3$
d.  HBrO

e.  HI
f.  $HNO_2$

### Solution

a.  hydrofluoric acid   (monatomic anion)
b.  acetic acid   (polyatomic anion)
c.  bromic acid   (polyatomic anion)
d.  hypobromous acid   (the anion is hypobromate)
e.  hydroiodic acid   (monatomic anion)
f.  nitrous acid   (the anion is nitrate)

Let's tie everything together with the following example, which makes use of all the nomenclature you've learned so far.

## Example 2.8 G - Tying it all Together

a.  $PCl_5$
b.  $HClO_2$
c.  $Ni(NO_3)_2$

d.  $Sb_2S_3$
e.  $XeF_4$
f.  $NH_4OH$

g.  $NaC_2H_3O_2$
h.  $NaHCO_3$
i.  LiH

### Solution

a.  phosphorous pentachloride
b.  chlorous acid   (The anion is chlorite.)
c.  nickel (II) nitrate   (The nickel is a transition metal.)
d.  antimony (III) sulfide (3 sulfurs x - 2 each = -6; each antimony must have +6/2 = +3 oxidation state.)
e.  xenon (IV) fluoride **or** xenon tetrafluoride
f.  ammonium hydroxide   (two polyatomic ions)
g.  sodium acetate
h.  sodium bicarbonate
i.  lithium hydride   (Hydrogen is an anion here.)

## B. Getting Formulas from Names

The key to getting **formulas** from **names** is to remember that chemical names contain both a **cation** and an **anion** part.  In just about every case, one of the two parts will unequivocally tell you what its oxidation state is.  Remembering that the compound is neutral, and given the oxidation state on one part of the compound, dictates the oxidation state on the other part.

Let's work together to determine the formula of **magnesium nitride**.  Magnesium is in group 2A, and as such, can only form the $Mg^{2+}$ ion.

Nitrogen can occur in many oxidation states.  However, when it exists as an <u>anion</u> in an ionic binary compound, we need to use our rule that the oxidation state equals the group number minus eight, or:

$$\text{oxidation state} = 5 - 8 = -3.$$

Nitrogen is present as $N^{3-}$ ion.

The simplest way for a +2 ion to combine with a -3 ion to maintain charge neutrality is to have <u>3 $Mg^{2+}$</u> and 2 <u>$N^{3-}$</u>, or $Mg_3N_2$.

Let's look at **nitric acid**.  An **acid** implies that the cation is **hydrogen**.  Recall the rules for **anions of acids**.  The endings "<u>ic acid</u>" came from an anion with an "**ate**" ending.  The anion was **nitrate**, or $NO_3^-$.

Therefore, the formula is $HNO_3$.

As a final exercise in this section, let's work out the formulas of compounds from their names.

---

## Example 2.8 H - Formulas from Names

Write the formulas for each compound:

|   |   |   |   |
|---|---|---|---|
| a. | sodium chloride | e. | hypoiodous acid |
| b. | calcium fluoride | f. | tin (IV) oxide |
| c. | iron (III) nitrate | g. | dinitrogen tetroxide |
| d. | copper (I) chloride | h. | ammonium acetate |

## <u>Solution</u>

|   |   |   |   |
|---|---|---|---|
| a. | NaCl | e. | HIO   ($IO^-$ is the hypoiodite ion) |
| b. | $CaF_2$ | f. | $SnO_2$ |
| c. | $Fe(NO_3)_3$ | g. | $N_2O_4$ |
| d. | CuCl | h. | $NH_4C_2H_3O_2$ |

---

# Problems 15-28 at the end of this study chapter cover material from this section

---

**END OF CHAPTER TWO**

# SELF-TEST QUESTIONS

1. In an exothermic (heat producting) reaction, chlorine reacts with 2.02 g of hydrogen to form 72.926 g of hydrogen chloride gas. How many grams of chlorine reacted with hydrogen?

2. Sulfur and oxygen can react to form both sulfur dioxide and sulfur trioxide. In sulfer dioxide, there are 32.06 g of sulfur and 32.00 g of oxygen. In sulfur trioxide, 32.06 g of sulfur are combined with 48.00 g of oxygen.

   a. What is the ratio of the weights of oxygen that combine with 32.06 g of sulfur?
   b. How do these data illustrate the law of multiple proportions?

3. Describe what part of Dalton's Atomic Theory each chemical statement relates to.

   a. $H_2 + Cl_2 \longrightarrow 2HCl$
   b. There are $3.01 \times 10^{23}$ atoms in 20.04 g of calcium.
   c. Lead does not change to chromium when it forms lead hydroxide.

4. Identify each of the following elements:

   a. $^{91}_{40}X$                    d. $^{85}_{36}X$

   b. $^{108}_{47}X$                    e. $^{51}_{23}X$

   c. $^{33}_{16}X$                    f. $^{133}_{55}X$

5. Identify each of the following elements:

   a. $^{98}_{43}X$                    d. $^{14}_{6}X$

   b. $^{186}_{75}X$                    e. $^{40}_{19}X$

   c. $^{75}_{33}X$                    f. $^{131}_{54}X$

6. How many protons and neutrons are in each of the following elements?

   a. $^{89}_{39}Y$    50    d. $^{238}_{92}U$    1 4 6
                39                      9 2

   b. $^{73}_{32}Ge$   4 1    e. $^{35}_{17}Cl^-$    18
                32                      17

   c. $^{24}_{12}Mg^{+2}$   12    f. $^{65}_{30}Zn$    35
                12                      3 0

7. How many protons and neutrons are in each of the following elements?

   a. $^{227}Ac$             d. $^{251}Cf$
   b. $^{70}Ga$             e. $^{239}Pu$
   c. $^{11}B$             f  $^{64}Cu$

8.  How many protons, neutrons and electrons are in each of the following ions?

    a.  $^{56}_{26}$ Fe$^{3+}$          d.  $^{31}_{15}$ P$^{3-}$

    b.  $^{40}_{20}$ Ca$^{2+}$          e.  $^{127}_{53}$ I$^-$

    c.  $^{19}_{9}$ F$^-$          f.  $^{127}_{53}$ I$^{7+}$

9.  How many protons, neutrons, and electrons are in each of the following?

    a.  $^{195}_{78}$ Pt$^+$          f.  $^{56}_{26}$ Fe$^{3+}$

    b.  $^{93}_{41}$ Nb          g.  $^{184}_{74}$ W

    c.  $^{40}_{18}$ Ar$^-$          h.  $^{133}_{55}$ Cs$^+$

    d.  $^{16}_{8}$ O$^{2-}$          i.  $^{28}_{14}$ Si$^{3-}$

    e.  $^{122}_{51}$ Sb$^{2+}$

10. Fill in the missing information in the following table:

| Symbol | Protons | Neutrons | Electrons | Charge |
|---|---|---|---|---|
| $^{80}_{35}$ Br$^-$ | 35 | 45 | 36 | −1 |
| $^{55}_{35}$ —$^{5+}$ | 35 | 55 | 30 | +5 |
| $^{137}_{56}$ —$^{2+}$ | 56 | 81 | 54 | +2 |
| $^{108}_{47}$ Ag$^+$ | 47 | 61 | 46 | +1 |
| $^{\ \ }_{23}$ —$^{5+}$ | 23 | | | +5 |
| — Co$^{2+}$ | | | | +2 |

11. Fill in the missing information in the following table:

| Symbol | Protons | Neutrons | Electrons | Charge |
|---|---|---|---|---|
| $_{13}$ —$^{3+}$ | 13 | | 10 | +3 |
| $^{88}_{38}$ —$^-$ | 38 | 50 | 37 | +1 |
| $^{65}_{30}$ —$^+$ | 30 | 35 | | |
| $^{35}_{17}$ —$^{-2}$ | 17 | 18 | 19 | −2 |
| — Te$^{3-}$ | | 76 | | |
| — Rb — | | | | +1 |

12. Name the family to which each of the following elements belong:

    a.  Fe  T
    b.  Cl  G
    c.  Ar  NG

    d.  Sr  AE
    e.  Rb  AE
    f.  Nd T

13. Are the following elements metals or nonmetals?

    a.  Mg  M
    b.  Si  G
    c.  Ge
    d.  Br
    e.  O

    f.  Bi
    g.  Co
    h.  Mo
    i.  Xe

14. Name the family to which each of the following elements belong:

    a.  Es
    b.  I
    c.  Au
    d.  Yb

    e.  Kr
    f.  Fr
    g.  Ca

15. Given the position in the periodic table, what is the most likely oxidation state that each element will have when forming an ion?

    a.  Cs
    b.  N
    c.  Br

    d.  K
    e.  Al
    f.  S

16. Would you expect the following atoms to gain or lose electrons when forming an ion?  If so, how many would be gained or lost?

    a.  Be
    b.  Cl
    c.  Al
    d.  O

    e.  F
    f.  Li
    g.  P

17. Name each of the following compounds:

    a.  $PbI_2$
    b.  $NH_4Cl$
    c.  $Fe_2O_3$
    d.  LiH
    e.  CsCl
    f.  $OsO_4$

    g.  $Cr(OH)_3$
    h.  $NaC_2H_3O_2$
    i.  $K_2Cr_2O_7$
    j.  $Na_2SO_4$
    k.  $KH_2PO_4$

18. Name each of the following compounds:

    a.  $MgSO_4$
    b.  $N_2O_3$
    c.  $Ce_2O_3$
    d.  $KMnO_4$
    e.  NiO

    f.  $BaSO_4$
    g.  $Fe(IO_4)_3$
    h.  $SO_3$
    i.  $KClO_4$

19. Name each of the following compounds:

    a.  $NI_3$
    b.  $PCl_5$
    c.  Co

    d.  $P_4O_{10}$
    e.  $N_2O_4$
    f.  $NH_3$

20. Name each of the following compounds:

a. $P_4O_6$          e. $BF_3$
b. KOH             f. AgCl
c. $N_2$             g. $KHCO_3$
d. $PH_3$            h. $AgNO_3$

21. Name each of the following compounds:

a. $HIO_3$           e. $NaNO_2$
b. HBr             f. $K_2SO_3$
c. $HNO_2$           g. $NaHSO_3$
d. HCN

22. Name each of the following compounds:

a. $UF_6$            e. $N_2H_4$
b. $Cu(NO_3)_2$        f. $Mg(OH)_2$
c. $H_3PO_4$          g. $SnCl_2$
d. $SF_6$            h. $NaCO_3$

23. Write formulas for each of the following compounds:

a. sodium cyanide         e. iron (III) oxide
b. tin (II) floride        f. calcium phosphate
c. sodium hydrogen sulfate  g. sodium bromate
d. lead (II) nitrate       h. hydrogen iodide

24. Write formulas for each of the following compounds:

a. sodium sulfate         e. lithium aluminum hydride
b. manganese dioxide      f. barium chloride
c. potassium chlorate      g. magnesium oxide
d. potassium hypochlorite   h. copper (I) oxide

25. Write formulas for each of the following compounds:

a. potassium carbonate     i. carbon tetrachloride
b. magnesium hydroxide     j. sodium iodate
c. dinitrogen tetroxide     k. potassium permanganate
d. hypoiodous acid        l. sulfurous acid
e. iron (III) chloride      m. potassium hydrogen phosphate
f. tin (IV) oxide         n. ammonium acetate
g. rubidium nitrate        o. ammonium dichromate
h. potassium chlorate      p. hydroiodic acid

26. Give the names of the following acids:

a. $H_2SO_3$          d. $HNO_2$
b. HI              e. $H_3PO_4$
c. HBr             f. HCl

27. Give formulas for the following acids:

a. nitric acid          d. hydrocyanic acid
b. hydrofluoric acid      e. hydrosulfuric acid
c. sulfuric acid        f. acetic acid

28. Give the alternate or common name for each of the following compounds or cations:

    a.   sodium hydrogen carbonate ($NaHCO_3$)

    b.   dinitrogen monoxide ($N_2O$)

    c.   nitrogen monoxide (NO)

    d.   iron (II) [$Fe^{2+}$]

    e.   tin (IV) [$Sn^{4+}$]

    f.   lead (II) [$Pb^{2+}$]

# ANSWERS TO SELF-TEST QUESTIONS

1. 70.906 g

2. a. The ratio is 1.5 to 1, or 3 to 2.
   b. The ratios are 2 whole numbers.

3. a. Chemical reactions involve reorganization of the atoms.
   b. Each element is made up of tiny particles called atoms.
   c. Chemical reactions involve reorganization of the atoms.

4. a. Zr          d. Kr
   b. Ag          e. V
   c. S           f. Cs

5. a. Tc          d. C
   b. Re          e. K
   c. As          f. Xe

6. a. Y = 39 p, 50 n          d. U = 92 p, 146 n
   b. Ge = 32 p, 41 n         e. $Cl^-$ = 17 p, 18 n
   c. $Mg^{2+}$ = 12 p, 12 n   f. Zn = 30 p, 35 n

7. a. Ac = 89 p, 138 n        d. Cf = 98 p, 153 n
   b. Ga = 31 p, 39 n         e. Pu = 94 p, 145 n
   c. B = 5 p, 6 n            f. Cu = 29 p, 35 n

8.

| | protons | neutrons | electrons |
|---|---|---|---|
| a. | 26 | 30 | 23 |
| b. | 20 | 20 | 18 |
| c. | 9 | 10 | 10 |
| d. | 15 | 16 | 18 |
| e. | 53 | 74 | 54 |
| f. | 53 | 74 | 46 |

9.

| | protons | neutrons | electrons |
|---|---|---|---|
| a. | 78 | 117 | 77 |
| b. | 41 | 52 | 41 |
| c. | 18 | 22 | 19 |
| d. | 8 | 8 | 10 |
| e. | 51 | 71 | 49 |
| f. | 26 | 30 | 23 |
| g. | 74 | 110 | 74 |
| h. | 55 | 78 | 54 |
| i. | 14 | 14 | 17 |

10.

| Symbol | Protons | Neutrons | Electrons | Charge |
|---|---|---|---|---|
| $^{80}_{35}Br^-$ | 35 | 45 | 36 | -1 |
| $^{80}_{35}Br^{5+}$ | 35 | 45 | 30 | +5 |
| $^{137}_{56}Ba^{2+}$ | 56 | 81 | 54 | +2 |
| $^{108}_{47}Ag^+$ | 47 | 61 | 46 | +1 |
| $^{51}_{23}V^{5+}$ | 23 | 28 | 18 | +5 |
| $^{59}_{27}Co^{2+}$ | 27 | 32 | 25 | +2 |

11.

| Symbol | Protons | Neutrons | Electrons | Charge |
|---|---|---|---|---|
| $^{27}_{13}Al^{3+}$ | 13 | 14 | 10 | +3 |
| $^{88}_{38}Sr^+$ | 38 | 50 | 37 | +1 |
| $^{65}_{30}Zn^+$ | 30 | 35 | 29 | +1 |
| $^{35}_{17}Cl^{2-}$ | 17 | 18 | 19 | -2 |
| $^{128}_{52}Te^{3-}$ | 52 | 76 | 55 | -3 |
| $^{85}_{37}Rb^+$ | 37 | 48 | 36 | +1 |

12. a. transition metal
    b. halogen
    c. noble gas
    d. alkaline earth
    e. alkali
    f. lanthanide

13. a. metal
    b. nonmetal
    c. metal
    d. nonmetal
    e. nonmetal
    f. metal
    g. metal
    h. metal
    i. nonmetal

14. a. Actinides
    b. Halogens
    c. Transition metals
    d. Lanthanide
    e. Noble gases
    f. Alkali metals
    g. Alkaline earth metal

15. a. 1+
    b. 3-
    c. 1-
    d. 1+
    e. 3+
    f. 2-

16. a. lose 2
    b. gain 1
    c. lose 3
    d. gain 2
    e. gain 1
    f. lose 1
    g. gain 3

17. a. lead (II) iodide
    b. ammonium chloride
    c. iron (III) oxide
    d. lithium hydride
    e. cesium chloride
    f. osmium tetroxide
    g. chromium (III) hydroxide
    h. sodium acetate
    i. potassium dichromate
    j. sodium sulfate
    k. potassium dihydrogen phosphate

18. a. magnesium sulfate
    b. dinitrogen trioxide
    c. cerium (III) oxide
    d. potassium permanganate
    e. nickel (II) oxide
    f. barium sulfate
    g. iron (III) periodate
    h. sulfur trioxide
    i. potassium perchlorate

19. a. nitrogen triiodide
    b. phosphorous pentachloride
    c. carbon monoxide
    d. phosphorous (V) oxide
    e. dinitrogen tetroxide
    f. ammonia

20. a. tetraphosphorus hexoxide
    b. potassium hydroxide
    c. molecular nitrogen
    d. phosphine
    e. boron trifluoride
    f. silver chloride
    g. potassium hydrogen carbonate
    h. silver nitrate

21. a. iodic acid
    b. hydrobromic acid
    c. nitrous acid
    d. hydrocyanic acid
    e. sodium nitrite
    f. potassium sulfite
    g. sodium bisulfite

22. a. uranium (VI) fluoride
    b. copper (II) nitrate
    c. phosphoric acid
    d. sulfur hexafluoride
    e. hydrazine
    f. magnesium hydroxide
    g. tin (II) chloride
    h. sodium carbonate

23. a. $NaCN$
    b. $SnF_2$
    c. $NaHSO_4$
    d. $Pb(NO_3)_2$
    e. $Fe_2O_3$
    f. $Ca_3(PO_4)_2$
    g. $NaBrO_3$
    h. $HI$

24. a. $Na_2SO_4$
    b. $MnO_2$
    c. $KClO_3$
    d. $KClO$
    e. $LiAlH_4$
    f. $BaCl_2$
    g. $MgO$
    h. $Cu_2O$

25. a. $K_2CO_3$
    b. $Mg(OH)_2$
    c. $N_2O_4$
    d. $HIO$
    e. $FeCl_3$
    f. $SnO_2$
    g. $RbNO_3$
    h. $KClO_3$
    i. $CCl_4$
    j. $NaIO_3$
    k. $KMnO_4$
    l. $H_2SO_3$
    m. $K_2HPO_4$
    n. $NH_4C_2H_3O_2$
    o. $(NH_4)_2Cr_2O_7$
    p. $HI$

26. a. sulfurous acid
    b. hydroiodic acid
    c. hydrobromic acid
    d. nitrous acid
    e. phosphoric acid
    f. hydrochloric acid

27. a. $HNO_3$
    b. $HF$
    c. $H_2SO_4$
    d. $HCN$
    e. $H_2S$
    f. $HC_2H_3O_2$

28. a.  sodium bicarbonate
    b.  nitrous oxide
    c.  nitric oxide
    d.  ferrous ion
    e.  stannic ion
    f.  plumbous ion

# CHAPTER 3

# Stoichiometry

| Section | Contents | Page in this Study Guide |
|---|---|---|
| 3.1 | Atomic Masses | 46 |
| 3.2 | The Mole | 49 |
| 3.3 | Molar Mass | 51 |
| 3.4 | Percent Composition of Compounds | 53 |
| 3.5 | Determining the Formula of a Compound | 55 |
| 3.6 | Chemical Equations | 59 |
| 3.7 | Balancing Chemical Equations | 60 |
| 3.8 | Stoichiometric Calculations | 62 |
| 3.9 | Calculations Involving a Limiting Reactant | 64 |

This chapter will help guide your study of the **mathematics of chemical reactions**.

## 3.1 - Atomic Masses

When you finish this section you will be able to:

- Calculate average atomic masses from isotopic data.
- Calculate relative isotope abundances from pertinent mass spectral data.

The **Mass Spectrometer** is the best instrument for measuring the masses of atoms and ions (charged particles). The essential operation of the mass spectrometer is given <u>in your textbook</u>.

---

## Example 3.1 A - Calculation of Average Atomic Masses from Isotopic Data

A sample of metal "M" is vaporized and injected into a mass spectrometer. The mass spectrum tells us that 60.10% of the metal is present as $^{69}M$ and 39.90% is present as $^{71}M$. The mass values for $^{69}M$ and $^{71}M$ are 68.93 amu and 70.92 amu, respectively.

- What is the **average atomic mass** of the element?
- What is the element?

## Strategy

The basic question here is "how much does each isotope contribute to the overall atomic mass of the element?"

$^{69}$M contributes 60.10%
$^{71}$M contributes 39.90%

For every 100 atoms of M, on average

60.10 are $^{69}$M
39.90 are $^{71}$M

The total mass of 100 atoms of M is therefore a weighted average of the atomic mass of $^{69}$M and $^{71}$M.

$^{69}$M has a mass of 68.93 amu's
$^{71}$M has a mass of 70.92 amu's

The mass of **one atom** is the total mass of 100 atoms divided by 100.

## Solution

Mass of 100 atoms of M = Mass of $^{69}$M + Mass of $^{71}$M

$$= 60.10 \text{ atoms} \times 68.93 \ \frac{\text{amu}}{\text{atom}} + 39.90 \text{ atoms} \times 70.92 \ \frac{\text{amu}}{\text{atom}}$$

$$= 6972 \text{ amu}$$

Average atomic mass of M = Mass of 1 atom of M

$$= \frac{\text{Mass of 100 atoms of M}}{100 \text{ atoms}}$$

$$= \frac{6972 \text{ amu}}{100 \text{ atoms}}$$

$$= \textbf{69.72 amu/atom}$$

Look at your periodic table. The element with an average atomic mass of 69.72 amu/atom is **Ga**.

### Does the Answer Make Sense?

Many more of the isotopes are present as $^{69}$M than $^{71}$M. We would therefore expect the average atomic mass to be closer to 69 than to 71.

---

## Example 3.1 B - Calculation of Relative Isotope Abundances

The element indium exists naturally as two isotopes. $^{113}$In has a mass of 112.9043 amu, and $^{115}$In has a mass of 114.9041 amu. The average atomic mass of indium is 114.82 amu. Calculate the **percent relative abundance** of the two isotopes of indium.

## Strategy

We have 2 unknowns:

Unknown 1 = relative abundance of $^{113}$In (expressed as a fraction between 0 and 1).

Unknown 2 = relative abundance of $^{115}$In (also expressed as a fraction between 0 and 1).

Because we have two unknowns, we need two equations to solve the problem. The first equation expresses the "weighted average" of the two isotopes, as in Example 3.1 A.

**Equation #1:**

atomic mass of $^{113}$In × relative abundance of $^{113}$In
+ atomic mass of $^{115}$In × relative abundance of $^{115}$In  =  **average atomic mass of In**

The second equation indicates that the sum of the relative abundances of the isotopes of indium equals 1. (The total of the two isotopic abundances must be 100%, because all the isotopes present must be accounted for. This total, expressed as a whole number, rather than a percentage, is equal to 1.)

**Equation #2:**

relative abundance of $^{113}$In  +  relative abundance of $^{115}$In = 1

The hard part of this problem is coming up with the proper equations.

## Solution

Let the relative abundance of $^{113}$In = X
Let the relative abundance of $^{115}$In = Y

The two simultaneous equations for two unknowns:

Equation #1:  112.9043(X) + 114.9041(Y) = 114.82
Equation #2:  X + Y = 1

Make the coefficients in front of Y in the two equations equal by multiplying equation #2 by 114.9041, so we can solve for X. Our two simultaneous equations become:

Equation #1:  112.9043(X) + 114.9041(Y) = 114.82
Equation #2:  114.9041(X) + 114.9041(Y) = 114.9041

Subtract equation #2 from equation #1 and solve for X.

$$-1.9998(X) + 0(Y) = -0.0841$$

$$X = \frac{0.0841}{1.9998}$$

$$= 0.042$$

To solve for Y, plug the answer for X back into the original equation #2.

$$0.042 + Y = 1$$

$$Y = 1 - 0.042$$

$$= 0.958$$

To calculate the percent relative abundance of each isotope, multiply X and Y by 100%. Thus the answer to the problem is:

**% Relative abundance of $^{113}$In = 4.2%**
**% Relative abundance of $^{115}$In = 95.8%**

## Does the Answer Make Sense?

Since the average atomic mass of indium is very close to the mass of $^{115}$In, we would expect most of the indium to be in that form. Be careful not to accidentally put the wrong percentage with the wrong isotope.

---

Problems 1-9 at the end of this study chapter cover material from this section.

---

## 3.2 - The Mole

When you finish this section, you will be able to interconvert between **moles, mass,** and the **number of particles** of a given element.

The **mole** is the key to many chemical calculations. A **mole** is defined as **the number equal to the number of carbon atoms in exactly 12 grams of pure $^{12}$C.**

**1 mole of anything = 6.022 × 10$^{23}$ units of that thing**
1 mole of klingons = 6.022 × 10$^{23}$ klingons
1 mole of compact discs = 6.022 × 10$^{23}$ compact discs

One mole of a substance contains **Avogadro's number** of particles.

**Avogadro's number = 6.022 × 10$^{23}$   particles**

One mole of klingons are heavier than 1 mole of compact discs.

<u>BUT</u>  1 mole of klingons contain the **same** (Avogadro's) number of units as 1 mole of compact discs.

The masses of 1 mole of various elements are given in <u>Table 3.1</u> in your textbook.

**Key Problem Solving Relationship:** The average mass of **one atom** of a substance expressed in **amu's** is the same number as the mass of **one mole** of a substance expressed in **grams.**

1 atom of $^{20}$Ne = 20.18 **amu**
1 mole of $^{20}$Ne = 20.18 **grams**
1 mole of $^{20}$Ne = 6.022 × 10$^{23}$ **atoms** of Ne
1 gram (exactly) = 6.022 × 10$^{23}$ **amu**

---

## Examples 3.2 A - Conversion Between Grams, Atoms, and AMU's

How many grams does a sample containing 34 atoms of neon weigh?

## Strategy

A good way to approach this problem is with a "**flowchart**":

$$
\begin{array}{ccc}
\text{Convert} & & \text{Convert} \\
\text{to} & & \text{to} \\
\text{34 Ne atoms} \longrightarrow \text{amu's in 34 Ne atoms} & \longrightarrow & \text{grams in 34 Ne atoms} \\
\uparrow & & \uparrow \\
\text{Start} & & \text{Finish}
\end{array}
$$

As we get to more difficult problems, the skillful use of **conversion factors** will become more and more important.

## Solution

$$
34 \text{ Ne atoms} \times \frac{20.18 \text{ amu}}{1 \text{ Ne atom}} \times \frac{1 \text{ g}}{6.022 \times 10^{23} \text{ amu}} = 1.1 \times 10^{-21} \text{ g}
$$

## Does the Answer Make Sense?

Having so few atoms of a substance should lead to a **very** low mass. Be careful not to accidently **invert** conversion factors so that your answer comes out to be unfortunately large!

---

# Example 3.2 B - Conversions Between Atoms, Moles and Mass

A sample of elemental silver (Ag) has a mass of 21.46 g.

- How many **moles** of silver are in the sample?
- How many **atoms** of silver are in the sample?

## Strategy

We can find the number of moles of silver by multiplying mass by atomic weight. If you are not certain of this relationship, the **units** in the problem tell you that this must be so.

Once we calculate the number of moles of Ag, we can use **Avogadro's number** to find the number of atoms. (Make sure your units cancel properly!)

## Solution

$$
21.46 \text{ g Ag} \times \frac{1 \text{ mole Ag}}{107.9 \text{ g Ag}} = 0.1989 \text{ moles Ag}
$$

$$
0.1989 \text{ moles Ag} \times \frac{6.022 \times 10^{23} \text{ atoms Ag}}{1 \text{ mole Ag}} = 1.198 \times 10^{23} \text{ atoms Ag}
$$

---

# Example 3.2 C - Practice with Conversions

What is the weight of $7.81 \times 10^{22}$ atoms of calcium?

## Strategy

You want **grams**. You are given **atoms**. The central relationship between the mass and the number of particles is the **mole**. We will go through moles to get to grams.

atoms Ca $\xrightarrow{\text{Avogadro's number}}$ moles Ca $\xrightarrow{\text{atomic mass}}$ grams Ca

CAUTION: MAKE SURE YOUR UNITS CANCEL PROPERLY!!!!

## Solution

$$\text{g Ca} \ \bigg\| \ 7.81 \times 10^{22} \text{ atoms Ca} \times \frac{1 \text{ mole Ca}}{6.022 \times 10^{23} \text{ atoms Ca}} \times \frac{40.08 \text{ g Ca}}{1 \text{ mole Ca}} = \mathbf{5.20 \text{ g Ca}}$$

## Does the Answer Make Sense?

You have $7.81 \times 10^{22}$ atoms of Ca, which is a little over 10% of a mole. Therefore, we would expect a mass of a little over 10% of the atomic weight.

Problems 10-16 at the end of this study chapter cover material from this section.

## 3.3 - Molar Mass

When you finish this section, you will be able to:

- Calculate molar masses.
- Interconvert among molar mass, moles, mass, and number of particles in a given sample.

Recall that a **molecule** is a **covalently bonded collection of atoms**. Carbon dioxide ($CO_2$), benzene ($C_6H_6$) and ethanol ($C_2H_6O$) are all examples of molecules.

**MOLAR MASS OF A MOLECULE = The sum of the masses of the individual atoms in the molecule**

In most chemical calculations, we express molar masses in **grams**, using prefixes as necessary (mg, μg, kg, etc.).

## Example 3.3 A - Calculation of the Molar Mass of a Compound

Calculate the molar mass of potassium dichromate, $K_2Cr_2O_7$.

## Strategy

To calculate the molar mass of a compound, you **total** the atomic weights of **EVERY ATOM** in the compound. If an atom appears more than once (as all do in our example), you must multiply its atomic weight by the number of times it appears in the compound (represented by its **subscript**; if there is no subscript, it is present only once).

## Solution

| atom | subscript | | atomic mass | | total mass |
|------|-----------|---|-------------|---|------------|
| K | 2 | × | 39.10 | = | 78.20 |
| Cr | 2 | × | 52.00 | = | 104.00 |
| O | 7 | × | 16.00 | = | 112.00 |
| | **Molar Mass of K$_2$Cr$_2$O$_7$** | | | = | **294.20 g/mole** |

---

Once you are comfortable calculating molar masses, you can use the conversion factor of **grams/mole** (in our previous example (294.20 g K$_2$Cr$_2$O$_7$)/(1 mole K$_2$Cr$_2$O$_7$)) to solve a host of fundamental chemistry problems.

---

## Example 3.3 B - Converting Among Molar Mass, Moles, Mass, and Number of Particles

How many μg are there in $3.82 \times 10^{-7}$ moles of pyrogallol, C$_6$H$_6$O$_3$?

## Strategy

This is a two-step conversion:

$$\text{moles pyrogallol} \xrightarrow[\text{mass}]{\text{molar}} \text{g pyrogallol} \xrightarrow[\text{conversion}]{\text{unit}} \text{μg pyrogallol}$$

(Be careful to use the proper unit conversion from g to μg!)

## Solution

The molar mass of pyrogallol (pyrgl) is:

C: 6 × 12.01 = 72.06 g
H: 6 × 1.008 = 6.048 g
O: 3 × 16.00 = 48.00 g
   **molar mass = 126.11 g/mole pyrogallol**

$$\text{μg pyrgl} \quad \bigg|\bigg|\quad 3.82 \times 10^{-7} \text{ moles pyrgl} \times \frac{126.11 \text{ g pyrgl}}{1 \text{ mole pyrgl}} \times \frac{1 \times 10^{6} \text{ μg pyrgl}}{1 \text{ g pyrgl}} = \textbf{48.2 μg pyrogallol}$$

(exactly)

---

## Example 3.3 C - Practice with Conversions

Freon-12, which has the formula CCl$_2$F$_2$, is used as a refrigerant in air conditioners and as a propellant in aerosol cans.  Given a 5.56 mg sample of Freon-12:

    a.   Calculate the number of molecules of Freon-12 in that sample.
    b.   How many mg of chlorine are in the sample?

## Strategy

Use **moles** as your **bridge** between milligrams and molecules.  Also, always keep in mind that a conversion factor is valid in two ways, for example:

$$\frac{6.022 \times 10^{23} \text{ molecules}}{1 \text{ mole}} \text{ is the same as } \frac{1 \text{ mole}}{6.022 \times 10^{23} \text{ molecules}}$$

You must be careful to use the conversion factor in the correct fashion so that the units cancel.

$$\text{mg Freon-12} \xrightarrow[\text{molar mass}]{\text{mg to grams, and}} \text{moles Freon-12} \xrightarrow{\text{Avogadro's number}} \text{molecules Freon-12}$$

The mass of 2 chlorine atoms **divided by** the molar mass of Freon-12 gives the **fraction** of the total mass that is chlorine.  Multiplying the **chlorine fraction** by the **total sample size** gives the mg chlorine in the sample.

## Solution

The molar mass of Freon-12 (Fr-12) is:

C :  $1 \times 12.01 = 12.01$ g
Cl: $2 \times 35.45 = 70.90$ g
F :  $2 \times 19.00 = \underline{38.00 \text{ g}}$
 **molar mass $= 120.91$ g/mole Freon-12**

$$5.56 \text{ mg Fr-12} \times \frac{1 \text{ g Fr-12}}{1000 \text{ mg Fr-12}} \times \frac{1 \text{ mole Fr-12}}{120.91 \text{ g Fr-12}} \times \frac{6.022 \times 10^{23} \text{ molecules Fr-12}}{1 \text{ mole Fr-12}}$$

$$= \textbf{2.77} \times \textbf{10}^{19} \textbf{ molecules Fr-12}$$

From the molar mass calculation above, Cl $= 70.90$ g of the molar mass of Freon-12.

$$\text{fraction of Freon-12 that is Cl} = \frac{70.90 \text{ g Cl}}{120.91 \text{ g Fr-12}} = \textbf{0.586}$$

$$\text{mg Cl in 5.56 mg of Freon-12} = 0.586 \times 5.56 \text{ mg} = \textbf{3.26 mg Cl}$$

---

## Problems 17-27 at the end of this study chapter cover material from this section.

---

## 3.4 - Percent Composition of Compounds

When you finish this section, you will be able to calculate the mass percent of each element in a compound.

## General Strategy

There are 3 steps to calculating the **mass percent** of each element in a compound.

a.  Compute the **molecular mass** of the compound.
b.  Calculate how much of the molecular mass comes from each element.
c.  Divide each element's mass contribution by the total molecular mass and multiply by 100 to convert to percent.

## Example 3.4 A - Determination of the Mass Percent in a Compound

Calculate the mass percent of each element in glucose, $C_6H_{12}O_6$.

### Solution

Using the **general strategy** outlined above,

a.  Molar mass:

$$C: \ 6 \times 12.01 \ = \ 72.06 \text{ g}$$
$$H: \ 12 \times 1.008 \ = \ 12.096 \text{ g}$$
$$O: \ 6 \times 16.00 \ = \ \underline{96.00 \text{ g}}$$
$$\textbf{molar mass} = \textbf{180.156 g glucose}$$

**(Note:  We retain all figures in the molar mass because this is an intermediate calculation)

b.  Contribution of each element:

As shown in the molar mass calculation,

C contributes  **72.06 g**
H contributes  **12.096 g**
O contributes  **96.00 g**

c.  Mass percent of each element:

$$\text{Mass percent of C} = \frac{72.06 \text{ g C}}{180.156 \text{ g glucose}} \times \ 100\% \ = \ \textbf{40.00\% C}$$

$$\text{Mass percent of H} = \frac{12.096 \text{ g H}}{180.156 \text{ g glucose}} \times \ 100\% \ = \ \textbf{6.71\% H}$$

$$\text{Mass percent of O} = \frac{96.00 \text{ g O}}{180.156 \text{ g glucose}} \times \ 100\% \ = \ \textbf{53.29\% O}$$

### Does the Answer Make Sense?

The best way to check your answer is to add up the percentages.  If they total 100.00% (as they do here) then your answers are fine (assuming your molar mass is correct).

**COMMENT:**  Although it would have been correct to calculate the mass percent of O by subracting (%C + %H) from 100%, in the previous problem, this eliminates your ability to double check your answer, because you have, in effect, already "guaranteed" that the mass percents must equal 100%.  We therefore recommend against this procedure.

## Example 3.4 B - Practice with Mass Percents

Calculate the mass percent of each element in potassium ferricyanide, $K_3Fe(CN)_6$.

## Solution

Using the same strategy as in the previous example,

a. Molar Mass:

$$
\begin{array}{ll}
K: & 3 \times 39.10 = 117.30 \text{ g} \\
Fe: & 1 \times 55.85 = \phantom{0}55.85 \text{ g} \\
C: & 6 \times 12.01 = \phantom{0}72.06 \text{ g} \\
N: & 6 \times 14.01 = \underline{\phantom{0}84.06 \text{ g}} \\
\textbf{molar mass} & = \textbf{329.27 g/mole } \mathbf{K_3Fe(CN)_6}
\end{array}
$$

b. Contribution of each element:

$$K = 117.30 \text{ g};\ \ Fe = 55.85 \text{ g};\ \ C = 72.06 \text{ g};\ \ N = 84.06 \text{ g}$$

c. Mass percent of each element:

$$\text{Mass percent of K} \ = \ \frac{117.30 \text{ g K}}{329.27 \text{ g } K_3Fe(CN)_6} \times 100\% = \mathbf{35.62\% \ K}$$

$$\text{Mass percent of Fe} \ = \ \frac{55.85 \text{ g Fe}}{329.27 \text{ g } K_3Fe(CN)_6} \times 100\% = \mathbf{16.96\% \ Fe}$$

$$\text{Mass percent of C} \ = \ \frac{72.06 \text{ g C}}{329.27 \text{ g } K_3Fe(CN)_6} \times 100\% = \mathbf{21.88\% \ C}$$

$$\text{Mass percent of N} \ = \ \frac{84.06 \text{ g N}}{329.27 \text{ g } K_3Fe(CN)_6} \times 100\% = \mathbf{25.53\% \ N}$$

## Checking the Figures

Adding up the percentages gives **99.99%**, not 100.00%. This is an example of the kind you will see from time to time, where rounding off to 2 significant figures after the decimal point leads to a **loss** of .01%. In such cases, we say that the answers are correct "within round-off error."

---

Problems 28-33 at the end of this study chapter cover material from this section.

---

## 3.5 - Determining the Formula of a Compound

When you finish this section you will be able to:

- Determine the empirical formula of a compound.
- Calculate the molecular formula of a compound.

The **empirical formula** is represented by the **simplest whole number ratio** of atoms in a compound. Examples of empirical formulas are:

$$CH, CH_4, CH_2O, K_2Cr_2O_7, K_3Fe(CN)_6.$$

The **molecular formula** is the **actual ratio** of atoms in a compound.  Examples of molecular formulas are:

$$C_6H_6, CH_4, C_6H_{12}O_6, K_2Cr_2O_7, K_3Fe(CN)_6.$$

The empirical and molecular formulas **can** be the same.

The common methods for experimentally determining the empirical and molecular formulas are given in your textbook.

---

## Example 3.5 A - Determination of the Empirical Formula of a Compound

The analysis of a rocket fuel showed that it contained 87.4% nitrogen and 12.6% hydrogen by weight.  Mass spectral analysis showed the fuel to have a molar mass of 32.05 grams.  What are the empirical and molecular formulas of the fuel?

### Strategy

The problems in this section involve just the opposite strategy as those in section 3.4.  In that section, you found what percentage of each element made up a known compound.  Here, you **know** the percentages, and you **need** the compound.

**To solve empirical formula problems that do not involve combustion (added oxygen):**

1.  Assume the compound has a mass of exactly 100 grams.  You can therefore convert percentage to grams.
2.  Calculate moles of each kind of atom present.
3.  Determine simplest whole number ratios by dividing the moles of each compound by the smallest calculated mole value.

**To determine the molecular formula**, divide the molar mass by the empirical formula mass.  This will give the number of empirical formula units in the actual molecule.  For example, if the **empirical formula** is CH and you determine that the **molar mass is 6 times** the empirical formula mass, the **molecular formula** is $C_6H_6$.

### Solution

Assuming the compound to have a mass of 100 grams (we will drop this assumption after calculating the empirical formula),

$$N = 87.4 \text{ g } (87.4\% \text{ of } 100)$$
$$H = 12.6 \text{ g } (12.6\% \text{ of } 100)$$

Moles of N $\left|\left|\right.\right.$ $87.4 \text{ g N} \times \dfrac{1 \text{ mole N}}{14.01 \text{ g H}} = \textbf{6.24 moles N}$

Moles of H $\left|\left|\right.\right.$ $12.6 \text{ g H} \times \dfrac{1 \text{ mole H}}{1.008 \text{ g H}} = \textbf{12.5 moles H}$

Simplest whole number ratio $= N = \dfrac{6.24}{6.24} = \textbf{1.0}$

$$H = \dfrac{12.5}{6.24} = \textbf{2.0}$$

The **empirical formula** is $\textbf{NH}_2$

To find the molecular formula,

$$\text{mass of empirical formula} = (NH_2) = 16.026 \text{ g}$$
$$\text{mass of compound} = 32.05 \text{ g}$$

$$\text{units of empirical formula in compound} = \frac{32.05 \text{ g}}{16.026 \text{ g}} = \textbf{2.00 units.}$$

Therefore, **the molecular formula is $N_2H_4$** (hydrazine).

**(A note on Precision:  Empirical formula determinations are not among the most precise experiments in chemistry.  You may expect variations of up to about 5% in your values.  Be flexible!)

---

## Example 3.5 B - Determination of the Empirical Formula of a Compound

Determine the empirical and molecular formulas for a deadly nerve gas that gives the following mass percent analysis:

$$C = 39.10\% \qquad H = 7.67\% \qquad O = 26.11\% \qquad P = 16.82\% \qquad F = 10.30\%$$

The molar mass is known to be 184.1 grams.

## Solution

We will proceed as we did with the previous example that is, by assuming a molar mass of exactly 100 grams (**for purposes of determining the empirical formula only!**)

$$C = 39.10 \text{ g} \qquad H = 7.67 \text{ g} \qquad O = 26.11 \text{ g} \qquad P = 16.82 \text{ g} \qquad F = 10.30 \text{ g}$$

$$\text{moles of C} \quad \Big\| \quad 39.10 \text{ g C} \times \frac{1 \text{ mole C}}{12.01 \text{ g C}} = \textbf{3.26 moles C}$$

$$\text{moles of H} \quad \Big\| \quad 7.67 \text{ g H} \times \frac{1 \text{ mole H}}{1.008 \text{ g H}} = \textbf{7.61 moles H}$$

$$\text{moles of O} \quad \Big\| \quad 26.11 \text{ g O} \times \frac{1 \text{ mole O}}{16.00 \text{ g O}} = \textbf{1.63 moles O}$$

$$\text{moles of P} \quad \Big\| \quad 16.82 \text{ g P} \times \frac{1 \text{ mole P}}{30.97 \text{ g P}} = \textbf{0.543 moles P}$$

$$\text{moles of F} \quad \Big\| \quad 10.32 \text{ g F} \times \frac{1 \text{ mole F}}{19.00 \text{ g F}} = \textbf{0.543 moles F}$$

Dividing through by smallest number:

$$C_{\frac{3.26}{0.543}} \quad H_{\frac{7.61}{0.543}} \quad O_{\frac{1.63}{0.543}} \quad P_{\frac{0.543}{0.543}} \quad F_{\frac{0.543}{0.543}}$$

Gives an empirical formula of

$$C_{6.00} \quad H_{14.01} \quad O_{3.00} \quad P_1 \quad F_1 = C_6H_{14}O_3PF$$

The empirical formula mass = **184.09 grams.  Therefore, the molecular formula and the empirical formula are the same.**

## Example 3.5 C - Determining the Empirical Formula from Combustion Data

A combustion device was used to determine the empirical formula of a compound containing **ONLY carbon, hydrogen and oxygen.** A 0.6349 g sample of the unknown produced 1.603 g of $CO_2$ and 0.2810 g of $H_2O$. Determine the empirical formula of the compound.

### Strategy

The primary assumption is that the **combustion is complete.** Therefore, **all** of the carbon was converted to $CO_2$. All of the hydrogen was converted to $H_2O$. Then, if you calculate the **grams of carbon in 1.603 g $CO_2$** and the **grams of hydrogen in 0.2810 g $H_2O$**, you know C and H in your original compound. You can calculate grams of oxygen in the original compound as:

$$\text{grams O} = \text{total grams} - (\text{grams C} + \text{grams H}).$$

### Solution

a.   Molar mass of $CO_2$ = 44.01 grams.

Mass fraction of C in $CO_2$ = $\dfrac{12.01}{44.01}$ = 0.2729  (0.2729 g of every g of $CO_2$ is C)

**g C in 1.603 g $CO_2$** = 0.2729 × 1.603 g = **0.4374 g C** in $CO_2$ (and therefore in our original compound).

b.   Molar mass of $H_2O$ = 18.016 grams

Mass fraction of H in $H_2O$ = $\dfrac{2.016}{18.016}$ = 0.1119  (0.1119 g of every g of $H_2O$ is H)

**g H in 0.2810 g $H_2O$** = 0.1119 × 0.2810 g = **0.0314 g H** in $H_2O$ (and therefore in our original compound).

c.   grams O in original compound

= 0.6349 g - (0.4374 g C + 0.0314 g H) = **0.1661 g O** in original compound.

d.   Converting each to moles,

**moles C**  $\left\|\right.$  0.4374 g C × $\dfrac{1 \text{ mole C}}{12.01 \text{ g C}}$ = **0.0364 mole C**

**moles H**  $\left\|\right.$  0.0314 g H × $\dfrac{1 \text{ mole H}}{1.008 \text{ g H}}$ = **0.0312 mole H**

**moles O**  $\left\|\right.$  0.1661 g O × $\dfrac{1 \text{ mole O}}{16.00 \text{ g O}}$ = **0.0104 mole O**

e.   dividing by smallest number

$C_{\frac{0.0364}{0.0104}}$      $H_{\frac{0.0312}{0.0104}}$      $O_{\frac{0.0104}{0.0104}}$      $= C_{3.5}H_3O_1$

**The empirical formula** (simplest WHOLE number ratio) = $C_7H_6O_2$. Two step-by-step methods for obtaining the molecular formula of a compound are summarized at the end of Section 3.5 of your textbook.

Problems 34-38 at the end of this study chapter cover material from this section

## 3.6 - Chemical Equations

When you finish this section you will be able to:

- State the meaning of each part of a chemical reaction.
- State a variety of different relationships that are inferred from a chemical equation.

Chemical equations are all of the form

Reactants --------> Products

Chemical equations describe chemical changes that occur in a reaction. The physical states that can be present (with their symbols) in a reaction are described in your textbook. The general information that can be gotten from chemical reactions is given in Table 3.2 in your textbook.

## Example 3.6 A - Interpreting a Chemical Equation

State, completely, what is happening in the reaction given below. Include in your answer the physical states of each component of the reaction.

$$Na_2CO_3(aq) + 2HCl(aq) -----> CO_2(g) + H_2O(l) + 2NaCl(aq)$$

## Solution

In this reaction, one mole of aqueous sodium carbonate is reacting with two moles of aqueous hydrochloric acid to produce one mole of gaseous carbon dioxide, one mole of water, and two moles of aqueous sodium chloride.

In this problem you will note that the equation is **balanced.** That is, there the same number of atoms of each element on each side of the equation.

## Example 3.6 B - Relating Reactions and Products of a Chemical Equation

Show, by means of molar masses, that matter is neither created nor destroyed in the equation given below.

$$C_2H_6O(l) + 3O_2(g) -----> 2CO_2(g) + 3H_2O(g)$$

## Strategy

The law of conservation of matter says that matter is neither created nor destroyed in a chemical reaction. Therefore,

**number of grams of reactant = number of grams of product**

Remember to take the **number of moles** of each compound into account when doing your calculations!

## Solution

| Compound | Molar Mass | Number of moles in equation | Total grams in equation |
|---|---|---|---|
| $C_2H_6O$ | 46.068 g | 1 | 46.068 g |
| $O_2$ | 32.00 g | 3 | 96.00 g |
| $CO_2$ | 44.01 g | 2 | 88.02 g |
| $H_2O$ | 18.016 g | 3 | 54.048 g |

So, the **total grams of reactant** = 46.068 + 96.00 = **142.07 g**
the **total grams of product** = 88.02 + 54.048 = **142.07 g**
**grams of reactant = grams of product.**

This relationship must hold in every chemical reaction.

---

Problems 39-41 at the end of this study chapter cover material from this section.

---

## 3.7 - Balancing Chemical Equations

When you finish this section you will be able to balance many chemical equations.

**A balanced chemical equation means that the number of atoms of each element on the reactants side equal the number of atoms of each element on the product side.**

---

## Example 3.7 A - Balancing Chemical Equations

Balance the following chemical equation

$$NaOH(aq) + H_3PO_4(aq) \text{ -----> } Na_3PO_4(aq) + H_2O(l)$$

## Strategy

Chemists have many different strategies for balancing chemical equations. There is no one best way. There is an excellent strategy given just before Exercise 3.14 in your textbook.

An alternate strategy is:

1.  Never change a molecular structure. Only use coefficients.
2.  Find the atoms that are in only one compound on the reactant side. Balance those first.
3.  In general, leave oxygen and hydrogen until the very end. They usually appear many times, and balancing other atoms will often force O and H to become balanced.
4.  Always double check **AFTER** balancing to make sure that the atoms of each element are equal on the reactant and product side.

## Solution

1.  Multiply **NaOH by 3** to balance sodium atoms, as there are 3 sodium atoms in $Na_3PO_4$.

$$3NaOH + H_3PO_4 \text{-----}> Na_3PO_4 + H_2O$$

Totals:

|  |  |
|---|---|
| 3 Na atoms | 3 Na atoms |
| 6 H atoms | 2 H atoms |
| 1 P atom | 1 P atom |
| 7 O atoms | 5 O atoms |

2.  Phosphorus is already balanced.  There are 6 hydrogen atoms on the left side and 2 on the right.  Multiply $H_2O$ by 3 to balance hydrogen.

$$3NaOH + H_3PO_4 \text{-----}> Na_3PO_4 + 3H_2O$$

Totals:

|  |  |
|---|---|
| 3 Na atoms | 3 Na atoms |
| 6 H atoms | 6 H atoms |
| 1 P atom | 1 P atom |
| 7 O atoms | 7 O atoms |

**We are now balanced.**


## Does the Answer Make Sense?

Yes, because the law of conservation of mass holds here.  The number of atoms of each element on the left side of the reaction equals the number of atoms of each element on the right side.

---

# Example 3.7 B - Practice with Balancing

Balance the following combustion reaction:

$$C_4H_{10}(g) + O_2(g) \text{-----}> CO_2(g) + H_2O(g)$$

## Strategy

Combustion represents a very common class of reactions.  We will proceed as before, balancing carbon first and saving hydrogen and oxygen for later.  See if you can keep a running tally of atoms as we proceed.

## Solution

1.  Multiply $CO_2$ by **4** to balance carbons.
2.  Multiply $H_2O$ by **5** to balance hydrogens.

    This leaves us with 4 carbons and 10 hydrogens on both sides.  We have **2 oxygens on the left side** and **13 oxygens on the right.**

3.  Multiply $O_2$ by **13/2** to balance oxygens.

Our balanced equation becomes:

$$C_4H_{10}(g) + 13/2 O_2(g) \text{-----}> 4CO_2(g) + 5H_2O(g)$$

we have:

$$4C, 10H, 13O \text{ -----> } 4C, 10H, 13O.$$

Fractional coefficients **are** used from time to time in chemistry. For clarity, however, we prefer to use simple **whole number ratios.** Therefore:

4. Multiply both sides of the reaction by 2. This converts the 13/2 to 13, a whole number. The number of atoms of each element on both sides of the equation are **still equal**, because we have multiplied both sides by 2.

**Final Answer:**

$$2C_4H_{10}(g) + 13O_2(g) \text{ -----> } 8CO_2(g) + 10H_2O(g)$$

---

Problems 42-45 at the end of this study chapter cover material from this section.

---

## 3.8 - Stoichiometric Calculations

When you finish this section you will be able to:

- Calculate the amount of products obtained from a given amount of reactant.
- Calculate the amount of reactants required to generate a desired amount of product.

The most important reason for learning to balance chemical equations is (as in section 3.8) to establish reactant-product mole (and therefore mass) relationships. For instance, in **example 3.7 B**, we found that:

**2 moles of $C_4H_{10}$ + 13 moles of $O_2$ -----> 8 moles of $CO_2$ + 10 moles of $H_2O$**

In terms of grams of $O_2$ to grams of $CO_2$ relationship,

$$13 \text{ moles } O_2 \times \frac{32.00 \text{ g } O_2}{1 \text{ mole } O_2} = \textbf{416 g } O_2 \text{ reacts with two moles of } C_4H_{10} \text{ to yield 10 moles of } H_2O$$

$$\text{and } 8 \text{ moles } CO_2 \times \frac{44.01 \text{ g } CO_2}{1 \text{ mole } CO_2} = \textbf{352 g } CO_2.$$

The **key relationship** in problems such as these is the **moles of reactant to moles of product** conversion factor. Note That

$$\frac{\textbf{13 moles } O_2}{\textbf{8 moles } CO_2} = \frac{\textbf{8 moles } CO_2}{\textbf{13 moles } O_2}$$

---

## Example 3.8 A - Amount of Product from a Given Amount of Reactant

Given the following reaction:

$$Na_2S(aq) + AgNO_3(aq) \text{ -----> } Ag_2S(s) + NaNO_3(aq)$$

How many grams of $Ag_2S$ can be generated from the reaction of 3.94 g of $AgNO_3$ with excess $Na_2S$?

## Strategy

1. **Always balance the chemical equation!**
2. The **mole ratio conversion factor** acts as the **bridge** between $AgNO_3$ and $Ag_2S$. This leads to the following strategy:

$$g\ AgNO_3 \xrightarrow[\text{mass}]{\text{molar}} moles\ AgNO_3 \xrightarrow[\text{ratio}]{\text{mole}} moles\ Ag_2S \xrightarrow[\text{mass}]{\text{molar}} g\ Ag_2S$$

It is recommended that you solve such problems with one set of equations, rather than splitting the calculation up into separate steps. Many calculations may lead to confusion and errors.

## Solution

The balanced chemical equation is:

$$Na_2S(aq) + 2AgNO_3(aq) \longrightarrow Ag_2S(s) + 2NaNO_3(aq).$$

Then, use one equation:

$$g Ag_2S \quad \bigg\| \quad 3.94\ g\ AgNO_3 \times \frac{1\ mole\ AgNO_3}{169.88\ g\ AgNO_3} \times \frac{1\ mole\ Ag_2S}{2\ mole\ AgNO_3} \times \frac{247.8\ g\ Ag_2S}{1\ mole\ Ag} = 2.87\ g\ Ag_2S$$

## Does the Answer Make Sense?

Although the answer looks reasonable, the best way to check your answer is by double checking your use of units - if they cancel properly, than your answer is likely to be correct.

---

# Example 3.8 B - Amount of Reactant Needed to Produce a Product

Aspirin (acetylsalicylic acid) is prepared by the reaction of salicylic acid ($C_7H_6O_3$) and acetic anhydride ($C_4H_6O_3$). How many grams of salicylic acid (sal) are needed to make 500 aspirin tablets weighing 1.00 g each (assuming 100% yield)?

$$C_7H_6O_3 + C_4H_6O_3 \longrightarrow C_9H_8O_4 + HC_2H_3O_2$$

| salicylic | acetic | aspirin | acetic |
| acid | anhydride | | acid |

## Strategy

We will work this problem in basically the same fashion as we did the previous example. The only difference is that we are going from **product** to **reactant**.

**number of aspirin tablets** $\xrightarrow{\text{mass per tablet}}$ grams of aspirin $\xrightarrow{\text{molar mass}}$ moles of aspirin $\xrightarrow{\text{mole ratio}}$

(start)

**grams of salicylic acid** $\xleftarrow{\text{molar mass}}$ moles of salicylic acid $\longleftarrow$

(finish)

## Solution

$$\text{g salicylic acid} \left|\left| \; 500 \text{ aspirin tablets} \times \frac{1 \text{ gram aspirin}}{1 \text{ aspirin tablet}} \times \frac{1 \text{ mole aspirin}}{180.15 \text{ g aspirin}} \times \frac{1 \text{ mole Sal}}{1 \text{ mole aspirin}} \times \frac{138.12 \text{ g Sal}}{1 \text{ mole Sal}} \right.\right.$$

$$= 383 \text{ g salicylic acid}$$

*Remember that the most important relationships in all of chemistry are between moles of reactants and moles of products.*

## Problems 46-48 at the end of this study chapter cover material from this section.

## 3.9 - Calculations Involving a Limiting Reactant

When you finish this section, you will be able to:

- Identify the limiting reactant.
- Solve problems involving a limiting reactant.
- Calculate percent yield of a product.

In this previous section, we were given the amount of a reactant and asked to find the amount of product formed. The **assumption** then was that **we had as much of the other reactants as we required to make the reactant go completely.** This will not always be the case. For example, given the following reaction:

$$2H_2(g) + O_2(g) \text{-----} > 2H_2O(g),$$

we need **2 moles of $H_2$** to completely react with **1 mole of $O_2$**, forming **2 moles of $H_2O$.** If we only have **1.5 moles of $H_2$** only **0.75 moles of $O_2$** will react, forming **1.5 moles of $H_2O$.** That means that (1 mole $O_2$ - 0.75 mole $O_2$) **0.25 moles of $O_2$** will be left over, or **in excess.**

$H_2$ is said to be the **limiting reactant** in this reaction, because it **limits the amount of product that can form.**

## Example 3.9 A - Calculations Involving a Limiting Reactant

Sodium hydroxide reacts with phosphoric acid to give sodium phosphate and water. If 17.82 g of NaOH is mixed with 15.40 g of $H_3PO_4$,

a.   How many grams of $Na_3PO_4$ can be formed?
b.   How many grams of the excess reactant remains unreacted?
c.   If the <u>actual yield</u> of $Na_3PO_4$ was 15.00 g, what is the percent yield of $Na_3PO_4$?

## Strategy

We must first write a balanced chemical equation for this reaction:

$$3NaOH(aq) + H_3PO_4(aq) \text{-----} > Na_3PO_4(aq) + 3H_2O(l)$$

a. The basic question to be addressed in limiting reactant problems is **"which compound limits the amount of product formed?"** It is therefore recommended that you calculate the amount of product formed by **each** reactant. **The reactant that forms less product is limiting**, and that total amount of product will be formed:

$$\text{g of reactant} \xrightarrow{\text{molar mass}} \text{moles of reactant} \xrightarrow{\text{mole ratio}} \text{moles of product} \xrightarrow{\text{molar mass}} \text{g of product.}$$

b. To determine grams of excess reactant, you must calculate **how many moles of excess reactant were actually used.**

**# moles excess = # moles original - # moles used.**

Then convert moles to grams.

$$15.40g\ H_3PO_4 \times \frac{1\ mole\ H_3PO_4}{98g\ H_3PO_4} \times \frac{1\ mole\ Na_3PO_4}{1\ mole\ H_3PO_4} \times \frac{163.94}{1\ mole\ Na_3PO_4}$$

c. The percent yield = $\dfrac{\text{actual yield}}{\text{theoretical yield}} \times 100\%$.

## Solution

a. Determination of limiting reactant

$$\text{g } Na_3PO_4 \text{ from NaOH} \;\bigg|\bigg|\; 17.80 \text{ g NaOH} \times \frac{1 \text{ mole NaOH}}{40.00 \text{ g NaOH}} \times \frac{1 \text{ mole } Na_3PO_4}{3 \text{ moles NaOH}} \times \frac{163.94 \text{ g } Na_3PO_4}{1 \text{ mole } Na_3PO_4}$$

$$= \textbf{24.31 g } \mathbf{Na_3PO_4} \textbf{ from NaOH}$$

$$\text{g } Na_3PO_4 \text{ from } H_3PO_4 \;\bigg|\bigg|\; 15.40 \text{ g } H_3PO_4 \times \frac{1 \text{ mole } H_3PO_4}{98.00 \text{ g } H_3PO_4} \times \frac{1 \text{ mole } Na_3PO_4}{1 \text{ mole } H_3PO_4} \times \frac{163.94 \text{ g } Na_3PO_4}{1 \text{ mole } Na_3PO_4}$$

$$= \textbf{25.76 g } \mathbf{Na_3PO_4} \textbf{ from } \mathbf{H_3PO_4}$$

Therefore, **NaOH is the limiting reactant** and **24.31 g $Na_3PO_4$** are formed.

b. $H_3PO_4$ is in excess. If 24.31 $Na_3PO_4$ is formed, the **moles of $H_3PO_4$ used** is given by:

$$\text{moles } H_3PO_4 \text{ used} \;\bigg|\bigg|\; 24.31 \text{ g } Na_3PO_4 \times \frac{1 \text{ mole } Na_3PO_4}{163.94 \text{ g } Na_3PO_4} \times \frac{1 \text{ mole } H_3PO_4}{1 \text{ mole } Na_3PO_4}$$

$$= \textbf{0.1483 moles } \mathbf{H_3PO_4} \textbf{ used}$$

The number of **moles of $H_3PO_4$ originally present** is:

$$17.82g \times \frac{1\ mole\ NaOH}{40g\ NaOH} \times \frac{1\ mole\ Na_3PO_4}{3\ mole\ NaOH} \times \frac{163.94g\ Na_3PO_4}{1\ mole\ Na_3PO_4}$$

$$15.40 \text{ g NaOH} \times \frac{1 \text{ mole } H_3PO_4}{98.00 \text{ g } H_3PO_4} = \textbf{0.1571 moles } \mathbf{H_3PO_4} \textbf{ originally present}$$

moles $H_3PO_4$ excess = moles $H_3PO_4$ originally present - moles $H_3PO_4$ used
$$= 0.1571 \text{ moles} - 0.1483 \text{ moles}$$
$$= \textbf{0.0088 moles } \mathbf{H_3PO_4} \textbf{ excess}$$

$$\text{grams } H_3PO_4 \text{ excess} \;\bigg|\bigg|\; 0.0088 \text{ moles } H_3PO_4 \times \frac{98.00 \text{ g } H_3PO_4}{1 \text{ mole } H_3PO_4}$$

$$= \textbf{0.8624 g } \mathbf{H_3PO_4} \textbf{ excess}$$

c. percent yield $= \dfrac{\text{actual yield}}{\text{theoretical yield}} \times 100\% = \dfrac{15.00 \text{ g}}{24.31 \text{ g}} \times 100\% = \textbf{61.70\% yield}$

$\frac{17.82 \quad 15.04}{3\ NaOH + H_3PO_4 \longrightarrow Na_3PO_4 + 3H_2O}$

$14.53g\ H_3PO_4$

**Remember that the single most important relationships in chemistry are among moles of reactants and moles of products!**

---

Problems 49-59 at the end of this study chapter cover material from this section.

---

**END OF CHAPTER THREE**

# SELF-TEST QUESTIONS

1. An element "E" is present as $^{10}E$ with a mass value of 10.01 amu, and as $^{11}E$ with a mass value of 11.01 amu. The natural abundances of $^{10}E$ and $^{11}E$ are 19.78% and 80.22% respectively. What is the average atomic mass of the element? What is the element?

2. Naturally occurring sulfur consists of four isotopes, $^{32}S$ (95.0%), $^{33}S$ (0.76%), $^{34}S$ (4.22%), and $^{36}S$ (0.014%). Using these data, calculate the atomic weight of naturally occurring sulfur. The masses of the isotopes are given in the table below.

| Isotope | Atomic mass (amu's) |
|---------|---------------------|
| $^{32}S$ | 31.97 × 95% |
| $^{33}S$ | 32.97 × .76% |
| $^{34}S$ | 33.97 × 4.22% |
| $^{36}S$ | 35.97 × .014% |

3. An unknown sample of mystery element "T" is injected into the mass spectrometer. According to the mass spectrum, 7.42% of the element is present as $^{6}T$ and 92.58% is present as $^{7}T$. The mass value for $^{6}T$ is 6.02 amu and 7.02 amu for $^{7}T$. Calculate the average atomic mass and identify the mystery element.

4. A noble gas consists of three isotopes of masses 19.99 amu, 20.99 amu, and 21.99 amu. The relative abundance of these isotopes is 90.92%, 0.257%, and 8.82% respectively. What is the average atomic mass of this noble gas? What noble gas is this?

5. Chlorine has two stable isotopes. The mass of one isotope is 34.97 amu. It's relative abundance is 75.53%. What is the mass of the other stable isotope?

6. Complete the following table of isotopic information for the element neon (Ne).

| Isotope | Mass (amu) | Abundance |
|---------|------------|-----------|
| $^{20}Ne$ | 19.99 | _____ |
| $^{21}Ne$ | 20.99 | 0.257% |
| $^{22}Ne$ | 21.99 | _____ |

7. Silicon has three stable isotopes in nature as shown in the table below. Fill in the missing information.

| Isotope | Mass (amu) | Abundance |
|---------|------------|-----------|
| $^{28}Si$ | 27.98 | _____ |
| $^{29}Si$ | _____ | 4.70% |
| $^{32}Si$ | 29.97 | 3.09% |

8. Gallium has two stable isotopes of masses 68.93 amu ($^{69}Ga$) and 70.92 amu ($^{71}Ga$). What are the relative abundances of the two isotopes?

9. Magnesium exists as three isotopes in nature. One isotope ($^{25}Mg$) has a mass of 24.99 amu and a relative abundance of 10.13%. The other two isotopes have masses of 23.99 amu ($^{24}Mg$) and 25.98 amu ($^{26}Mg$). What are their relative abundances? (atomic mass Mg = 24.305 amu).

10. How many moles are in a sample of 300 atoms of Nitrogen (N)? How many grams?

11. How many atoms of gold (Au) does it take to make 1 gram of gold?

12. If you buy 38.9 moles of M&M's®, how many M&M's® do you have? (1 mole of M&M's® = $6.022 \times 10^{23}$ M&M's®)

13. A sample of sulfur (S) has a mass of 5.37 g. How many moles are in the sample? How many atoms?

14. Give the number of moles of each element present in 1.0 mol of each of the following substances:

    a. $Hg_2I_2$                    d. $Ba_3(AsO_4)_2$
    b. LiH                          e. $RbOH \cdot 2H_2O$
    c. $PbCO_3$                     f. $H_2SiF_6$

15. How many grams of zinc are in $1.16 \times 10^{22}$ atoms of zinc (Zn)?

16. How many amu are in 3.68 moles of iron (Fe)?

17. Calculate the molar masses of each of the following:

    a. $Cu_2SO_4$                   d. $Zr(SeO_3)_2$
    b. $NH_4OH$                     e. $Ca_2Fe(CN)_6 \cdot 12H_2O$
    c. $C_{10}H_{16}O$              f. $Cr_4(P_2O_7)_3$

18. What is the mass of $4.28 \times 10^{22}$ molecules of water ($H_2O$)?

19. How many milligrams of $Br_2$ are in $4.8 \times 10^{20}$ molecules of $Br_2$?

20. Determine the molar mass of $KAl(SO_4)_2 \cdot 12H_2O$.

21. How many moles of cadmium bromide ($CdBr_2$) are in a 39.25 g sample?

22. A sample of calcium chloride ($CaCl_2$) has a mass of 23.8 g. How many moles of calcium chloride is this?

23. If 1.72 moles of baking soda ($NaHCO_3$) were used to bake a chocolate cherry cake, how many grams of baking soda would the recipe call for?

24. How many moles are there in a sample of barium sulfate ($BaSO_4$) weighting $9.90 \times 10^7$ g?

25. How many grams are there in 0.36 moles of cobalt (III) acetate ($Co(C_2H_3O_2)_3$)? How many grams of cobalt are in this sample? How many atoms of cobalt?

26. How many milligrams of chlorine are there in a sample of $3.9 \times 10^{19}$ molecules of chlorine gas ($Cl_2$)? How many atoms of chlorine?

27. Bauxite, the principle ore used in the production of aluminum cans, has a molecular formula of $Al_2O_3 \cdot 2 H_2O$.

    a. Determine the molar mass of bauxite.
    b. How many grams of Al are in 0.58 moles of bauxite?
    c. How many atoms of Al are in 0.58 moles of bauxite?
    d. What is the mass in grams of $2.1 \times 10^{24}$ molecules of bauxite?

28. Calculate the mass percent of Cl in each of the following compounds:

    a. ClF
    b. $HClO_2$
    c. $CuCl_2$
    d. PuOCl

29. Calculate the mass percent of each element in $C_5H_{10}O$.

30. Calculate the mass percent of each element in barium sulfite ($BaSO_3$).

31. Calculate the mass percent of each element in natural lucite ($KAlSi_2O_6$).

32. Calculate the mass percent of silver (Ag) in each of the following compounds:

    a.  AgCl            b.  AgCN            c.  AgNO$_3$

33. Chlorophyll **a** is essential for photosynthesis. It contains 2.72% magnesium (Mg) by mass. What is the molar mass of chlorophyll **a** assuming there is one atom of magnesium in every molecule of chlorophyll **a**?

34. Which of the following formulas can be empirical?

    a.  CH$_4$            f.  NH$_4$Cl
    b.  CH$_2$            g.  Sb$_2$S$_3$
    c.  KMnO$_4$          h.  N$_2$O$_4$
    d.  N$_2$O$_5$        i.  CH$_2$O
    e.  B$_2$H$_6$

35. Determine the empirical and molecular formulas of a compound that has a mass of 31.04 g/mole and contains the following percentages of elements by mass:

    C = 38.66%,  H = 16.24%,  N = 45.10%

36. A compound is found, by mass spectral analysis, to contain the following percentages of elements by mass:

    C = 49.67%,  Cl = 48.92%,  H = 1.39%

    The molar mass of the compound is 289.9 g/mole. Determine the empirical and molecular formulas of the compound.

37. Vanillin, the pleasant smelling ingredient used to bake chocolate chip cookies (my favorite), is often used in the production of vanilla extract. Vanillin has a mass of 152.08 g/mol and contains the following percentages of elements by mass:

    C = 63.18%, H = 5.26%, O = 31.56%

    Determine the empirical and molecular formula of vanillin.

38. Determine the empirical formula of a compound that contains the following percentages of elements by mass:

    Mo = 43.95%, O = 7.33%, Cl = 48.72%

39. How many grams of product are formed in each of the following reactions?

    a.  Two moles of H$_2$ react with one mole of O$_2$.
    b.  One mole of silver nitrate reacts with one mole of sodium chloride.
    c.  Three moles of sodium hydroxide react with one mole of phosphoric acid.

40. Calculate the mass percent of each of the elements in Nicotine (C$_{10}$H$_{14}$N$_2$).

41. The following reaction was performed:

    Fe$_2$O$_3$(s) + 2X(s) ----> 2Fe(s) + X$_2$O$_3$(s)

    It was found that 79.847 g of Fe$_2$O$_3$ reacted with "X" to form 55.847 g of Fe and 50.982 g of X$_2$O$_3$. Identify element X.

42. Fill in the blanks to balance the following chemical equations:

   a. __AgI + __Na$_2$S ----> __Ag$_2$S + __NaI
   b. __(NH$_4$)$_2$Cr$_2$O$_7$ ----> __Cr$_2$O$_3$ + __N$_2$ + __H$_2$O
   c. __Na$_3$PO$_4$ + __HCl ----> __NaCl + __H$_3$PO$_4$
   d. __TiCl$_4$ + __H$_2$O ----> __TiO$_2$ + __HCl
   e. __Ba$_3$N$_2$ + __H$_2$O ----> __Ba(OH)$_2$ + NH$_3$
   f. __HNO$_2$ ----> __HNO$_3$ + __NO + __H$_2$O

43. Balance the following equation:

$$NH_4OH(l) + KAl(SO_4)_2 \cdot 12H_2O \rightarrow Al(OH)_3(s) + (NH_4)_2SO_4(aq) + KOH(aq) + H_2O(l)$$

44. Balance the following equation:

$$Fe(s) + HC_2H_3O_2(aq) \rightarrow Fe(C_2H_3O_2)_3(aq) + H_2(g)$$

45. Complete the following reactions:  (Make sure they are balanced.)

   a. HNO$_3$     +                →  H$_2$O     + KNO$_3$
   b.              + Na$_3$PO$_4$  →  Ca$_3$(PO$_4$)$_2$ + Na Cl
   c. Mg(OH)$_2$ + HCl        →  MgCl$_2$    +
   d.              + Cl$_2$        →  NaCl      + Br$_2$

46. How many grams of water vapor can be generated from the combustion of 18.74 g of ethanol?

$$C_2H_6O(g) + O_2(g) ----> CO_2(g) + H_2O(g) \text{ (unbalanced)}$$

47. How many grams of sodium hydroxide are required to form 51.63 g of lead hydroxide?

$$Pb(NO_3)_2(aq) + NaOH(aq) ----> Pb(OH)_2(s) + NaNO_3(aq) \text{ (unbalanced)}$$

48. How many grams of potassium iodide are necessary to completely react with 20.61 g of mercury (II) chloride?

$$HgCl_2(aq) + KI(aq) ----> HgI_2(s) + KCl(aq) \text{ (unbalanced)}$$

49. What mass of calcium carbonate (CaCO$_3$) would be formed if 248.6 g of carbon dioxide (CO$_2$) were exhaled into limewater, Ca(OH)$_2$?  How many grams of calcium would be needed to form that amount of calcium carbonate?  Assume 100% yield in each reaction.

50. The following reaction is used to form lead iodide crystals.  What mass of crystal (PbI$_2$) could be formed from $1.0 \times 10^3$ g of lead (II) acetate [Pb(C$_2$H$_3$O$_2$)$_2$]?

$$Pb(C_2H_3O_2)_2(aq) + 2KI(aq) \rightarrow PbI_2(s) + 2K(C_2H_3O_2)_2(aq)$$

51. How many grams of precipitate (Hg$_2$Cl$_2$) would be formed in the following reaction if you were given 102.9 g of mercury to begin with?

$$2Hg^+(aq) + 2Cl^-(aq) \rightarrow Hg_2Cl_2(s)$$

52. You were hired by a laboratory to recycle 6 moles of silver ions.  You were given 150 g of copper (Cu).  How many grams of silver (Ag) can you recover?  Is this enough copper to recycle 6 moles of silver ions?

$$2Ag^+ + Cu \rightarrow 2Ag + Cu^{2+}$$

53. Fermentation converts sugar into ethanol and carbon dioxide. If you were to ferment a bushel of apples containing 235 g of sugar, what is the maximum amount of ethanol in grams that would be produced?

$$C_6H_{12}O_6 \rightarrow 2C_2H_6O + 2CO_2$$

54. The reaction between potassium chlorate and red phosphorus is highly exothermic and takes place when you strike a match on a matchbox. If you were to react 52.9 g of potassium chlorate ($KClO_3$) with red phosphorus, how many grams of tetraphosphorus decaoxide ($P_4O_{10}$) would be produced?

$$KClO_3(s) + P_4(s) \rightarrow P_4O_{10}(s) + KCl(s) \ \underline{(unbalanced)}$$

55. A reaction combines 113.484 g of lead (II) nitrate with 45.010 g of sodium hydroxide (see problem 47).

    a.  How much lead (II) hydroxide is formed?
    b.  Which reactant is limiting? Which is in excess?
    c.  How much of the excess reactant is left over?
    d.  If the actual yield of lead (II) hydroxide were 80.02 g, what was the percent yield?

56. A reaction combines 64.81 grams of silver nitrate with 92.67 grams of potassium bromide.

$$AgNO_3(aq) + KBr(aq) ----> AgBr(s) + KNO_3(aq)$$

    a.  How much silver bromide is formed?
    b.  Which reactant is limiting? Which is in excess?
    c.  How much of the excess reactant is left over?
    d.  If the actual yield of silver bromide were 14.77 g, what was the percent yield?

57. A reaction proceeds between 94.6 g of $KClO_3$ and 65.3 g of $P_4$ (see problem #54).

    a.  How much potassium chloride is formed?
    b.  Which reactant is limiting? Which is in excess?
    c.  How much of the excess reactant is left over?
    d.  If the actual yield of potassium chloride were 21.0 g, what was the percent yield?

58. DDT, an insecticide harmful to fish, birds and humans, is produced by the following reaction:

$$2C_6H_5Cl + C_2HOCl_3 \rightarrow C_{14}H_9Cl_5 + H_2$$
$$\text{chlorobenzene} \quad \text{chloral} \quad\quad\quad \text{DDT}$$

In a government lab 1142 g of chlorobenzene were reacted with 485 g of chloral.

    a.  How much DDT is formed?
    b.  Which reactant is limiting? Which is in excess?
    c.  How much of the excess reactant is left over?
    d.  If the actual yield of DDT is 2000 g, what was the percent yield?

# ANSWERS TO SELF-TEST QUESTIONS

1.  The average atomic mass is 10.81 amu.  The element is boron.

2.  The molar mass of sulfur is 32.06 amu.

3.  The average atomic mass is 6.95 amu.  The element is lithium.

4.  The average atomic mass is 20.17 amu.  The gas is neon.

5.  The mass of the other isotope is 36.93 amu.

6.  $^{20}$Neon = 90.37%; $^{22}$Neon = 9.37%

7.  The abundance of silicon-28 is 92.21%.  The atomic mass of silicon-29 is 29.01.

8.  gallium-69 = 60.3%;  gallium-71 = 39.7%

9.  magnesium 24 = 79.13%;  magnesium-25 = 10.13%;  magnesium-26 = 10.74%

10. $4.98 \times 10^{-22}$ moles of nitrogen;  mass = $6.98 \times 10^{-21}$ g

11. $3.06 \times 10^{21}$ atoms of gold

12. $2.34 \times 10^{25}$ M & M's$^{©}$

13. 0.168 moles of sulfur;  $1.01 \times 10^{23}$ atoms in the sample

14. a.   2Hg/2I              d.   3Ba/2As/8O
    b.   1Li/1H              e.   1Rb/3O/5H
    c.   1Pb/1C/3O           f.   2H/1Si/6F

15. 1.26 g Zn

16. $1.23 \times 10^{26}$ amu of iron

17. a.   223.1 g/mol      c.   152.1 g/mol      e.   508.3 g/mol
    b.   35.0 g/mol       d.   345.1 g/mol      f.   729.8 g/mol

18. mass = 1.28 g

19. 127 mg $Br_2$

20. 474.2 g/mole

21. 0.144 moles of cadmium bromide

22. This is 0.214 moles of calcium chloride.

23. 144.5 g of $NaHCO_3$

24. $4.24 \times 10^{-4}$ moles

25. 85 g of cobalt (III) acetate.  There are 21 g of cobalt.  There are $2.2 \times 10^{23}$ atoms.

26. There are 4.6 mg of chlorine.  There are $7.8 \times 10^{19}$ atoms of chlorine.

27. a.  bauxite = 137.96 g/mole   c.  $6.99 \times 10^{23}$ atoms of Al
    b.  31.3 g Al                 d.  $4.81 \times 10^2$ g of bauxite

28. a.  65.1% Cl          c.  52.77% Cl
    b.  51.79% Cl         d.  12.0% Cl

29. Mass percent of C = 69.76%, H = 11.62%, O = 18.60%

30. mass percent of Ba = 63.2%, S = 14.7%, O = 22.1%

31. mass percent of K = 17.9%, Al = 12.4%, Si = 25.7%, O = 44.0%

32. a.  Ag = 75.3%          b.  Ag = 80.6%          c.  Ag = 63.5%

33. chlorophyll = 894 g/mol

34. Empirical formulas can be: a, b, c, d, f, g, i

35. Empirical formula = molecular formula = $CNH_5$

36. empirical formula = $C_3HCl$;  molecular formula = $C_{12}H_4Cl_4$

37. Empirical formula = molecular formula = $C_8H_8O_3$

38. Empirical formula = $Mo_3O_3Cl$

39. a.  36 grams
    b.  228 g of product, 143 g AgCl
    c.  165 grams of sodium phosphate, 54 grams of water

40. Percent composition of C = 74.1%, H = 8.64%, N = 17.3%

41. X = aluminum

42. a.  2, 1, 1, 2          c.  1, 3, 3, 1          e.  1, 6, 3, 2
    b.  1, 1, 1, 4          d.  1, 2, 1, 4          f.  3, 1, 2, 1

43. $4NH_4OH(l) + KAl(SO_4)_2 \cdot 12H_2O(s) \rightarrow Al(OH)_3(s) + 2(NH_4)_2SO_4(aq) + KOH(aq) + 12H_2O(l)$

44. $2Fe(s) + 6HC_2H_3O_2(aq) \rightarrow 2Fe(C_2H_3O_2)_3(aq) + 3H_2(g)$

45. a.  $HNO_3 + KOH \rightarrow H_2O + KNO_3$
    b.  $3CaCl_2 + 2Na_3PO_4 \rightarrow Ca_3(PO_4)_2 + 6NaCl$
    c.  $Mg(OH)_2\ 2HCl \rightarrow MgCl_2 + 2H_2O$

46. 22.0 g of water vapor

47. 17.12 g of NaOH

48. 25.20 g of KI

49. 565.38 g of $CaCO_3$; 226.4 g of $Ca^{2+}$

50. $1.42 \times 10^3$ g of $PbI_2$

51. 121.09 g of $Hg_2Cl_2$

52. 509.3 g of Ag recovered; no, you could recycle only 4.72 moles of $Ag^+$

53. 120.1 g of ethanol

54. 36.76 g of $P_4O_{10}$; (balanced equation: $10KClO_3(s) + 3P_4(s) \rightarrow 3P_4O_{10}(s) + 10KCl(s)$)

55. a.  82.54 g of lead hydroxide is formed
    b.  Lead nitrate is limiting.  Sodium hydroxide is in excess.
    c.  17.59 g of sodium hydroxide is left over.
    d.  The yield was 96.83%.

56. a.  71.62 g of silver bromide is formed.
    b.  Silver nitrate is limiting.  Potassium bromide is in excess.
    c.  47.29 g of potassium bromide is left over.
    d.  The yield was 20.62%.

57. a.  57.6 g of KCl is formed
    b.  Potassium chlorate is limiting.  Red phosphorus is in excess.
    c.  36.54 g of $P_4$ is left over.
    d.  The yield was 36.5%.

58. a.  1165 g of DDT is formed.
    b.  Chlorobenzene is limiting.  Chloral is in excess.
    c.  403 g of chloral is left over.
    d.  The yield was 17.2%.

# CHAPTER 4

# Types of Chemical Reactions and Solution Stoichiometry

| Section | Contents | Page in this Study Guide |
|---|---|---|
| 4.1 | Water, the Common Solvent | 75 |
| 4.2 | Strong and Weak Electrolytes | 77 |
| 4.3* | The Composition of Solutions - Dilution | 78 |
| 4.5 | Precipitation Reactions | 85 |
| 4.6 | Describing Reactions in Solution | 86 |
| 4.7 | Stochiometry of Precipitation Reactions | 87 |
| 4.8 | Acid - Base Reactions | 89 |
| 4.9 | Oxidation - Reduction Reactions | 92 |
| 4.10 | Balancing Oxidation - Reduction Equations | 96 |
| 4.11 | Simple Oxidation - Reduction Titrations | 99 |

In this review chapter, we discuss, from both a quantitative and qualitative standpoint, what happens in the preparation of aqueous solutions.

## 4.1 - Water, the Common Solvent

When you finish this section you will be able to:

- State why water acts as a common solvent.
- Draw the structure of water, including partial charges.
- Write equations for the dissolution of some ionic salts in water.

Your textbook introduces you to the nature of the interactions between atoms and electrons in the water molecule. Pay special attention to the following ideas:

a. **Water is not a linear molecule.** It is bent at an angle of about 105°. (This will be justified in chapter 8.)

b. Electrons are not evenly distributed around the atoms in water. Look carefully at <u>Figure 4.2 in your textbook</u>. Notice the position of the **partial charges** in the molecule. This is why the molecule is **polar**. The "+" and "-" charges are not distributed symmetrically.

c. Like dissolves like. The following classes of substances, in general, are miscible:

- polar and ionic
- polar and polar
- nonpolar and nonpolar.

---

*Section 4.4 in your textbook serves as an introduction to later sections. We will not review it in this study guide.

Ionic salts dissolve in water. Water is polar. Compounds that contain **only** carbon and hydrogen are nonpolar. Given that information, try the following exercise.

## Example 4.1 A - Will the Substances Mix?

Predict whether each pair of substances is miscible. State why or why not.

a. $NaNO_3$ and $H_2O$
b. $C_6H_{14}$ and $H_2O$
c. $I_2$ and $C_6H_{14}$
d. $I_2$ and $H_2O$

## Solution

a. Miscible. Sodium nitrate is ionic and water is polar.
b. Immiscible. Cyclohexane ($C_6H_{14}$), is nonpolar, while water is polar.
c. Miscible. The iodine molecule is composed of two identical atoms. Therefore, the electrons are distributed symmetrically. Iodine is therefore almost completely nonpolar. Cyclohexane is also nonpolar.
d. Immiscible. The iodine molecule is nonpolar, and water is polar.

The dissociation of simple ionic salts in water are often written as shown in the following equations:

$$NaI(s) \xrightarrow{H_2O(l)} \underset{\text{cation}}{Na^+(aq)} + \underset{\text{anion}}{I^-(aq)}$$

$$K_2Cr_2O_7(s) \xrightarrow{H_2O(l)} \underset{\text{cation}}{2K^+(aq)} + \underset{\text{anion}}{Cr_2O_7^{2-}(aq)}$$

$$Ba(OH)_2(s) \xrightarrow{H_2O(l)} \underset{\text{cation}}{Ba^{2+}(aq)} + \underset{\text{anion}}{2OH^-(aq)}$$

## Example 4.1 B - Practice with Equations

Complete each of the following dissociation equations:

a.    $CaCl_2(s) \xrightarrow{H_2O(l)}$

b.    $Fe(NO_3)_3(s) \xrightarrow{H_2O(l)}$

c.    $KBr(s) \xrightarrow{H_2O(l)}$

d.    $(NH_4)_2Cr_2O_7(s) \xrightarrow{H_2O(l)}$

## Helpful Hint

Recall your nomenclature rules, and the oxidation states that go along with common anions and cations.

## Solution

a.  $CaCl_2(s) \xrightarrow{\text{H}_2\text{O(l)}} Ca^{2+}(aq) + 2Cl^-(aq)$

b.  $Fe(NO_3)_3(s) \xrightarrow{\text{H}_2\text{O(l)}} Fe^{3+}(aq) + 3NO_3^-(aq)$

c.  $KBr(s) \xrightarrow{\text{H}_2\text{O(l)}} K^+(aq) + Br^-(aq)$

d.  $(NH_4)_2Cr_2O_7(s) \xrightarrow{\text{H}_2\text{O(l)}} 2NH_4^+(aq) + Cr_2O_7^{2-}(aq)$

---

Problems 1-3 at the end of this study chapter cover material from this section.

---

## 4.2 - Strong and Weak Electrolytes

When you finish this section, you will be able to classify many substances as strong, weak or nonelectrolytes.

As you read the textbook material in this section, pay particular attention to the **terms and definitions** given. You will use these for as long as you are involved with the study of chemistry.

Let's review, and add to, the definitions of the key terms of this section.

- **solute:**   1.  If it and the solvent are present in the *same phase*, it is the one in *lesser amount*.
            2.  If it and the solute are present in *different phases*, it is the one that *changes phase*.
            3.  Your textbook puts it in more general terms by saying that it is the one that *dissolves into the solvent*.
- **solvent:**   1.  If it and the solute are present in the *same phase*, it is the one in *greater amount*.
            2.  If it and the solute are present in *different phases*, it is the one that *retains its phase*.
            3.  Your textbook says that it is the one *into which the solute dissolves*.

An **aqueous solution** means that **water is the solvent.**

Notice Figure 4.4 in your textbook.  It shows the affect of strong, weak and nonelectrolytes on the ability to pass a current (conductivity) in an aqueous solution.

| Electrolyte | Conductivity | Degree of Dissociation | Examples |
|---|---|---|---|
| strong | high | total | strong acids, such as HCl; many salts such as NaCl and $Sr(NO_3)_3$; strong bases, such as NaOH, $Ba(OH)_2$, and other group I and II hydroxides. |
| weak | low to moderate | partial | weak organic acids, such as $HCO_2H$ and $HC_2H_3O_2$; weak bases, such as $NH_3$ |
| non | none | close to zero | sugar, AgCl, $Fe_2O_3$ |

## Example 4.2 - Strong, Weak, or Nonelectrolyte

List whether each of the following is a <u>strong</u>, <u>weak</u>, or <u>nonelectrolyte</u>.

    a.  $HClO_4$         d.  $NH_3$

    b.  $C_6H_{12}$        e.  $CaCl_2$

    c.  $LiOH$          f.  $HC_2H_3O_2$

## <u>Solution</u>

    a.  strong  ($HClO_4$ is a strong acid)

    b.  non  (cyclohexane contains only carbon and hydrogen, and is therefore <u>not soluble</u> in water!!)

    c.  strong

    d.  weak

    e.  strong

    f.  weak (acetic acid)

---

Problem 4 at the end of this study chapter covers material from this section.

---

## 4.3 - The Composition of Solutions - Dilution

When you finish this section you will be able to:

- Determine the molarity of a solution.
- Calculate the molarity of each ion in a solution.
- Determine the mass and/or volume of reagents necessary to prepare a solution of a given molarity.
- Solve problems related to dilution.
- Solve problems involving parts per million.

This section is very important to your understanding of chemistry, because we deal here with the *preparation of solutions*. The definition of **molarity** is in your textbook, and it says that **molarity (M) is defined as moles of solute per liter of solution.**

$$M = \frac{\textbf{moles of solute}}{\textbf{liter of solution}}.$$

Keep in mind that $\dfrac{\text{moles}}{\text{liter}} = \dfrac{\text{millimoles}}{\text{milliliter}} = \dfrac{\text{micromoles}}{\text{microliter}}.$

But $\dfrac{\text{moles}}{\text{liter}}$ **DOES NOT EQUAL** $\dfrac{\text{millimoles}}{\text{liter}}$ **OR** $\dfrac{\text{moles}}{\text{microliter}}.$

Be very careful with your units!!!

## Example 4.3 A - Calculating Molarity

Calculate the molarity of a solution prepared by dissolving 11.85 g of solid $KMnO_4$ in enough water to make 750 mL of solution.

### Strategy

Remember that our units are **moles per liter**. Therefore, you must first convert *grams of $KMnO_4$* to *moles of $KMnO_4$*, then divide by the volume. (Remember to convert volume from mL to L!)

### Solution

The molar mass of $KMnO_4$ = 158.04 g/mole.

### Method A

In this method, we can proceed in two steps:

1.  Convert from grams of $KMnO_4$ to moles of $KMnO_4$.

$$\text{moles } KMnO_4 \; \left\| \; \frac{1 \text{ mol } KMnO_4}{158.04 \text{ g } KMnO_4} \times 11.85 \text{ g } KMnO_4 = 0.07498 \text{ mol } KMnO_4 \right.$$

2.  Divide moles by volume (in liters!) to get molarity.

$$M = \frac{\text{mol}}{L} = \frac{0.07498 \text{ mol } KMnO_4}{0.750 \text{ L solution}} = 0.09997 \text{ M } KMnO_4 = \textbf{0.100 M } KMnO_4$$

### Method B

We can use dimensional analysis to solve the problem in one longer step:

$$\frac{\text{mol } KMnO_4}{L \text{ solution}} \; \left\| \; \frac{11.85g \text{ } KMnO_4}{0.750 \text{ L soln}} \times \frac{1 \text{ mol } KMnO_4}{158.04 \text{ g } KMnO_4} = \textbf{0.100 M } KMnO_4 \right.$$

We will often use dimensional analysis in this study guide. However, it is never the ONLY correct method. Let's change the problem around a bit. Instead of using the mass of the solute to calculate the solution molarity, let's begin with **molarity** to calculate **mass**.

## Example 4.3 B - Mass from Molarity

Calculate the mass of NaCl needed to prepare 175 mL of a 0.500 M NaCl solution.

### Strategy

From a *dimensional analysis* point of view, the question we have to answer is, "how can we go from molarity (mol/L) to mass (g)?" Note that this means that somewhere along the line, we will have to *cancel volume*.

### Solution

$$\text{g NaCl} \; \left\| \; \frac{0.500 \text{ mol NaCl}}{L \text{ soln}} \times \frac{58.44 \text{ g NaCl}}{1 \text{ mol NaCl}} \times 0.175 \text{ L soln} = \textbf{5.11 g NaCl} \right.$$

$$\uparrow \qquad\qquad\qquad \uparrow$$
$$\text{convert to grams} \qquad \text{cancel volume}$$

Our values for molarity and volume had 3 significant figures, so we rounded off accordingly.

Whether you want **mass, molarity,** or **volume,** the essential problem solving strategy is the same - you can use dimensional analysis to get your units in the proper place. Try the next problem.

## Example 4.3 C - Volume from Molarity

How many mL of solution are necessary if we are to have a 2.48 M NaOH solution that contains 31.52 g of the dissolved solid?

### Strategy

The units must cancel to give mL.  Starting off by inverting molarity will put volume in the numerator.  Remember to convert L to mL!

### Solution

$$\text{mL solution} \parallel \frac{1.00 \text{ L soln}}{2.48 \text{ mol NaOH}} \times \frac{1000 \text{ mL}}{\text{L}} \times \frac{1 \text{ mol NaOH}}{40.00 \text{g NaOH}} \times 31.52 \text{ g NaOH}$$

$$\underset{\substack{\uparrow \\ \text{convert from} \\ \text{L to mL}}}{} \qquad \underset{\substack{\uparrow \\ \text{cancel moles} \\ \text{NaOH}}}{} \qquad \underset{\substack{\uparrow \\ \text{cancel grams} \\ \text{NaOH}}}{}$$

$$= 318 \text{ mL solution}$$

### Does the Answer Make Sense?

The amount of NaOH, 31.52 g, is about 3/4 of a mole.  If the solution is 2.48 M, you have about 1/3 as much NaOH as you need to make 1 L, or enough to make about 300 mL, so the answer seems to make sense.

In the previous problems, we calculated the molarity of the solute.  However, we have neglected to take into account the fact that each of the solutes, $KMnO_4$, NaCl, and NaOH are **strong electrolytes, and completely dissociate** in aqueous solution (recall Section 4.2).  For example,

$$KMnO_4(s) \xrightarrow{\text{H}_2\text{O}(l)} K^+(aq) + MnO_4^-(aq)$$

This means that while it is generally acceptable to discuss your solution concentration as "molarity of $KMnO_4$," it is more chemically correct to discuss "molarity of $K^+$ ions and molarity of $MnO_4^-$ ions."

A solution that is 0.85 M in $KMnO_4$ is really 0.85 M in $K^+$ ion and 0.85 M in $MnO_4^-$ ion, because $KMnO_4$ completely dissociates, and the dissociation is a 1 to 1 to 1 ratio; that is, *one mole of $KMnO_4$ dissociates into one mole of $K^+$ ion and one mole of $MnO_4^-$ ion.*

Keeping the stoichiometry of the dissociation equation in mind, try the next problem.

## Example 4.3 D - Molarity of Ions in Solution

Calculate the molarity of all of the ions in each of the following solutions.

    a.    0.25 M $Ca(OCl)_2$                      b.    2 M $CrCl_3$

## Helpful Hint

Always write out the dissociation equation, so your "ion-to-solute" mole ratio is clear.

## Solution

a.   $Ca(OCl)_2(s) \xrightarrow{\ H_2O(l)\ } Ca^{2+}(aq) + 2OCl^-(aq)$

molarity of $Ca^{2+}$ = molarity of $Ca(OCl)_2$ = **0.25 M.**
molarity of $OCl^-$ = twice the molarity of $Ca(OCl)_2$ = **0.50 M.**

b.   $CrCl_3(s) \xrightarrow{\ H_2O(l)\ } Cr^{3+}(aq) + 3Cl^-(aq)$

molarity of $Cr^{3+}$ = molarity of $CrCl_3$ = **2 M.**
molarity of $Cl^-$ = three times the molarity of $CrCl_3$ = **6 M.**

---

# Example 4.3 E - Molarity of Ions in Solution

Determine the molarity of $Cl^-$ ion in a solution prepared by dissolving 9.82 g of $CuCl_2$ in enough water to make 600 mL of solution.

## Strategy

Let's finish a strategy that we began to discuss in the last problem.  One process to find the ion concentration (molarity of an ion) of a strong electrolyte is to:

1.   Calculate the solute concentration (molarity of the solute).
2.   Determine the ion-to-solute mole ratio by writing the dissociation equation.
3.   Use that mole ratio, along with the solute concentration, to calculate the ion concentration.

## Solution

1.   *Solute concentration:*

$$\frac{mol\ CuCl_2}{L\ soln} \ \left\| \ \frac{1\ mol\ CuCl_2}{134.55\ g\ CuCl_2} \times \frac{9.82\ g\ CuCl_2}{1} \times \frac{1}{0.600\ L\ soln} = \textbf{0.1216 M CuCl}_\textbf{2} \right.$$

We retain all of our figures because this is an *intermediate* calculation.

2.   *Ion to solute ratio:*

$$CuCl_2(s) \xrightarrow{\ H_2O\ (l)\ } Cu^{2+}(aq) + 2Cl^-(aq)$$

Therefore, the ratio $\dfrac{ion}{solute} = \dfrac{2\ moles\ Cl^-}{1\ mole\ CuCl_2}.$

3.   *The final molarity of $Cl^-$ in solution (also expressed as "$[Cl^-]$"):*

$$[Cl^-] = \frac{0.1216\ mol\ CuCl_2}{L\ soln} \times \frac{2\ mol\ Cl^-}{1\ mol\ CuCl_2} = \textbf{0.243 M}$$

We rounded off because we were at the end of the problem.  Three significant figures is correct because both mass and volume were given to three figures.  Note that if we had rounded off earlier, we would have calculated that $[Cl^-]$ = 0.122 M $\times$ 2 = 0.244 M, which is not strictly correct.

**ALTERNATIVELY**, we can use dimensional analysis to solve the entire problem with one equation.

$$\frac{\text{mol } Cl^-}{\text{L soln}} \left\| \underset{\underset{\text{equation}}{\text{from dissociation}}}{\frac{2 \text{ mol } Cl}{1 \text{ mol } CuCl_2}} \times \underset{\text{molar mass}}{\frac{1 \text{ mol } CuCl_2}{134.55 \text{ g } CuCl_2}} \times \underset{\text{given}}{\frac{9.82 \text{ g } CuCl_2}{1}} \times \underset{\text{given}}{\frac{1}{0.600 \text{ L soln}}} = \mathbf{0.243 \, M} = [Cl^-] \right.$$

---

# Example 4.3 F - Practice with Ion Concentration

Determine the molarity of $Fe^{3+}$ ions and $SO_4^{2-}$ ions in a solution prepared by dissolving 48.05 g of $Fe_2(SO_4)_3$ in enough water to make 800 mL of solution.

## Solution

The dissociation equation is:

$$Fe_2(SO_4)_3(s) \xrightarrow{H_2O(l)} 2Fe^{3+}(aq) + 3SO_4^{2-}(aq)$$

So the mole ratio of $\dfrac{\text{ion}}{\text{solute}} = \dfrac{2 \text{ mol } Fe^{3+}}{1 \text{ mol } Fe_2(SO_4)_3}$ and $\dfrac{3 \text{ mol } SO_4^{2-}}{1 \text{ mol } Fe_2(SO_4)_3}$

Let's work with $[Fe^{3+}]$.

$$\frac{\text{mol } Fe^3}{\text{L soln}} \left\| \frac{2 \text{ mol } Fe^{3+}}{1 \text{ mol } Fe_2(SO_4)_3} \times \frac{1 \text{ mol } Fe_2(SO_4)_3}{399.9 \text{ g } Fe_2(SO_4)_3} \times \frac{48.05 \text{ g } Fe_2(SO_4)_3}{1} \times \frac{1}{0.800 \text{ L soln}} = \mathbf{0.300 \, M} = [Fe^{3+}] \right.$$

Looking at the **mole ratios**, can you tell that the ratio of $\dfrac{[SO_4^{2-}]}{[Fe^{3+}]}$ must equal $\dfrac{3 \text{ moles}}{2 \text{ moles}}$ ?

Therefore, since the volume is constant, $[SO_4^{2-}] = \dfrac{3}{2}[Fe^{3+}]$, or $= \dfrac{3}{2}(0.300 \text{ M})$

$$[SO_4^{-2}] = \mathbf{0.450 \, M}$$

---

An important part of your chemistry experience is to be able to *prepare dilute solutions from more concentrated ("stock") solutions*. Your textbook points out that the most important idea in diluting solutions is that

**moles of solute after dilution = moles of solute before dilution**

If molarity = $\dfrac{\text{moles of solute}}{\text{liter of solution}}$ , then

moles of solute = $\dfrac{\text{moles of solute}}{\text{liter of soln}} \times$ liters of soln = M × V.

If the **moles of solute remain identical** before and after dilution (only the amount of water changes), then

$$M_i V_i = M_f V_f$$

where $M_i$ = molarity of concentrated (initial) solution
   $V_i$ = volume of concentrated solution that you add to water to dilute
        . . . this will often be your "unknown volume."
   $M_f$ = molarity of dilute (final) solution
   $V_f$ = total volume of your dilute solution

Try the following introductory example.

---

# Example 4.3 G - Preparation of a Dilute Solution

What volume of 12 M hydrochloric acid must be used to prepare 600 mL of a 0.30 M HCl solution?

## Strategy

Be aware that no matter how much you dilute your acid, **the number of moles of acid in the solution will remain the same.** Your molarity will change, but not the total number of moles. $M_iV_i = M_fV_f$ is your best bet in solving these types of problems.

## Solution

Let   $M_i$ = 12 M HCl         $V_i$ = unknown
      $M_f$ = 0.30 M HCl        $V_f$ = 600 mL (or 0.600 L) the units of molarity cancel, you may use any
                                       volume unit that you want.

$$12 \text{ M} \times V_i = 0.30 \text{ M (600 mL)} = \frac{0.30 \text{ M (600 mL)}}{12 \text{ M}} = V_i = \textbf{15 mL of 12 M HCl}$$

## Double Check

$$\frac{12 \text{ mol}}{\text{L}} \ (0.015 \text{ L}) = \frac{0.30 \text{ mol}}{\text{L}} \ (0.600 \text{ L})$$

$$0.18 \text{ mol} = 0.18 \text{ mol}$$

The number of moles, 0.18 mol, is the same on each side.

---

# Example 4.3 H - More Practice on Preparing Dilute Solutions

What volume of 9.0 M sodium hydroxide must be used to prepare 1.2 L of a 1.0 M NaOH solution?

## Solution

We proceed as before.

Let   $M_i$ = 9.0 M NaOH        $V_i$ = unknown
      $M_f$ = 1.0 M NaOH        $V_f$ = 1.2 L

$$9.0 \text{ M} \times V_i = 1.0 \text{ M (1.2 L)} = \frac{1.0 \text{ M (1.2 L)}}{9.0 \text{ M}} = V_i = \textbf{0.13 L of 9.0 M NaOH}$$

Upon checking, using the <u>non</u>-rounded value of 0.1333 L, the number of moles on each side is identical.

---

Although not discussed in Chapter Four of your text, some of the additional exercises in your text use the term **parts per million**.

$$1 \text{ part per million of "X" (ppm)} = \frac{1 \text{ part X}}{1 \times 10^6 \text{ parts solution}},$$

$$= \frac{1 \text{ g X}}{1 \times 10^6 \text{ g soln}} \times \frac{1 \text{ μg X}}{1 \text{ g soln}}$$

If the solution is water (whose density = 1.00 g/mL),

$$1 \text{ ppm X} = \frac{1 \text{ μg X}}{1 \text{ mL soln}} = \frac{1 \text{ mg X}}{1 \text{ L soln}}$$

Note that this differs from molarity in that the units are *mass per volume*, not moles per volume.

To set the stage for your work with the additional exercises, consider the following problem.

## Example 4.3 I - Parts Per Million

An aqueous solution with a total volume of 750 mL contains 14.38 mg of $Cu^{2+}$. What is the concentration of $Cu^{2+}$ in parts per million?

### Helpful Hint

The use of ppm DOES NOT necessarily require you to calculate moles on your way to your answer. The dimensional analysis unit of choice is **mg/L**.

### Solution

$$\frac{14.38 \text{ mg } Cu^{2+}}{0.750 \text{ L soln}} = \textbf{19.2 ppm } Cu^{2+}$$

The next problem requires that you deal with moles.

## Example 4.3 J - Molarity to Parts Per  Million

A solution is $3 \times 10^{-7}$ M in manganese (VII) ion. What is the $Mn^{7+}$ concentration in ppm?

### Solution

You want $\dfrac{\text{mg } Mn^{7+}}{\text{L soln}}$, and you have $\dfrac{\text{moles } Mn^{7+}}{\text{L soln}}$. Dimensional analysis is a good approach.

$$\frac{\text{mg } Mn^{7+}}{\text{L soln}} \;\|\; \frac{3 \times 10^{-7} \text{ mol } Mn^{7+}}{\text{L soln}} \times \frac{54.94 \text{ g } Mn^{7+}}{1 \text{ mol } Mn^{7+}} \times \frac{1000 \text{ mg } Mn^{7+}}{1 \text{ g } Mn^{7+}}$$

$$= \frac{0.0164 \text{ mg } Mn^{7+}}{\text{L soln}}, \text{ or } \textbf{0.02 ppm } Mn^{7+} \textbf{ (rounded off)}$$

Problems 5-20 at the end of this study chapter cover material from this section.

## 4.5 - Precipitation Reactions

When you finish this section you will be able to predict the products of many reactions that occur in solution.

The major ideas in this section are that:

1. Many salts dissociate into ions in aqueous solution.
2. If a solid forms from a combination of selected ions in solution, the solid must contain an anion part and cation part, and the net charge on the solid must be zero.
3. There are some simple solubility rules you can use that can help you predict the products of reactions in aqueous solutions.

Table 4.1 in your textbook lists solubility rules for salts in water. The information is important enough to reprint below:

Simple Rules for the Solubility of Salts in Water

1. Most nitrate ($NO_3^-$) salts are soluble.
2. Most salts containing the alkali metal ions ($Li^+$, $Na^+$, $K^+$, $Cs^+$, $Rb^+$) and the ammonium ion ($NH_4^+$) are soluble.
3. Most chloride, bromide, and iodide, salts are soluble. Notable exceptions are salts containing the ions $Ag^+$, $Pb^{2+}$, and $Hg_2^{2+}$.
4. Most sulfate salts are soluble. Notable exceptions are $BaSO_4$, $HgSO_4$, $PbSO_4$, and $CaSO_4$.
5. Most hydroxide salts are only slightly soluble. The important soluble hydroxides are $NaOH$ and $KOH$. The compounds $Ba(OH)_2$, $Sr(OH)_2$, and $Ca(OH)_2$ are marginally soluble.
6. Most sulfide ($S^{2-}$), carbonate ($CO_3^{2-}$), chromate ($CrO_4^{2-}$), and phosphate ($PO_4^{3-}$) salts are only slightly soluble.

If two soluble substances (call them AX and BZ), are combined, you can assume that the products will be AZ and BX.

$$AX(aq) + BZ(aq) ----> AZ + BX$$

Your goal is to determine, based on your knowledge of solubility rules, whether AZ or BX will form a solid (precipitate).

Let's look at the following reaction.

$$AgNO_3(aq) + Na_2S(aq) ----> \text{products.}$$

The reactants are electrolytes that will dissociate to form the ions

$$Ag^+(aq) + NO_3^-(aq) + 2Na^+(aq) + S^{2-}(aq).$$

If a solid forms, and if it is to have zero charge, it can be either **$Ag_2S$** or **$NaNO_3$**. According to solubility rule #1, $NaNO_3$ is soluble. According to solubility rule #6, $Ag_2S$ is insoluble, and will therefore precipitate. The correct (balanced) reaction, therefore, is

$$2AgNO_3(aq) + Na_2S(aq) ----> Ag_2S(s) + 2 NaNO_3(aq)$$

## Example 4.5 - Predicting Precipitates

Complete and balance the following reactions, determining, in each case, if a precipitate is formed.

    a.   $KCl(aq) + Pb(NO_3)_2(aq) \longrightarrow$
    b.   $AgNO_3(aq) + MgBr_2(aq) \longrightarrow$
    c.   $Ca(OH)_2(aq) + FeCl_3(aq) \longrightarrow$
    d.   $NaOH(aq) + HCl(aq) \longrightarrow$

## Solution

    a.   $2KCl(aq) + Pb(NO_3)_2(aq) \longrightarrow \mathbf{PbCl_2(s)} + 2KNO_3(aq)$
    b.   $2AgNO_3(aq) + MgBr_2(aq) \longrightarrow \mathbf{2AgBr(s)} + Mg(NO_3)_2(aq)$
    c.   $3Ca(OH)_2(aq) + 2FeCl_3(aq) \longrightarrow \mathbf{2Fe(OH)_3(s)} + 3CaCl_2(aq)$
    d.   $NaOH(aq) + HCl(aq) \longrightarrow H_2O(l) + NaCl(aq)$
                                   ↑
                        water is the
                 solvent, hence "(l)"

Problems 21-22 at the end of this study chapter cover material from this section.

## 4.6 - Describing Reactions In Solution

When you finish this chapter you will be able to write molecular, ionic and net ionic equations to describe reactions in solution.

This section describes the three kinds of equations that are used to describe reactions in aqueous solution. Specific definitions for the **molecular**, **complete ionic** and **net ionic equations** are given in your textbook. Let's look at how the aqueous reaction of <u>silver nitrate</u> with <u>sodium sulfide</u> can be expressed with each type of equation.

    a.  **Molecular:** This gives the overall reaction. While it does give information on stoichiometry it gives no information on whether or not compounds really exist as ions in solution. Molecular form:

$$2AgNO_3(aq) + Na_2S(aq) \longrightarrow Ag_2S(s) + 2NaNO_3(aq)$$

    b.  **Complete Ionic:** This gives the equation including all ions in solution. Because **all** compounds and ions are present, some information may be redundant. Complete ionic form:

$$2Ag^+(aq) + 2NO_3^-(aq) + 2Na^+(aq) + S^{2-}(aq) \longrightarrow Ag_2S(s) + 2Na^+(aq) + 2NO_3^-(aq)$$

    c.  **Net Ionic:** This only gives information on those species that undergo a chemical change. Ions that appear in the same form on both sides of the complete ionic equation are called **spectator ions**, and are not included in the net ionic equation. In our sample equation, $Na^+$ and $NO_3^-$ are unaltered during the reactions. They are therefore omitted from the net ionic equation. Net ionic form:

$$2Ag^+(aq) + S^{2-}(aq) \longrightarrow Ag_2S(s)$$

The **molecular** and **net ionic** forms of equations are the most commonly used. The **complete ionic** form helps us to determine the net ionic form.

# Example 4.6 - Molecular, Complete Ionic and Ionic Equations

Write the molecular, complete ionic and net ionic forms for each of the following equations.

a.  Aqueous nickel (II) chloride reacts with aqueous sodium hydroxide to give a nickel (II) hydroxide precipitate and aqueous sodium chloride.

b.  Solid potassium metal reacts with water to give aqueous potassium hydroxide and hydrogen gas.

c.  Aqueous sodium hydroxide reacts with aqueous phosphoric acid to give water and aqueous sodium phosphate.

## Keep in Mind

Remember to <u>balance</u> the **molecular** equation before you write the **complete and net ionic** forms.

## Solution

a.  **molecular:** $NiCl_2(aq) + 2NaOH(aq) \longrightarrow Ni(OH)_2(s) + 2NaCl(aq)$

   **complete ionic:** $Ni^{2+}(aq) + 2Cl^-(aq) + 2Na^+(aq) + OH^-(aq) \longrightarrow Ni(OH)_2(s) + 2Na^+(aq) + 2Cl^-(aq)$

   **net ionic:** $Ni^{2+}(aq) + 2OH^-(aq) \longrightarrow Ni(OH)_2(s)$

b.  **molecular:** $2K(s) + 2H_2O(l) \longrightarrow 2KOH(aq) + H_2(g)$

   **complete ionic:** $2K(s) + 2H_2O(l) \longrightarrow 2K^+(aq) + 2OH^-(aq) + H_2(g)$

   **net ionic:** $2K(s) + 2H_2O(l) \longrightarrow 2K^+(aq) + 2OH^-(aq) + H_2(g)$

Note that in <u>part b</u>, **every** reactant undergoes some chemical change.  Therefore, the complete and net ionic equations are the same.

c.  **molecular:** $3NaOH(aq) + H_3PO_4(aq) \longrightarrow 3H_2O(l) + Na_3PO_4(aq)$

   **complete ionic:** $3Na^+(aq) + 3OH^-(aq) + 3H^+(aq) + PO_4^{3-}(aq) \longrightarrow 3H_2O(l) + 3Na^+(aq) + PO_4^{3-}(aq)$

   **net ionic:** $3OH^-(aq) + 3H^+(aq) \longrightarrow 3H_2O(l)$

---

Problem 23 at the end of this study chapter covers material from this section.

---

# 4.7 - Quantitative Aspects of Precipatation Reactions

When you finish this section you will:

- have your first exposure to "railroad" problems in chemistry, and
- be able to solve a variety of problems involving the formation of precipitates.

The material in this section is very important because it combines most of the previous chemical ideas that you have learned up until now.  Solving problems involving precipitates from solution makes use of *molarity, solubility rules, balancing equations and limiting reactant calculations.*

The problems can be quite "wordy," but wordy problems are not necessarily hard.  They are just wordy.  Each sentence contains some chemical information, and it is useful to jot down information as it appears.  Keep in mind that although no two examples are exactly alike, the approach to solving the problems will be similar.

Your textbook suggests **SIX STEPS** to solving solution problems.  Let's use these ideas in the following example.

---

## Example 4.7 A - An Introduction to Problems Based on Precipitation Reactions

Calculate the mass of $Ag_2S$ produced when 125 mL of 0.200 M $AgNO_3$ is added to excess $Na_2S$ solution ("excess" $Na_2S$ means that you have more than you need - $AgNO_3$ is the limiting reactant.)

### Strategy and Solution

Before we do any calculations, we first have to determine what is happening in the solution.  This is best done by writing a **balanced molecular equation.**

The problem says that $AgNO_3$(aq) is being added to $Na_2S$(aq).  According to our solubility rules (section 4.5), insoluble silver sulfide will be formed.

$$2AgNO_3(s) + Na_2S(aq) \longrightarrow Ag_2S(s) + 2NaNO_3(aq)$$

The critical point here is that the *mole ratio* of $Ag_2S$ to $AgNO_3$ is

$$\frac{1 \text{ mol } Ag_2S}{2 \text{ mol } AgNO_3}$$

Now that we know what is going on in solution, we need to know **how many moles of reactant ($AgNO_3$)** we have.  You learned how to do such calculations in <u>Section 4.3 of your textbook and this study guide</u>.

$$\text{moles } AgNO_3 \quad \left\| \quad \frac{0.200 \text{ mol } AgNO_3}{\text{L solution}} \times 0.125 \text{ L solution} = \textbf{0.0250 mol } AgNO_3 \right.$$

**How many moles of product** ($Ag_2S$) can be formed from this many moles of reactant?  Given the mole ratio of 2:1 that we determined earlier,

$$\text{moles } Ag_2S \quad \left\| \quad 0.0250 \text{ mol } AgNO_3 \times \frac{1 \text{ mol } Ag_2S}{2 \text{ mol } AgNO_3} = \textbf{0.0125 mol } Ag_2S \textbf{ formed} \right.$$

To finish up, we need to convert **moles of $Ag_2S$ to grams of $Ag_2S$.**

$$\text{g } Ag_2S \quad \left\| \quad 0.0125 \text{ mol } Ag_2S \times \frac{247.9 \text{ g } Ag_2S}{1 \text{ mol } Ag_2S} = \textbf{3.10 g } Ag_2S \textbf{ formed} \right.$$

Note that early on, once we wrote down the correct equation, we could have solved the problem with one long dimensional analysis equation:

$$\text{g } Ag_2S \quad \left\| \quad \frac{247.9 \text{ g } Ag_2S}{1 \text{ mol } Ag_2S} \times \frac{1 \text{ mol } Ag_2S}{2 \text{ mol } AgNO_3} \times \frac{0.200 \text{ mol } AgNO_3}{1 \text{ L solution}} \times 0.125 \text{ L solution} \right.$$

$$= \textbf{3.10 g } Ag_2S \textbf{ formed}$$

However you choose to solve such problems, a stepwise approach is helpful.

## Example 4.7 B. - Practice with Precipitation Problems

What mass of $Fe(OH)_3$ is produced when 35 mL of a 0.250 M $Fe(NO_3)_3$ solution is mixed with 55 mL of a 0.180 M KOH solution?

### Helpful Hint

Remember your systematic procedure for solving precipitation problems.  The complication here is that by combining different amounts of reactant, you have a **limiting reactant** problem.  Also, remember to **properly balance your chemical equation!**

### Solution

The reaction of interest is:

$$Fe(NO_3)_3(aq) + 3KOH(aq) ----> Fe(OH)_3(s) + 3KNO_3(aq)$$

Using a stepwise approach,

$$\text{moles } Fe(NO_3)_3 \text{ before reaction} \quad \left\| \quad \frac{0.250 \text{ mol } Fe(NO_3)_3}{\text{L solution}} \times 0.035 \text{ L soln} = \textbf{0.00875 moles } Fe(NO_3)_3 \right.$$

$$\text{moles KOH before reaction} \quad \left\| \quad \frac{0.180 \text{ mol KOH}}{\text{L solution}} \times 0.055 \text{ L soln} = \textbf{0.00990 moles KOH} \right.$$

According to the balanced equation, 0.00875 moles of **$Fe(NO_3)_3$** can yield **0.00875 moles of $Fe(OH)_3$** (1:1 mole ratio), and 0.00990 moles of **KOH** can yield **0.00330 moles of $Fe(OH)_3$** (3:1 mole ratio).

Therefore, KOH is the <u>limiting reactant</u> (see section 3.9) and **0.00330 moles of $Fe(OH)_3$(s) will be produced**.
Converting from moles of $Fe(OH)_3$ to grams of $Fe(OH)_3$ (F.W. = 106.9 g/mol) gives **0.35 g $Fe(OH)_3$ produced**.

---

Problems 24-31 at the end of this study chapter cover material from this section.

---

## 4.8 - Acid-Base Reactions

When you finish this chapter you will be able to solve a variety of problems related to acid-base neutralizations.

The key to solving acid-base problems is to know that they **require the same strategy** as most of the other types of problems in this chapter.  **Writing down a balanced chemical equation is always your first, and most important, step.**

### AN ACID IS A PROTON DONOR.

### A BASE IS A PROTON ACCEPTOR.

<table>
<tr><td align="center"><u>Some STRONG ACIDS</u></td><td align="center"><u>Some STRONG BASES</u></td></tr>
</table>

| <u>Some STRONG ACIDS</u> | <u>Some STRONG BASES</u> |
|:---:|:---:|
| HCl | NaOH |
| $HNO_3$ | KOH |
| $H_2SO_4$ | $NH_2^-$ |
| $HClO_4$ | |

You may assume that the acid-base reactions used in this section go to completion.

The steps for solving acid-base problems are given just before <u>Sample Exercise 4.14 in your textbook</u>. Let's solve the following problem together to show how these steps are used.

---

# Example 4.8 A - Neutralization of a Strong Acid

How many mL of a 0.800 M NaOH solution is needed to just neutralize 40.00 mL of a 0.600 M HCl solution?

## Looking at the Terms

You are beginning to realize that terms such as **molarity** are used over and over again in chemistry. **Neutralization** is often used in acid-base chemistry. Neutralization of an acid implies stoichiometric addition (ie; "just enough") of a strong base, so that no acid remains. The analogous definition applies to the neutralization of a base by a strong acid.

## Solution

Let's go step-by-step.

1.   *List the species present in solution before reaction.*

$$H^+, Cl^- \qquad Na^+, OH^- \quad \text{and } H_2O$$
$$\text{from HCl(aq)} \qquad \text{from NaOH(aq)}$$

2.   *Write the balanced net ionic reaction.*

$$H^+(aq) + OH^-(aq) \text{----> } H_2O(l)$$

3.   *Find out the number of moles of acid we need to neutralize.*

$$\text{moles } H^+ \left\|\ \frac{0.600 \text{ moles } H^+}{1 \text{ L soln}} \times 0.04000 \text{ L soln} = \textbf{0.0240 moles } H^+ \right.$$
$$\text{change mL to L}$$

4.   *Because the stoichiometry of the reaction is 1 mole of base to 1 mole of acid, 0.0240 moles OH⁻ are required to neutralize the acid.*

5.   *Determine the volume of OH⁻ (from NaOH) needed to give that many moles.*

$$\text{L NaOH soln} \left\|\ \frac{1 \text{ L NaOH soln}}{0.800 \text{ moles NaOH}} \times 0.0240 \text{ moles NaOH} = 0.0300 \text{ L soln} = \textbf{300 mL NaOH solution} \right.$$

You should <u>double check</u> to make sure that the **moles of acid** equals the **moles of base**.

---

<u>Sample Exercise 4.15 in your textbook</u> is really a limiting reactant problem, such as we did in Section 3.9 of this study guide and your textbook. Review that material, and remember, when doing limiting reactant problems, **convert everything to moles.**

The remainder of this section deals with **VOLUMETRIC ANALYSIS**. This kind of analysis uses **precisely measured amounts of liquid** to carry out an analysis. There are several new terms introduced such as **titration, buret, equivalence point, indicator,** and **endpoint**. See your textbook to get definitions for each of these terms. When solving volumetric analysis problems, the same chemical rules apply as with most other acid-base problems:

- Write down the reaction.
- Convert to moles, and relate moles of acid to moles of base.
- Don't be frightened by "wordy" railroad problems.
- Ignore superfluous information.

See if you can do the next example on your own.

---

# Example 4.8 B - Acid-Base Titration

You wish to determine the molarity of a solution of sodium hydroxide. To do this, you titrate a 25.00 mL aliquot of your sample, which has had 3 drops of phenolphthalein indicator added so that it is pink, with 0.1067 M HCl. The sample turns clear (indicating that the NaOH(aq) has been precisely neutralized by the HCl solution) after the addition of 42.95 mL of the HCl. Calculate the molarity of your NaOH solution.

## Solution

1. The net ionic reaction is:

$$H^+(aq) + OH^-(aq) \longrightarrow H_2O(l)$$

2. The total number of moles of $H^+$ added to neutralize the NaOH equals:

$$\text{moles } H^+ \parallel \frac{0.1067 \text{ mol H}}{\text{L soln}} \times 0.04295 \text{ L soln} = \mathbf{4.583 \times 10^{-3} \text{ mol } H^+}$$

3. The net ionic reaction tells us that **the number of moles of $H^+$ equals the number of moles of $OH^-$.** Therefore, our $OH^-$ sample of unknown molarity contains **$4.583 \times 10^{-3}$ mol $OH^-$.**

4. We had a 25.00 mL (0.02500 L) aliquot of our NaOH solution. The molarity of the solution is:

$$\frac{\text{mol } OH^-}{\text{L soln}} = \frac{4.583 \times 10^{-3} \text{ mol } OH^-}{0.02500 \text{ L soln}} = \mathbf{0.1833 \text{ M NaOH}}$$

## Does the Answer Make Sense?

You had 25.00 mL of NaOH solution. It took almost twice the volume of 0.1067 M HCl to neutralize. Because you have the **same number of moles in about half the solution volume**, it makes sense that the NaOH solution would be almost twice the concentration.

---

Your textbook does a number of problems that call for you to determine the molar mass of an acid using an acid-base titration. The basic steps (writing the equation, calculating moles, etc.) are the same. Let's try such a problem.

---

## Example 4.8 C - Molar Mass of an Acid

You want to determine the molar mass of an acid. The acid contains one acidic hydrogen per molecule. You weigh out a 2.879 g sample of the pure acid and dissolve it, along with 3 drops of phenolphthalein indicator, in distilled water. You titrate the sample with 0.1704 M NaOH. The pink endpoint is reached after addition of 42.55 mL of the base. Calculate the molar mass of the acid.

### Solution

As always, we write the equation first. Let "HA" be your acid. It has one acidic hydrogen. Therefore, it will react with base in a 1:1 ratio.

$$HA + NaOH ----> H_2O + NaA,$$
$$or$$
$$H^+(aq) + OH^-(aq) ----> H_2O(l)$$

We have the number of grams of our acid, **2.879 g**. In order to calculate the molar mass (g/mol) of HA, we need to know **how many moles** of HA we have.

$$moles\ NaOH = moles\ HA$$

Our task, then, is to find moles of NaOH (present as OH⁻ in solution).

$$moles\ OH^-\ \left|\left|\ \frac{0.1704\ moles\ OH^-}{L\ soln} \times 0.04255\ L\ soln = 7.251 \times 10^{-3}\ mol\ OH^-\right.\right.$$

The reaction stoichiometry is 1 mole of HA per mole of NaOH. Therefore, there are $7.251 \times 10^{-3}$ moles HA in 2.879 g HA.

$$molar\ mass\ of\ the\ acid = \frac{2.879\ g}{7.251 \times 10^{-3}\ mol} = 397.1\ g/mole$$

Problems 32-42 at the end of this study chapter cover material from this section.

## 4.9 - Oxidation-Reduction Reactions

When you finish this section you will be able to:

- Assign **oxidation states** to atoms in a compound.
- Define **oxidation** and **reduction**.

Both this section and Section 4.10 seem, at first reading, to be unrelated to the general topic of solution chemistry. This section deals with definitions and assignments relating to **electron exchange**, and the next deals with balancing such equations. **You need these skills** before you do electrochemical solution problems, because you need to understand how to deal with a problem with electron exchange before you can manipulate values to solve problems.

Your textbook defines **oxidation-reduction reactions** as being those in which **one or more electrons are transferred**.

$$H_2(g) + Cl_2(g) ----> 2HCl(g)$$

In this case, electrons are transferred from the hydrogen to the chlorine.

From the last section, we saw that hydrogen can form $H^+$, or hydrogen with a "+1" oxidation state. It has one more proton than electron. Chlorine is very happy forming $Cl^-$, or chlorine with a "-1" oxidation state. Many elements are happy gaining or losing electrons so that they have a **"noble gas electronic configuration."** (We will learn why in Chapter 7.) Sodium forms $Na^+$, calcium forms $Ca^{2+}$, and sulfur can form $S^{2-}$. All these are common **oxidation states.**

There are a set of 6 rules for **assigning oxidation states** that are given in Table 4.2 of your textbook. You must, for now, memorize these rules. They will become second nature as you do more and more "redox" problems. Use those rules in the following problem.

Remember that the **sum of the oxidation states in a neutral compound must equal zero** and must be equal to the **overall charge** in an **ionic compound.**

---

## Example 4.9 A - Assigning Oxidation States

Assign oxidation states to each of the atoms in the following compounds:

a. $CaF_2$      c. $H_2O$      e. $KMnO_4$
b. $C_2H_6$     d. $ICl_5$     f. $SO_4^{2-}$

## Solution

a.  Fluorine has a greater attraction for electrons than calcium, and it has a charge of **1- (rule #5)**. There-
    fore, calcium must be 2+, so that the overall charge balances at zero. Notice that calcium is in group 2A.
    Metals in groups 1A and 2A always have an oxidation state equal to their group number when in com-
    pounds.

$$CaF_2$$
$$\uparrow \quad \uparrow$$
$$+2 \quad -1$$

b.  Hydrogen is **assigned** an oxidation state of **+1** in covalent compounds (rule #4). Because there are 6 H's,
    the total charge due to H is $6^+$. In order to balance, each of the two carbons must have an oxidation state
    of -3.

$$C_2H_6$$
$$\uparrow \quad \uparrow$$
$$-3 \quad +1$$

c.  **Rules #3 and 4** dictate that each H has a charge of $1^+$ and the oxygen has a charge of $2^-$. The water
    molecule therefore comes out to be electronically neutral.

$$H_2O$$
$$\uparrow \quad \uparrow$$
$$+1 \quad -2$$

d.  In this compound, Cl is the more "electronegative" (has greater attraction for electrons), and therefore
    gets the negative oxidation state, as $Cl^-$. To electrically balance, iodine, which acts as the **cation**, must
    be $I^{5+}$.

$$ICl^{5+}$$
$$\uparrow \quad \uparrow$$
$$+5 \quad -1$$

e.  Although this compound is not totally covalent, oxygen is the **most electronegative** element here, and is therefore assigned an oxidation state of -2. With four oxygens, **the total charge due to oxygen is 8⁻.** Potassium is in group 1A. According to our discussion in "part a" of this example, it must have an **oxidation state of +1**. Mathematically then, manganese must have an **oxidation state of +7** to allow the compound to electrically balance.

$$KMnO_4$$
$$\uparrow \quad \uparrow \quad \uparrow$$
$$+1 \quad +7 \quad -2$$

f.  According to **rule #2**, each oxygen is assigned a -2 oxidation state in this covalent ion, for a total of 8⁻ for four oxygens. According to **rule #6**, the sum of the oxidation states in this ion must add up to -2. Therefore, the sulfur must have an oxidation state of +6.

$$SO_4{}^{2-}$$
$$\uparrow \quad \uparrow$$
$$+6 \quad -2$$

## Additional Notes:

1.  In Sample Exercise 4.18 in your textbook, you will notice that Dr. Zumdahl **checks** the math at the end of each part, to make sure that the number of atoms of a given oxidation state, when added to all other atoms multiplied by their respective oxidation states, all add up to the total charge on the compound or ion. You should follow this practice with this example and other related examples.

2.  Oxidation states can be arbitrary. Don't be frightened by noninteger or awkward states. These things can happen.

---

## Example 4.9 B - Practice with Oxidation States

Assign oxidation states to each atom in the equation (treat each compound separately).

$$Fe_2O_3 + 2Al \longrightarrow Al_2O_3 + 2Fe$$

## Solution

$$Fe_2O_3(s) + 2Al(s) \longrightarrow Al_2O_3(s) + 2Fe(l)$$

| +3 per | 0 (rule #1) | +3 per | 0 per Fe |
| Fe | | Al | (rule #1) |

Let's look at what happened to each element as a result of the reaction.

- Fe went from $Fe^{3+}$ to $Fe°$  (gained 3 electrons)
- Al went from $Al°$ to $Al^{3+}$  (lost 3 electrons)
- O remained the same

THE IRON GAINED ELECTRONS.  IT HAS BEEN REDUCED.

THE ALUMINUM LOST ELECTRONS.  IT HAS BEEN OXIDIZED.

Remember **OIL RIG**; **O**xidation **I**nvolves **L**oss (of electrons)
                              **R**eduction **I**nvolves **G**ain (of electrons)

This next idea is tricky:  Something that is **reduced** is called an **oxidizing agent** (it causes something else to be oxidized).  Something that is **oxidized** is called a **reducing agent** (it causes something else to be reduced).

In the previous example:

- Iron was reduced.
- Aluminum was oxidized.
- Iron (III) oxide was the oxidizing agent.
- Aluminum was the reducing agent.

---

# Example 4.9 C - Which Atoms Undergo Redox?

For each reaction, identify the atoms that undergo reduction, or oxidation. Also, list the oxidizing and reducing agents.

a. $2H_2(g) + O_2(g) ----> 2 H_2O(g)$
b. $Zn(s) + Cu^{2+}(aq) ----> Zn^{2+}(aq) + Cu(s)$
c. $2AgCl(s) + H_2(g) ----> 2H^+(aq) + 2Ag(s) + 2Cl^-(aq)$
d. $2MnO_4^-(aq) + 16H^+(aq) + 5C_2O_4^{2-}(aq) ----> 2Mn^{2+}(aq) + 10CO_2(g) + 8H_2O(l)$

## Helpful Hint

For purposes of assigning oxidation states, treat each compound or ion by itself.

## Solution

a.  $2H_2(g) + O_2(g) ----> 2H_2O(g)$
       0       0     +1 per H; -2 per O

    oxidized: hydrogen (0 to +1)     oxidizing agent: molecular oxygen
    reduced: oxygen (0 to -2)       reducing agent: molecular hydrogen

b.  $Zn(s) + Cu^{2+}(aq) ----> Zn^{2+}(aq) + Cu(s)$
       0     +2        +2      0

    oxidized: zinc (0 to +2)       oxidizing agent: copper
    reduced: copper (+2 to 0)     reducing agent: zinc

c.  $2AgCl(s) + H_2(g) ----> 2H^+(aq) + 2Ag(s) + 2Cl^-(aq)$
      +1  -1     0       +1      0     -1

    oxidized: hydrogen (0 to +1)     oxidizing agent: silver chloride
    reduced: silver (+1 to 0)       reducing agent: molecular hydrogen

d.  $2MnO_4^-(aq) + 16H^+(aq) + 5C_2O_4^{2-}(aq) ----> 2Mn^{2+}(aq) + 10CO_2(g) + 8H_2O(l)$

    oxidized: carbon (+3 to +4)
    reduced: manganese (+7 to +2)
    oxidizing agent: permanganate ion ($MnO_4^-$)
    reducing agent: oxalate ion ($C_2O_4^{2-}$)

---

Problems 43-45 at the end of this study chapter cover material from this section.

---

## 4.10 - Balancing Oxidation-Reduction Equations

When you finish this section you will be able to balance redox equations using the half-reaction method in acidic or basic solutions.

In Section 4.9, you learned how to assign oxidation states to atoms in a compound. You also learned that in redox reactions, <u>oxidation states on some atoms change</u>. We will use this information now to learn how to **balance redox equations using the half-reaction method.**

When you learn about electrochemistry later in the year, you will find that splitting redox reactions into oxidation and reduction halves is very convenient for describing what happens at different electrodes. The **half-reaction method** is therefore very important to master.

The steps involved in balancing redox reactions by this method are given in your textbook. When you work redox problems, always remember that **you can only add substances to your equation that are already in solution** (such as $H_2O$, $H^+$ in acid solution and $OH^-$ in basic solution).

Let's try to balance the following equation together.

## Example 4.10 A - Balancing Redox Equations by the Half-Reaction Method

Balance the following equation **in acid solution** using the half-reaction method.

$$Cu(s) + HNO_3(aq) ----> Cu^{2+}(aq) + NO(g)$$

## Solution

*Step 1: Identify and write equations for the half-reactions.*

$$Cu(s) + HNO_3(aq) ----> Cu^{2+}(aq) + NO(g)$$
$$\quad 0 \qquad +5 \qquad\qquad +2 \qquad +2$$

Copper is being oxidized:  $Cu ----> Cu^{2+}$
Nitrogen is being reduced:  $HNO_3 ----> NO$    ($N^{5+} ----> N^{2+}$)

*Step 2: Balance each half-reaction.*

The oxidation is balanced atomically. We need to add two electrons to the right-hand side to balance electronically.

    i.   (oxidation)  $Cu ----> Cu^{2+} + 2e^-$

    ii.  (reduction)  $HNO_3 ----> NO$

    a.  **Balance all atoms that are neither oxygen nor hydrogen.** (Nitrogens are already balanced).

    b.  **Balance oxygens** by adding water to the side that needs oxygen. (The left-hand side has 3 oxygens, the right-hand side has one, so **2** waters must be added to the right-hand side.)

$$HNO_3 ----> NO + 2H_2O$$

    c.  **Balance hydrogens** by adding $H^+$ to the side that needs hydrogen. (The left-hand side has 1 hydrogen, the right-hand side has 4, so **3** hydrogen ions must be added to the left-hand side.)

$$HNO_3 + 3H^+ ----> NO + 2H_2O$$

The half-reaction is now balanced atomically, but not electronically.

d.  **Balance charges** by adding **electrons** to the side that is more positive.  (The left-hand side has 3 positives, the right-hand side is neutral.  Therefore, we need to add **3** electrons to the left-hand side.)

$$HNO_3 + 3H^+ + 3e^- ----> NO + 2H_2O$$

Both half-reactions are now balanced.

### Step 3:  Equalize electron transfer.

The same number of electrons must be gained as are lost in the reaction.  Therefore, we must multiply each reaction by numbers that will allow both reactions to have **the same** number of electrons exchanged.

With our reactions, the lowest common denominator of electrons is **6**.  Therefore, we would <u>multiply the oxidation by 3</u>, and <u>multiply the reduction by 2</u>.

$$3Cu ----> 3Cu^{2+} + 6e^-$$
$$2HNO_3 + 6H^+ + 6e^- ----> 2NO + 4H_2O$$

### Step 4:  Add the half-reactions and cancel appropriately to get a complete redox reaction.

$$3Cu ----> 3Cu^{2+} + 6e^-$$
$$\underline{2HNO_3 + 6H^+ + 6e^- ----> 2NO + 4H_2O}$$
$$3Cu + 2HNO_3 + 6H^+ + 6e^- ----> 3Cu^{2+} + 2NO + 4H_2O + 6e^-$$

Canceling electrons, we get our final result:

$$3Cu + 2HNO_3 + 6H^+ ----> 3Cu^{2+} + 2NO + 4H_2O$$

## Double Check

Do we have the same number of each kind of atom on both sides?  Yes, there are 3 coppers, 2 nitrogens, 6 oxygens and 8 hydrogens on each side.  Are the charges the same on both sides?  Yes, +6.  The equation is balanced.

---

# Example 4.10 B - Practice with the Half-Reaction Method

Balance the following equation in acidic solution:

$$Cr_2O_7{}^{2-}(aq) + NO(g) ----> Cr^{3+}(aq) + NO_3{}^-(aq)$$

## Solution

Step 1:  $Cr_2O_7{}^{2-} + NO ----> Cr^{3+} + NO_3{}^-$

Nitrogen is being oxidized:  $N^{2+}$ to $N^{5+}$
Chromium is being reduced:  $Cr^{6+}$ to $Cr^{3+}$

Step 2:  $Cr_2O_7{}^{2-} ----> Cr^{3+}$
   $Cr_2O_7{}^{2-} ----> 2Cr^{3+}$
   $Cr_2O_7{}^{2-} ----> 2Cr^{3+} + 7H_2O$
   $Cr_2O_7{}^{2-} + 14H^+ ----> 2Cr^{3+} + 7H_2O$
   $Cr_2O_7{}^{2-} + 14H^+ + 6e^- ----> 2Cr^{3+} + 7H_2O$    (balanced)

$$NO \longrightarrow NO_3^-$$
$$NO + 2H_2O \longrightarrow NO_3^-$$
$$NO + 2H_2O \longrightarrow NO_3^- + 4H^+$$
$$NO + 2H_2O \longrightarrow NO_3^- + 4H^+ + 3e^- \quad \text{(balanced)}$$

Step 3:  $2[\ NO + 2H_2O \longrightarrow NO_3^- + 4H^+ + 3e^-]$

$$2NO + 4H_2O \longrightarrow 2NO_3^- + 8H^+ + 6e^- \quad \text{(oxidation)}$$
$$\underline{Cr_2O_7^{2-} + 14H^+ + 6e^- \longrightarrow 2Cr^{3+} + 7H_2O} \quad \text{(reduction)}$$

$$Cr_2O_7^{2-}(aq) + 2NO(g) + 6H^+(aq) \longrightarrow 2Cr^{3+}(aq) + 2NO_3^-(aq) + 3H_2O(l)$$

## Double Check

There are 2 chromiums, 2 nitrogens, 9 oxygens and 6 hydrogens on each side.  Also, each side has a total charge of +4.  The equation is balanced.

---

## C.  Balancing in Basic Solutions

One of the conditions of balancing equations is that you can only add to equations what is actually in the solution. In **acid** solutions, you have a lot of $H^+$ ions.  In **basic** solutions however, you have a lot of $OH^-$ ions.  **We need to add $OH^-$ ions to balance for hydrogen when balancing a reaction that takes place in basic solution.**

However, adding $OH^-$ to balance hydrogens puts oxygens out of balance.  To get around that,

1.  Balance basic solutions as if they were acidic.

2.  Add a number of $OH^-$ ions to both sides so that you just balance excess $H^+$ ions.

3.  $H^+ + OH^-$ will form $H_2O$ on the side with excess $H^+$.  Free $OH^-$ will appear on one side of the equation.

4.  Double check atoms and charges, as always.

Try this technique on the equation in the previous example.

---

# Example 4.10 C - Balancing Redox Equations in Basic Solution

Balance the following equation (which <u>IS ALREADY</u> balanced in acid) assuming it is now in basic solution.

$$Cr_2O_7^{2-}(aq) + 2NO(g) + 6H^+(aq) \longrightarrow 2Cr^{3+}(aq) + 2NO_3^-(aq) + 3H_2O(l)$$

## Solution

We need to get rid of the excess $H^+$, because $OH^-$ is the dominant acid-base related species.  Therefore, **add $6OH^-$ to both sides.**  (Whatever is done to the left side, must be done to the right, if the equation is already balanced.)

$$Cr_2O_7^{2-} + 2NO + 6H^+ + 6OH^- \longrightarrow 2Cr^{3+} + 2NO_3^- + 3H_2O + 6OH^-$$

The $H^+$ will combine with the $OH^-$, giving $H_2O$.

$$Cr_2O_7^{2-} + 2NO + 6H_2O \longrightarrow 2Cr^{3+} + 2NO_3^- + 3H_2O + 6OH^-$$

The 3 $H_2O$'s on the right-hand side can be cancelled (with 3 of the waters on the left-hand side), which gives the final balanced equation.

$$Cr_2O_7^{2-}(aq) + 2NO(g) + 3H_2O(l) ----> 2Cr^{3+}(aq) + 2NO_3^-(aq) + 6OH^-(aq)$$

---

Problems 46-50 at the end of this study chapter cover material from this section.

---

## 4.11 - Simple Oxidation-Reduction Titrations

When you finish this section you will be able to solve many problems relating to redox titrations.

Solving problems involving redox titrations involves the same general strategy as any other titration problem. That is, you must:

  a.  Balance the redox equation.
  b.  Determine the moles of titrant.
  c.  Use the balanced redox equation to determine the moles of unknown.
  d.  Convert from moles of unknown to grams, percent, molarity, or whatever.

Though specific strategies may vary, the importance of **balancing the redox equation** and converting data to **moles** will always be the most important steps.

The following problem illustrates redox titration problem solving strategy.

---

## Example 4.11 - A Redox Titration

The use of potassium permanganate ($KMnO_4$) as an oxidizing agent is described in your textbook. A 0.0483 M $KMnO_4$ solution was used to titrate a solution containing 0.8329 g of impure calcium oxalate, $CaC_2O_4$. (The ionic form of the equation is given below.) If 30.25 mL of the $KMnO_4$ solution was required to reach the titration endpoint, calculate the **percent purity** of the $CaC_2O_4$.

$$MnO_4^-(aq) + C_2O_4^{2-}(aq) ----> Mn^{2+}(aq) + CO_2(g)    \text{(unbalanced)}$$

### Strategy

You must learn to separate the wheat from the chaff in such problems. List only the important information. Also, draw a "mole map" outlining your strategy.

Balance equation ---->   Determine moles     critical     Determine moles
                         of $MnO_4^-$ used  ------------->  of $C_2O_4^{2-}$ used
                                           relationship                 ¦
                                                                        ¦
            Calculate              Determine grams                      ¦
          % $CaC_2O_4^{2-}$ <----------    of $C_2O_4^{2-}$   <------------
            in sample                in sample

### Solution

The balanced equation is:

$$2MnO_4^-(aq) + 5C_2O_4^{2-}(aq) + 16H^+(aq) ----> 2Mn^{2+}(aq) + 10CO_2(g) + 8H_2O(l)$$

moles of $MnO_4^-$ used $\left|\right|$ $\dfrac{0.0483 \text{ moles } MnO_4^-}{\text{L solution}} \times 0.03025 \text{ L soln} = \mathbf{1.46_1 \times 10^{-3} \text{ moles } MnO_4^- \text{ used}}$

moles of $C_2O_4^{2-}$ used $\left|\right|$ $1.46_1 \times 10^{-3}$ moles of $MnO_4^-$ used $\times \dfrac{5 \text{ moles } C_2O_4^{2-}}{2 \text{ moles } MnO_4^-}$

$$= \mathbf{0.00365_3 \text{ moles of } C_2O_4^{2-} \text{ used}}$$

g of $CaC_2O_4$ in the sample $\left|\right|$ $0.00365_3$ moles $C_2O_4^{2-} \times \dfrac{1 \text{ mol } CaC_2O_4}{1 \text{ mol } C_2O_4^{2-}} \times \dfrac{128.10 \text{ g } CaC_2O_4}{1 \text{ mol } CaC_2O_4}$

$$= \mathbf{0.467_9 \text{ g } CaC_2O_4 \text{ in the sample}}$$

$$\% \text{ } CaC_2O_4 \text{ in the sample} = \dfrac{0.467_9 \text{ g } CaC_2O_4}{0.8329 \text{ g sample}} \times 100\% = \mathbf{56.2\% \text{ } CaC_2O_4}$$

(Note that you should keep one extra (insignificant) figure all the way through.  Drop it only at the very end.)

We could have done most of this problem using one equation, which contains most of our mole map:

g $CaC_2O_4$ $\left|\right|$ $\dfrac{0.0483 \text{ mol } MnO_4^-}{\text{L solution}} \times 0.03025 \text{ L soln} \times \dfrac{5 \text{ mol } C_2O_4^{2-}}{2 \text{ mol } MnO_4^-} \times \dfrac{1 \text{ mol } CaC_2O_4}{1 \text{ mol } C_2O_4^{2-}} \times \dfrac{128.10 \text{ g } CaC_2O_4}{1 \text{ mol } CaC_2O_4}$

$$= \mathbf{0.4679 \text{ g } CaC_2O_4}$$

---

Problems 51-53 at the end of this study chapter cover material from this section.

---

**END OF CHAPTER FOUR**

# SELF-TEST QUESTIONS

1.  Write dissociation equations for the following when they are dissolved in water:

    a.  HF
    b.  $SrBr_2$
    c.  $MgBr_2$

    d.  $NH_4Cl$
    e.  $NaNO_3$
    f.  $Al_2(SO_4)_3$

2.  Write dissociation equations for the following when they are dissolved in water:

    a.  $BaSO_4(s)$
    b.  $KCl(s)$
    c.  $NaOH(s)$

    d.  $CaCrO_4(s)$
    e.  $Mg(OH)_2(s)$
    f.  $HCOOH(l)$

3.  Which of the following pairs of substances are miscible and which are not?  Give a reason:

    a.  $CH_3CH_2OH$ and $H_2O$
    b.  $C_6H_6$ and $C_6H_{12}$

    c.  $C_6H_6$ and $H_2O$
    d.  LiBr and $H_2O$

4.  Classify the following as strong, weak or nonelectrolyte:

    a.  $CH_3CH_2OH$
    b.  $C_{12}H_{22}O_{11}$ (sugar)
    c.  HCl

    d.  $NH_3$
    e.  $C_6H_{12}$

5.  Calculate the molarity of the following solutions:

    a.  49.73 g $H_2SO_4$ in enough water to make 500 mL of solution
    b.  4.739 g $RuCl_3$ in enough water to make 1.00 L of solution
    c.  5.035 g $FeCl_3$ in enough water to make 250 mL of solution
    d.  27.74 g $C_{12}H_{22}O_{11}$ in enough water to make 750 mL of solution
    e.  218.7 g HCl in enough water to make 500 mL of solution

6.  Calculate the molarity of the following solutions:

    a.  18.92 g of $HNO_3$ in enough water to make 500 mL of solution
    b.  5.761 g of KOH in enough water to make 350 mL of solution
    c.  31.44 g of $Fe_2O_3$ in enough water to make 1.0 L of solution
    d.  72.06 g of $BaCl_2$ in enough water to make 800 mL of solution

7.  Calculate the concentrations of each of the ions in the following solutions:

    a.  0.25 M $Na_3PO_4$
    b.  0.15 M $Al_2(SO_4)_3$
    c.  0.87 M $Na_2CO_3$

8.  Calculate the concentrations of each of the ions in the following solutions:

    a.  0.62 M $K_2Cr_2O_7$
    b.  0.35 M NaOH
    c.  0.14 M $Co(NO_3)_2$

    d.  0.07 M $Na_3PO_4$
    e.  0.23 M $(NH_4)_2Cr_2O_7$
    f.  0.49 M $Al_2(SO_3)_3$

9.  Describe how you would prepare the following solutions:

    a.  100 mL of 1 M NaCl.
    b.  250 mL of 1.0 M $Na_2SO_4$.
    c.  1.5 L of 0.5 M $K_2Cr_2O_7$.

10. Calculate the amount of solute you would need to add to make the following solutions:

    a.  400 mL of 0.1 M HCl                     c.  1 L of 1.5 M $KMnO_4$
    b.  750 mL of 0.35 M $BaSO_4$              d.  250 mL of 0.2 M $AgNO_3$

11. Describe how you would prepare the following solutions:

    a.  500 mL of 1.0 M $H_2SO_4$ from 17.8 M $H_2SO_4$
    b.  1.5 L of 0.25 M $KMnO_4$ from 1.0 M stock solution
    c.  1.0 L of 0.15 M $KBrO_3$ from solid $KBrO_3$.
    d.  100 mL of 0.01 M $AgNO_3$ from 0.5 M stock solution
    e.  1 L of 0.5 M $AgNO_3$ from solid $AgNO_3$

12. Describe how you would prepare the following solutions:

    a.  250 mL of 0.1 M HCl from 12.5 M HCl
    b.  500 mL of 1.5 M NaCl from 7.3 M NaCl
    c.  800 mL of 0.2 M $NiCl_2$ from 4.6 M $NiCl_2$
    d.  750 mL of 0.05 M $FeSO_4$ from 0.1 M $FeSO_4$

13. A standard solution of KHP($C_8H_5O_4K$) was made by dissolving 3.697 g of KHP in enough water to make 100.0 mL of solution. Calculate the KHP concentration.

14. How many moles of KHP are contained in 30.00 mL of the solution in problem 8?

15. A solution of ammonium acetate ($NH_4C_2H_3O_2$) was made by dissolving 3.85 g of ammonium acetate in enough water to make 500 mL of solution. Calculate the solute concentration.

16. How many moles of ammonium acetate are contained in 17 mL of the solution in problem 15?

17. How many milliliters of 0.136 M NaOH is required to react with the $H_2SO_4$ in 10 mL of a 0.202 M solution? The reaction of the two is

$$2NaOH + H_2SO_4 \text{----}> Na_2SO_4 + 2H_2O$$

18. How many milliliters of 0.5 M $Ca(OH)_2$ are required to react with the HCl in 30 mL of a 0.12 M solution. The reaction of interest is:  $2HCl + Ca(OH)_2 \text{----}> CaCl_2 + 2H_2O$.

19. What is the concentration of the following in ppm?

    a.  $1.0 \times 10^{-2}$ g $Cu^{2+}$ in 2.0 L of solution.
    b.  concentration of $Pb^{2+}$ in $2.1 \times 10^{-5}$ M $Pb(NO_3)_2$.

20. What is the concentration of the following in ppm?

    a.  $6.2 \times 10^{-3}$ g $Be^{+2}$ in 750 mL of solution
    b.  255 mg $NaIO_3$ in 1.5 L of solution

21. Balance the following reactions:

    a.  $C_3H_8 + O_2 \text{----}> CO_2 + H_2O$          d.  $Ca_3(PO_4)_2 + H_2SO_4 \text{----}> CaSO_4 + H_3PO_4$
    b.  $K_2CO_3 + Al_2Cl_6 \text{----}> KCl + Al_2(CO_3)_3$     e.  $KOH + H_3PO_4 \text{----}> K_3PO_4 + H_2O$
    c.  $Mg_3N_2 + H_2O \text{----}> MgO + NH_3$             f.  $KClO_3 + C_{12}H_{22}O_{11} \text{----}> KCl + CO_2 + H_2O$

22. Complete and balance the following reactions:

    a.   $NaCl(aq) + HgNO_3(aq) \longrightarrow$
    b.   $Ca(OH)_2(aq) + Na_2CO_3(aq) \longrightarrow$
    c.   $Na_2S(aq) + FeCl_3(aq) \longrightarrow$

23. Write molecular, complete ionic and net ionic equations for the following reactions:

    a.   aqueous sodium sulfide reacts with aqueous copper(II) nitrate
    b.   aqueous hydrogen fluoride reacts with aqueous potassium hydroxide to give water and aqueous potassium
         fluoride

24. A solution contains $Ag^+$, $Pb^{+2}$, and $Fe^{+3}$. If you want to precipitate the $Pb^{+2}$ selectively, what anion would
    you choose?

25. What mass of $Mg(OH)_2$ is produced when 100 mL of 0.42 M $Mg(NO_3)_2$ is added to excess NaOH solution?

26. What mass of $BaSO_4$ is produced when 15 mL of 3M $H_2SO_4$ is added to 20 mL of 0.1M $BaCl_2$?

27. Calculate the mass of $CaSO_4$ produced when 10 mL of 6.0 M $H_2SO_4$ is added to 100 mL of 0.52 M
    $Ca(NO_3)_2$.

28. What mass of $CaCO_3$ is produced when 250 mL of 6 M $H_2CO_3$ is added to 750 mL of 1 M $CaF_2$?

29. Calculate the mass of $Al_2S_3$ produced when 100 mL of 0.50 M $AlCl_3$ is added to 100 mL of 0.50 M $Na_2S$.

30. You are given the equation: $AgBr + 2S_2O_3^{-2} \rightarrow Ag(S_2O_3)_2^{-3} + Br^-$. What mass of AgBr can be dissolved
    by 750 mL of 0.3M $Na_2S_2O_3$?

31. Given the following chemical equation, determine the theoretical yield of $B_2H_6$ if exactly 100.0 g of $LiAlH_4$
    was allowed to react with 225 g of $BF_3$:

$$3LiAlH_4 + 4BF_3 \rightarrow 3LiF + 3AlF_3 + 2B_2H_6$$

32. What volume of 0.1379 M HCl is required to neutralize 10.0 mL of 0.2789 M NaOH solution?

33. How many mL of 1.50 M NaOH is required to neutralize 275 mL of 0.5 M $H_2SO_4$?

34. What is the molarity of a solution of HCl if it requires 29.31 mL of a 0.0923 M NaOH solution to reach a
    phenolphthalein endpoint for the titration of a 10.0 mL aliquot of the HCl solution?

35. Given the following two equations, determine the number of grams of magnesium dioxide required to prepare
    enough chlorine gas to produce 25.0 g of potassium hypochlorite:

$$MnO_2 + 4HCl \rightarrow MnCl_2 + Cl_2 + 2H_2O$$
$$Cl_2 + 2KOH \rightarrow KCl + KClO + H_2O$$

36. A titration is done using 0.1302 M NaOH to determine the molar mass of an acid. The acid contains one
    acidic hydrogen per molecule. If 1.863 g of the acid require 70.11 mL of the NaOH solution, what is the
    molar mass of the acid?

37. A 2.0 g sample of silver alloy was dissolved in nitric acid and then precipitated as AgBr. After drying, the
    sample of silver bromide weighed 2.0 g. Calculate the percentage of silver in the alloy.

38. Complete and balance each acid-base equation:

    a.   $H_2SO_4 + NaOH \rightarrow$       c.   $H_2SO_3 + NaOH \rightarrow$
    b.   $H_3PO_4 + Mg(OH)_2 \rightarrow$    d.   $HC_2H_3O_2 + Ba(OH)_2 \rightarrow$

39. What volume of 0.2 M NaOH is required to neutralize 50 mL of 0.1 M $H_2SO_3$?

40. The neutralization of a 25 mL sample of an unknown base requires 18.34 mL of 0.1 M HCl. Assuming that the acid-base stoichiometry is 1:1, what is the concentration of the unknown base?

41. Assuming that the stoichiometry is 1:1, what is the concentration of an unknown acid if a 20 mL sample of it is neutralized with precisely 33.4 mL of 0.25 M base?

42. A 25 mL sample of an ammonia solution is analyzed by titration with HCl. The reaction is: $NH_3 + H^+ \rightarrow NH_4^+$. It took 18.96 mL of 0.15 M HCl to titrate the ammonia. What is the concentration of the original ammonia solution?

43. Determine the oxidation number for Mn in each of the following:

   a.  $KMnO_4$          d.  $K_2MnCl_4$
   b.  $LiMnO_2$         e.  $Mn_2O_7$
   c.  $MnO_2$

44. Determine the oxidation number for each atom in the following compounds or ions:

   a.  $H_3O^+$          d.  $H_2CO$
   b.  $P_4O_{10}$       e.  $NH_4ClO_4$
   c.  $S_8$

45. Determine the oxidation numbers for each atom in the following compounds:

   a.  $MgB_2$           c.  $Cr_2O_7^{-2}$         e.  $NaClF_4$
   b.  $Na_2SO_4$        d.  $CaCO_3$               f.  $HNO_3$

46. Balance the following oxidation reduction reactions. Which species in each is the oxidizing agent, reducing agent?

   a.  $P + Cl_2 ----> PCl_5$
   b.  $Sn^{2+} + Cu^{2+} ----> Sn^{4+} + Cu^+$
   c.  $Cu + H^+ + NO_3^- ----> Cu^{2+} + NO_2 + H_2O$
   d.  $Br_2 + SO_2 + H_2O ----> H^+ + Br^- + SO_4^{2-}$
   e.  $H_2SO_4 + HBr ----> SO_2 + Br_2 + H_2O$

47. Balance the following redox reactions. Identify the oxidizing agent and the reducing agent.

   a.  $ClO^- + H^+ + Cu \rightarrow Cl- + H_2O + Cu^{+2}$
   b.  $H_2O + Cr^{+3} + XeF_6 \rightarrow Cr_2O_7^{-2} + H^+ + Xe + F^-$
   c.  $H_2O + SO_3^{-2} + Fe^{+3} \rightarrow SO_4^{-2} + Fe^{+2} + H^+$

48. Balance the following in basic solution:

   a.  $P_4 ----> PH_3 + HPO_3^{2-}$
   b.  $Cl_2 + OH^- ----> Cl^- + ClO_3^-$
   c.  $Zn + NO_3^- ----> Zn^{2+} + NH_3$

49. Balance the following redox reactions which take place in an acid solution:

   a.  $H_3AsO_4 + Zn \rightarrow AsH_3 + Zn^{+2}$
   b.  $HS_2O_3^- \rightarrow S + HSO_4^-$
   c.  $Cr_2O_7^{-2} + Cl^{-2} \rightarrow Cr^{+3} + Cl_2$
   d.  $MnO_2 + Hg + Cl^- \rightarrow Mn^{+2} + Hg_2Cl_2$

50  Balance the following redox reactions that take place in a basic solution:

   a.  $HXeO_4^- + Pb \rightarrow He + HPbO_2^-$
   b.  $ClO_4^- + I- \rightarrow ClO_3^- + IO_3^-$
   c.  $NO_2 + Al \rightarrow NH_3 + AlO_2$
   d.  $Co(OH)_3 + Sn \rightarrow Co(OH)_2 + HSnO_2^-$

51. Iron(II) can be determined by redox titration with a cerium (IV) solution. The oxidation reduction reaction is

$$Fe^{2+} + Ce^{4+} ----> Fe^{3+} + Ce^{3+}$$

What is the concentration of $Fe^{2+}$ if it requires 21.35 mL of 0.3136 M $Ce^{4+}$ to titrate a 10.00 mL aliquot to the endpoint?

52. A solution containing an unknown concentration of $Sn^{2+}$ ions was titrated with a solution containing $Ce^{4+}$ ions.  One liter of the unknown solution required 22.9 mL of a 0.27 M $Ce^{4+}$ solution to reach the titration endpoint.  Balance the following equation and calculate the concentration of $Sn^{2+}$ ions in the unknown solution.

$$Ce^{4+} + Sn^{2+} \rightarrow Ce^{3+} + Sn^{4+}$$

53. A 0.0621 M $KMnO_4$ solution was used to titrate a solution containing $CaC_2O_4$.  If 22.76 mL of $KMnO_4$ solution was required to reach the titration endpoint, calculate the number of grams of $CaC_2O_4$ in the sample.

$$MnO_4^- + C_2O_4^{2-} \rightarrow Mn^{2+} + CO_2$$

## ANSWERS TO SELF-TEST QUESTIONS

1. a. $HF(g) \xrightarrow{H_2O} H^+(aq) + F^-(aq)$

   b. $SrBr_2(s) \xrightarrow{H_2O} Sr^{2+}(aq) + 2Br^-(aq)$

   c. $MgBr_2(s) \xrightarrow{H_2O} Sr^{2+}(aq) + 2Br^-(aq)$

   d. $NH_4Cl(s) \xrightarrow{H_2O} NH_4^+(aq) + Cl^-(aq)$

   e. $NaNO_3(s) \xrightarrow{H_2O} Na^+(aq) + NO_3^-(aq)$

   f. $Al_2(SO_4)_3 \xrightarrow{H_2O} 2Al^{3+}(aq) + 3SO_4^{2-}(aq)$

2. a. $BaSO_4(s) \xrightarrow{H_2O} Ba^{2+}(aq) + SO_4^{2-}(aq)$

   b. $KCl(s) \xrightarrow{H_2O} K^+(aq) + Cl^-(aq)$

   c. $NaOH(s) \xrightarrow{H_2O} Na^+(aq) + OH^-(aq)$

   d. $CaCrO_4(s) \xrightarrow{H_2O} Ca^{2+}(aq) + CrO_4^{2-}(aq)$

   e. $Mg(OH)_2(s) \xrightarrow{H_2O} Mg^{2+}(aq) + 2OH^-(aq)$

   f. $HCOOH(l) \xrightarrow{H_2O} H^+(aq) + COOH^-(aq)$

3. a. miscible;  polar O-H bonds in both.
   b. miscible;  both nonpolar.
   c. immiscible;  $C_6H_6$ nonpolar, water is polar.
   d. miscible;  LiBr is ionic and $H_2O$ is polar.

4. a. nonelectrolyte        c. strong          e. nonelectrolyte
   b. nonelectrolyte        d. weak

5. a. 1.01 M          c. 0.124 M         e. 12.0 M
   b. 0.0228 M        d. 0.108 M

6. a. 0.601 M         c. 0.197 M
   b. 0.293 M         d. 0.433 M

7. a. 0.75 M $Na^+$, 0.25 M $PO_4^{3-}$
   b. 0.3 M $Al^{3+}$, 0.45 M $SO_4^{2-}$
   c. 1.74 M $Na^+$, 0.87 M $CO_3^{2-}$

8. a. 1.24 M $K^+$, 0.62 M $Cr_2O_7^{2-}$
   b. 0.35 $Na^+$, 0.35 M $OH^-$
   c. 0.14 M $Co^{2+}$, 0.28 M $NO_3^{2-}$
   d. 0.21 M $Na^+$, 0.07 M $PO_4^{3-}$
   e. 0.46 M $NH_4^+$, 0.23 M $Cr_2O_7^{2-}$
   f. 0.98 M $Al^{3+}$, 1.47 M $SO_3^{2-}$

9. a. 5.844 g NaCl in enough water to make 100 mL.
   b. 35.5 g of $Na_2SO_4$ in enough water to make 250 mL.
   c. 221 g of $K_2Cr_2O_7$ in enough water to make 1.5 L.

10. a. 1.46 g of HCl in enough water to make 400 mL.
    b. 61.26 g of $BaSO_4$ in enough water to make 750 mL.
    c. 273 g of $KMnO_4$ in enough water to make 1 L.
    d. 8.5 g of $AgNO_3$ in enough water to make 250 mL.

11. a. 28 mL of 17.8 M $H_2SO_4$ diluted to 500 mL.
    b. 375 mL of 1 M $KMnO_4$ diluted to 1.5 L
    c. 25 g of $KBrO_3$ with enough water to make 1.0 L
    d. 2 mL of 0.5 M $AgNO_3$ diluted to 100 mL.
    e. 85 g $AgNO_3$ with enough water to make 1 L.

12. a.   2 mL of 12.5 M HCl diluted to 250 mL
    b.   103 mL of 7.3 M NaCl diluted to 500 mL
    c.   35 mL of 4.6 M $NiCl_2$ diluted to 800 mL
    d.   375 mL of 0.1 M $FeSO_4$ diluted to 750 mL

13. 0.1811 M

14. $5.433 \times 10^{-3}$ mole

15. 0.1 M

16. $1.7 \times 10^{-3}$ moles

17. 29.7 mL

18. 3.6 mL

19. a.   5 ppm
    b.   4.4 ppm

20. a.   8.3 ppm
    b.   170 ppm

21. a.   $C_3H_8 + 5O_2 \rightarrow 3CO_2 + 4H_2O$
    b.   $3K_2CO_3 + Al_2Cl_6 \rightarrow 6KCl + Al_2(CO_3)_3$
    c.   $Mg_3N_2 + 3H_2O \rightarrow 3MgO + 2NH_3$
    d.   $Ca_3(PO_4)_2 + 3H_2SO_4 \rightarrow 3CaSO_4 + 2H_3PO_4$
    e.   $3KOH + H_3PO_4 \rightarrow K_3PO_4 + 3H_2O$
    f.   $8KClO_3 + C_{12}H_{22}O_{11} \rightarrow 8KCl + 12CO_2 + 11H_2O$

22. a.   $2NaCl(aq) + 2HgNO_3(aq) \longrightarrow Hg_2Cl_2(s) + 2NaNO_3(aq)$
    b.   $Ca(OH)_2(aq) + Na_2CO_3(aq) \longrightarrow CaCO_3(s) + 2NaOH(aq)$
    c.   $3Na_2S(aq) + 2FeCl_3(aq) \longrightarrow Fe_2S_3(s) + 6NaCl(aq)$

23. a.   molecular:  $Na_2S(aq) + Cu(NO_3)_2(aq) \longrightarrow CuS(s) + 2NaNO_3(aq)$
        complete ionic:  $2Na^+(aq) + S^{2-}(aq) + Cu^{2+}(aq) + 2NO_3^-(aq) \longrightarrow CuS(s) + 2Na^+(aq) + 2NO_3^-(aq)$
        net ionic:  $S^{2-}(aq) + Cu^{2+}(aq) \longrightarrow CuS(s)$
    b.   molecular:  $HF(aq) + KOH(aq) \longrightarrow H_2O(l) + KF(aq)$
        complete ionic:  $H^+(aq) + F^-(aq) + K^+(aq) + OH^-(aq) \longrightarrow H_2O(l) + K^+(aq) + F^-(aq)$
        net ionic:  $H^+(aq) + OH^-(aq) \longrightarrow H_2O(l)$

24. $SO_4^{2-}$

25. 2.4 g

26. 0.467 g

27. 7.1 g

28. 75 g

29. 2.5 g

30. 21.15 g

31. 45.9 g $B_2H_6$

32. 20.2 mL

33. 183 mL

34. 0.271 M

35. 15.5 g $MnO_2$

36. 204.1 g/mole

37. 57.45% Ag in the alloy

38. a.  $H_2SO_4 + 2NaOH \rightarrow 2H_2O + Na_2SO_4$
    b.  $2H_3PO_4 + 3Mg(OH)_2 \rightarrow 6H_2O + Mg_3(PO_4)_2$
    c.  $H_2SO_3 + 2NaOH \rightarrow 2H_2O + Na_2SO_3$
    d.  $2HC_2H_3O_2 + Ba(OH)_2 \rightarrow 2H_2O + Ba(C_2H_3O_2)_2$

39. 50 mL

40. $7.34 \times 10^{-2}$ M

41. 0.418 M

42. 0.114 M

43. a.  +7          c.  +2          e.  +7
    b.  +3          d.  +2

44. a.  H, +1; O, -2          c.  S, O          e.  N, -3; H, +1; Cl, +7; O, -2
    b.  P, +5; O, -2          d.  H, +1; C, O; O,-2

45. a.  Mg, +2; B, -1          d.  Ca, +2; C, +4; Oc -2
    b.  Na, +1; S, +6; O, -2          e.  Na, +1; Cl, +3; F, -1
    c.  Cr, +6; O, -2          f.  H, +1; N, +5; O, -2

46. a.  $2P + 5Cl_2 \dashrightarrow 2PCl_5$: ox. agent $Cl_2$; red. agent P
    b.  $Sn^{2+} + 2Cu^{2+} \dashrightarrow Sn^{4+} + 2Cu^+$: ox. agent $Cu^{2+}$; red. agent $Sn^{2+}$
    c.  $Cu + 4H^+ + 2NO_3^- \dashrightarrow Cu^{2+} + 2NO_2 + 2H_2O$: ox. agent $NO_3^-$, red. agent Cu.
    d.  $Br_2 + SO_2 + 2H_2O \dashrightarrow 4H^+ + 2Br^- + SO_4^{2-}$: ox. agent $Br_2$; red. agent $SO_2$
    e.  $H_2SO_4 + 2HBr \dashrightarrow SO_2 + Br_2 + 2H_2O$: ox. agent $H_2SO_4$; red. agent HBr.

47. a.  $ClO^- + 2H^+ + Cu \rightarrow Cl^- + H_2O + Cu^{2+}$; ox. agent = $ClO^-$, red. agent = Cu
    b.  $7H_2O + 2Cr^{3+} + XeF_6 \rightarrow Cr_2O_7^{2-} + 14H^+ + Xe + 6F^-$; ox. agent = $XeF_6$, red. agent = $Cr^{3+}$
    c.  $H_2O + SO_3^{2-} + 2Fe^{3+} \rightarrow SO_4^{2-} + 2Fe^{2+} + 2H^+$; ox. agent = $Fe^{3+}$, red. agent = $SO_3^-$

48. a.  $2H_2O + 4OH^- + P_4 \dashrightarrow 2PH_3 + 2HPO_3^{2-}$
    b.  $3Cl_2 + 6OH- \dashrightarrow 5Cl^- + ClO_3^- + 3H_2O$
    c.  $6H_2O + 4Zn + NO_3^- \dashrightarrow 4Zn^{2+} + NH_3 + 9OH^-$

49. a.  $H_3AsO_4 + 4Zn + 8 H^+ \rightarrow AsH_3 + 4Zn^{+2} + 4H_2O$
    b.  $3HS_2O_3^- + H^+ \rightarrow 4S + 2HSO_4^- + H_2O$
    c.  $Cr_2O_7^{2-} + 6Cl^- + 14H^+ \rightarrow 2Cr^{3+} + 3Cl_2 + 7H_2O$
    d.  $MnO_2 + 2Hg + 2Cl^- + 4H^+ \rightarrow Mn^{2+} + Hg_2Cl_2 + 2H_2O$

50. a.  $HXeO_4^- + 3Pb + 2OH^- \rightarrow Xe + 3HPbO_2^-$
    b.  $3ClO_4^- + I^- \rightarrow 3ClO_3^- + IO_3^-$
    c.  $NO_2^- + 2Al + H_2O + OH^- \rightarrow NH_3 + 2AlO_2^-$
    d.  $2Co(OH)_3 + Sn + OH^- \rightarrow 2Co(OH)_2 + HSnO_2^- + H_2O$

51.  0.6695 M

52.  $2Ce^{4+} + Sn^{2+} \rightarrow 2Ce^{3+} + Sn^{4+}$; $3.09 \times 10^{-3}$ M

53.  0.4526 g $CaC_2O_4$

# CHAPTER 5

# Gases

| Section | Contents | Page in this Study Guide |
|---|---|---|
| 5.1 | Pressure | 110 |
| 5.2 | The Gas Laws of Boyle, Charles, and Avogadro | 112 |
| 5.3 | The Ideal Gas Law | 115 |
| 5.4 | Gas Stoichiometry - Molar Mass of a Gas | 118 |
| 5.5 | Dalton's Law of Partial Pressures | 121 |
| 5.6 | The Kinetic Molecular Theory of Gases | 123 |
| 5.7 | Effusion and Diffusion | 124 |
| 5.8 | Real Gases | 126 |
| 5.9 | Chemistry in the Atmosphere | 128 |

## 5.1 - Pressure

When you finish this section you will be able to:

- Define "pressure."
- State how pressure is different from force.
- Convert between various units of pressure.

Your textbook introduces you to the **manometer** and the **barometer**. In order to really know how they work, you must understand force, pressure and the units of both.

$$\text{Force} = \text{mass} \times \text{acceleration}$$

When you weigh yourself on a scale, you are measuring the **force** that your body exerts on the scale. Let's say that your body mass is 68 kilograms. Also, the **acceleration** due to **gravity** that the earth exerts on your constant 9.8 meters/second$^2$. The **total force** exerted by your body on the scale is

$$F = m \times a$$

$$F = 68 \text{ kg} \times \frac{9.8 \text{ m}}{s^2} = \frac{670 \text{ kg m}}{s^2}$$

The S.I. unit of force is the **Newton (N)**.

$$1 \text{ N} = \frac{1 \text{ kg m}}{s^2}$$

Therefore, your force on the scale is 670 N.  This corresponds to your weight of 150 pounds (4.47 N/pound).

---

## Example 5.1 A - The Units of Force

Let's say that you (with your mass of 68 kg) are on the moon, where the acceleration due to gravity is about 1.6 $m/s^2$.  What force would you exert on a scale ("how much would you weigh") in Newtons?  In pounds?

### Solution

$$F = m \times a = 68 \text{ kg} \times \frac{1.6 \text{ m}}{s^2} = 108.8 \frac{\text{kg m}}{s^2} = \textbf{110 N}$$

$$\text{weight in pounds} = 108.8 \text{ N} \times \frac{1 \text{ pound}}{4.47 \text{ N}} = \textbf{24 pounds}$$

---

$$\textbf{Pressure} = \frac{\textbf{Force}}{\textbf{Area}}$$

If you weigh 670 N (670 kg $m/s^2$) on earth, and you are standing on a 0.5 m × 0.5 m square, your **area = $(0.5 \text{ m})^2$ = 0.25 m$^2$**, the pressure you exert is

$$P = \frac{F}{A} = \frac{670 \text{ kg m}/s^2}{0.25 \text{ m}^2} = \frac{\textbf{2700 kg}}{\textbf{m}/s^2} \quad \text{or} \quad \frac{\textbf{2700 N}}{\textbf{m}^2}$$

The SI unit of pressure is the **Pascal (Pa)**,

$$1 \text{ Pa} = \frac{1 \text{ kg}}{m/s^2} = \frac{1 \text{ N}}{m^2}$$

Therefore, the pressure you exert is 2700 Pa.

---

## Example 5.1 B - The Units of Pressure

Let's say that you are wearing high heels with a total area (for both heels) of $1 \times 10^{-4}$ m$^2$.  (Men, make believe that you are doing this just for fun.)  Your force is **670 N**.  Calculate the pressure that you exert.

$$P = \frac{F}{A} = \frac{670 \text{ kg m}/s^2}{1 \times 10^{-4} \text{ m}^2} = \frac{6.7 \times 10^6 \text{ N}}{m^2} = \textbf{6.7} \times \textbf{10}^6 \text{ Pa}$$

This pressure is quite high!  That is why it hurts if someone wearing a stiletto heel stands on your foot.

---

Air exerts pressure as well.  **The "standard atmosphere" is = 101,325 Pa**, but this is an awkwardly large number.  We commonly express air pressure in units of **atmospheres (atm)**, **torr**, or **mm Hg**.

$$1 \text{ atm} = 101,325 \text{ Pa}$$

$$1 \text{ atm} = 760 \text{ torr} = 760 \text{ mm Hg}$$

---

## Example 5.1 C - Interconverting the Units of Pressure

The pressure exerted by a gas is measured to be 0.985 atm.  Convert this pressure to **torr** and **pascals**.

### Solution

$$0.985 \text{ atm} \times \frac{760 \text{ torr}}{1 \text{ atm}} = \textbf{749 torr (749 mm Hg)}$$

$$0.985 \text{ atm} \times \frac{101,325 \text{ Pa}}{1 \text{ atm}} = \textbf{98,300 Pa}$$

---

Problems 1-6 at the end of this study chapter cover material from this section.

---

## 5.2 - The Gas Laws of Boyle, Charles, and Avogadro

When you finish this section you will be able to solve problems relating to the fundamental gas laws.

There are **three gas laws** that you need to know about and be able to use.  In the next section, these laws will be combined to give a more general law governing the behavior of gases.

A.  *Boyle's Law*

This law says that **the pressure exerted by a gas is inversely proportional to the volume that the gas occupies.**  In other words, as you squeeze the gas, it exerts more pressure.  As an example, fill a zip-lock freezer bag with air, and then seal it.  Try to squeeze it (**reduce the volume**).  It is very difficult to squeeze because the **pressure of the gas is increasing** as you reduce the volume of the freezer bag.

**Pressure × Volume = Constant**

**PV = K**

If PV is a constant for a gas at constant temperature, then

$$P_1 V_1 = P_2 V_2 \text{ (at constant temperature)}$$

where   $P_1$ = initial pressure
$V_1$ = initial volume
$P_2$ = final pressure
$V_2$ = final volume.

Your textbook notes that you should make sure your problem answer makes physical sense.  This is especially important when doing gas law problems.

## Example 5.2 A - Boyle's Law

A gas which has a pressure of 1.3 atm occupies a volume of 27 L.  What volume will the gas occupy if the pressure is increased to 3.9 atm at constant temperature?

### Predicting the Answer

If the pressure on the gas is **increased** by a factor of 3 (3.9 atm/1.3 atm) (you are squeezing the gas), we would expect the volume to **decrease** proportionately.   Therefore, the volume should be 1/3 of its original value.

### Solution

$$P_1 = 1.3 \text{ atm} \qquad P_2 = 3.9 \text{ atm}$$
$$V_1 = 27 \text{ L} \qquad V_2 = ?$$

$$P_1V_1 = P_2V_2$$

$$1.3 \text{ atm } (27 \text{ L}) = 3.9 \text{ atm } (V_2)$$

$$V_2 = \frac{1.3 \text{ atm } (27 \text{ L})}{3.9 \text{ atm}} = 9 \text{ L}$$

The final volume is 9 L  (1/3 of 27), the answer we predicted.   Always consider, "does my answer make sense?"

---

## Example 5.2 B - The Units of P × V

The PV constant in the previous example was 35.1 L atm (1.3 atm × 27 L).  Convert this constant to

   a.   L Pa, and
   b.   fundamental SI units.

### Solution

   a.   $35.1 \text{ L atm } \times \dfrac{101{,}325 \text{ Pa}}{\text{atm}} = \mathbf{3.56 \times 10^6 \ L \ Pa}$

   b.   We know from section 5.1 that **1 Pascal = 1 kg m$^{-1}$ s$^{-2}$**.  1 L = 1000 mL, or 1000 cm$^3$ , or (10 cm)$^3$. This equals (0.1 m)$^3$, or, **1 L = 1 × 10$^{-3}$ m$^3$**. Therefore,

$$3.56 \times 10^6 \text{ L Pa} = 3.56 \times 10^6 \text{ L Pa } \times \frac{1 \times 10^{-3} \text{ m}^3}{1 \text{ L}} \times \frac{1 \text{ kg m}^{-1} \text{ s}^{-2}}{1 \text{ Pa}}$$

$$= \mathbf{3.56 \times 10^3 \ kg \ m^2 \ s^{-2}}$$

The unit kg m$^2$ s$^{-2}$ is the SI unit of **energy** called the **Joule**. *PV, then, is really a unit of energy.*  We will use this information in later chapters.

---

B. *Charles's Law*

   Charles's Law says that **at a constant pressure, the volume of a gas is directly proportional to the temperature (in Kelvins) of the gas.**

$$\frac{\text{Volume}}{\text{Temperature}} = \text{Constant}$$

$$\frac{V}{T} = K$$

Using the same reasoning as we did in Boyle's Law, if we change the temperature (at constant pressure), the volume will change so that the ratio of volume to temperature will remain constant.

$$\frac{V_1}{T_1} = \frac{V_2}{T_2} \quad \text{(at constant pressure)}$$

$V_1$ = initial volume
$T_1$ = initial temperature
$V_2$ = final volume
$T_2$ = final temperature

---

## Example 5.2 C - Charles's Law

A gas at 30°C and 1 atm occupies a volume of 0.842 L. What volume will the gas occupy at 60°C and 1 atm?

### Predicting the Answer

The temperature is increased, therefore, according to Charles's Law, the volume occupied should increase as well. We predict that our **final volume will be larger**. (Remember to convert temperature to Kelvins!)

### Solution

$V_1 = 0.842$ L                          $V_2 = ?$
$T_1 = 30°C + 273 = 303$ K          $T_2 = 60°C + 273 = 333$ K

$$\frac{V_1}{T_1} = \frac{V_2}{T_2}$$

$$\frac{0.842 \text{ L}}{303 \text{ K}} = \frac{V_2}{333 \text{ K}}$$

$$V_2 = \frac{0.842 \text{ L } (333 \text{ K})}{303 \text{ K}} = \textbf{0.925 L}$$

The final volume agrees with our prediction, so it "makes sense."

---

C.  *Avogadro's Law*

Avogadro's Law says that **for a gas at constant temperature and pressure the volume is directly proportional to the number of moles of gas.**

$$\textbf{Volume = Constant} \times \textbf{number of moles} \quad \textbf{(constant T, P)}$$

$$\textbf{V = kn (constant T, P)}$$

If you triple the number of moles of a gas (at constant temperature and pressure), the volume will also triple.

$$\frac{V_1}{n_1} = \frac{V_2}{n_2} \quad \text{(constant T, P)}$$

$V_1$ = initial volume
$n_1$ = initial number of moles
$V_2$ = final volume
$n_2$ = final number of moles

## Example 5.2 D - Avogadro's Law

A 5.20 L sample at 18°C and 2 atm pressure contains 0.436 moles of a gas. If we add an **additional** 1.27 moles of the gas at the same temperature and pressure, what will the **total volume** occupied by the gas be?

### Predicting the Answer

According to Avogadro's Law, as we increase the number of moles of a gas, the volume should increase proportionately. Therefore, we would predict that our volume would increase.

### Solution

$$V_1 = 5.20 \text{ L} \qquad\qquad V_2 = ?$$
$$n_1 = 0.436 \text{ moles} \qquad n_2 = 0.436 + 1.27 = 1.706 \text{ moles}$$

$$\frac{V_1}{n_1} = \frac{V_2}{n_2}$$

$$\frac{5.20 \text{ L}}{0.436 \text{ mol}} = \frac{V_2}{1.70_6 \text{ mol}}$$

$$V_2 = \frac{5.20 \text{ L} \, (1.70_6 \text{ mol})}{0.436 \text{ mol}} = \mathbf{20.3 \text{ L}}$$

The answer agrees with our prediction.

Keep in mind that we have assumed that our gases behave **ideally**. That is not always a valid assumption, as your textbook points out.

Problems 7-14 at the end of this study chapter cover material from this section.

## 5.3 - The Ideal Gas Law

When you finish this section you will be able to solve a variety of problems relating to the ideal gas law.

**The ideal gas law** is a combination of Boyle's, Charles's, and Avogadro's laws. It relates <u>pressure</u>, <u>temperature</u>, <u>volume</u> and the <u>number of moles</u> of a gas. The derivation of the ideal gas law is given in your textbook. The equation of interest is

**Pressure × Volume = # of moles of the gas × a constant × temperature**

$$PV = nRT$$

> **P** = pressure in **atm**
> **V** = volume in **L**
> **n** = number of **moles**
> **R** = 0.08206 **L·atm/K·mol**
> **T** = temperature in **Kelvins**

Please keep in mind that

1. This relationship assumes that the gas behaves **ideally**. As we will see in section 5.8, there are certain conditions under which a gas will not behave ideally, and correction factors must be added to the ideal gas law.

2. You need to keep track of your dimensions. Many ideal gas law problems are best solved using **dimensional analysis**.

3. Always list what you are given. You may be able to simplify the problem.

Let's try a few examples.

---

# Example 5.3 A - Ideal Gas Law

A sample containing 0.614 moles of a gas at 12°C occupies a volume of 12.9 L. What pressure does the gas exert?

## Strategy

List those things that you are given, including constants. Once you see what you have, and what you are being asked to find, you can decide how to manipulate your equation.

## Solution

> P = ?
> V = 12.9 L
> T = 285 K
> n = 0.614 moles
> R = 0.08206 L·atm/K·mol

$$PV = nRT$$

or

$$P = \frac{nRT}{V}$$

$$P = \frac{0.614 \text{ moles } (0.08206 \text{ L·atm/K·mol})(285 \text{ K})}{12.9 \text{ L}} = 1.11 \text{ atm}$$

You can check your answer by using dimensional analysis as shown below.

$$\text{atm} \left\| \frac{0.08206 \text{ L·atm}}{\text{K·mol}} \times 0.614 \text{ mol} \times 285 \text{ K} \times \frac{1}{12.9 \text{ L}} = 1.11 \text{ atm} \right.$$

↑
start with "R" with
the units where you
want them, then cancel

## Example 5.3 B - Practice With Gas Laws

A sample of methane gas ($CH_4$) at 0.848 atm and 4°C occupies a volume of 7.0 L.  What volume will the gas occupy if the pressure is increased to 1.52 atm and the temperature increased to 11°C?

### Predicting the Answer

According to Section 5.2 in your textbook, if the **pressure is increased**, the **volume should decrease** (Boyle's Law).  If the **temperature is increased**, the **volume should also increase** (Charles's Law).  However, the pressure almost doubles while the temperature increase is relatively small.  Therefore the pressure effect will be dominant.  Overall then, we would expect the **volume to decrease**.

### Setting Up the Equations

You are given two sets of data.

$$P_1 = 0.848 \text{ atm} \qquad P_2 = 1.52 \text{ atm}$$
$$V_1 = 7.0 \text{ L} \qquad V_2 = ?$$
$$T_1 = 277 \text{ K} \qquad T_2 = 284 \text{ K}$$

$$n_1 = n_2$$
$$R = R$$

$$P_1V_1 = n_1RT_1 \qquad \text{and} \qquad P_2V_2 = n_2RT_2$$

$$\frac{P_1V_1}{T_1} = n_1R \qquad \text{and} \qquad \frac{P_2V_2}{T_2} = n_2R$$

$$n_1R = n_2R \text{ (because } n_1 = n_2\text{), so}$$

$$\frac{P_1V_1}{T_1} = \frac{P_2V_2}{T_2}$$

This is often called the **combined gas law**.  The keys to setting up this equation were to **list what you are given** and to **know what you are being asked to solve for.**

### Solution

By converting the combined gas law to solve for $V_2$, we find that

$$V_2 = \frac{P_1V_1T_2}{P_2T_1} = \frac{0.848 \text{ atm } (7.0 \text{ L})(284 \text{ K})}{1.52 \text{ atm } (277 \text{ K})} = \textbf{4.0 L}$$

This agrees with our prediction.

## Example 5.3 C - More Practice with Gas Laws

How many moles of a gas at 104°C would occupy a volume of 6.8 L at a pressure of 270 mm Hg?

### Solution

$$P = 270 \text{ mm Hg} \times \frac{1 \text{ atm}}{760 \text{ mm Hg}} = 0.355 \text{ atm}$$

$$V = 6.8 \text{ L} \qquad\qquad T = 377 \text{ K} \qquad\qquad n = \text{?}$$

$$n = \frac{PV}{RT} = \frac{0.355 \text{ atm } (6.8 \text{ L})}{0.08206 \text{ L·atm/K·mol } (377 \text{ K})} = \textbf{0.078 moles}$$

Note that the units are in moles.  You can use dimensional analysis as an additional check.

$$\text{moles} \quad \Big\| \quad \frac{1 \text{ K·mol}}{0.08206 \text{ L·atm}} \times 0.355 \text{ atm } \times 6.8 \text{ L} \times \frac{1}{377 \text{ K}} = \textbf{0.078 moles}$$

Problems 15-21 at the end of this study chapter cover material from this section.

## 5.4 - Gas Stoichiometry - Molar Mass of a Gas

When you finish this section you will be able to:

- define STP conditions, and
- do a variety of calculations regarding molar mass, density, and stoichiometry of gases.

This section uses the ideal gas law to perform a variety of calculations.  Your textbook defines the volume occupied under **standard temperature and pressure (STP)**.  One mole of an ideal gas at 0°C (standard temperature) and 1.000 atm (standard pressure) will occupy **22.4 liters**.

**STP MEANS 0°C and 1.000 atm.**

## Example 5.4 A - The Ideal Gas Law and STP

What volume will 1.18 moles of $O_2$ occupy at STP?

## Solution

**Method 1:**

T = 273 K
P = 1.000 atm
R = 0.08206 L·atm/K·mol
n = 1.18 mol
V = ?

$$V = \frac{nRT}{P} = \frac{1.18 \text{ mol } (0.08206 \text{ L·atm/K·mol}) \ (273 \text{ K})}{1.000 \text{ atm}} = \textbf{26.4 L}$$

**Method 2:**

At STP, 1 mol occupies 22.4 L, therefore,

$$\frac{22.4 \text{ L}}{1 \text{ mol}} = \frac{V}{1.18 \text{ mol}}$$

$$V = \frac{22.4 \text{ L} \times 1.18 \text{ mol}}{1 \text{ mol}} = \textbf{26.4 L}$$

As you can see, problems are made much simpler when you are dealing with STP conditions.

---

Many gas law problems involve calculating the **volume** of a gas produced by the reaction of volumes of other gases.  The problem solving strategy that we have used throughout your chemistry course is still the same.  That is, you want to relate **moles of reactants to moles of products**.  The ideal gas law will allow you to use the following strategy:

$$\text{volume of reactants} \xrightarrow{\text{ideal gas law}} \text{moles of reactants} \xrightarrow{\text{stoichiometry}} \text{moles of products}$$

$$\text{volume of products} \xleftarrow{\text{ideal gas law}}$$

In the next problem, we are given moles of reactant and asked to find the volume of product.

---

## Example 5.4 B - Reactions and the Ideal Gas Law

A sample containing 15.0 g of dry ice ($CO_2(s)$) is put into a balloon and allowed to sublime according to the following reaction:

$$CO_2(s) \longrightarrow CO_2(g)$$

How big will the balloon be (ie; what is the volume of the balloon), at 22°C and 1.04 atm , after all of the dry ice has sublimed?

### Strategy

You are given *moles of $CO_2(s)$*.  You want *volume of $CO_2(g)$*.  The mole ratio of the solid to the gas is 1 to 1.  Given the mass of $CO_2$, we can go from

$$\text{g } CO_2 \xrightarrow{\text{molar mass}} \text{moles } CO_2(g) \xrightarrow{\text{ideal gas law}} \text{volume } CO_2$$

### Solution

$$\text{moles } CO_2 \,\Big|\!\Big|\; 15.0 \text{ g} \times \frac{1 \text{ mol } CO_2}{44.0 \text{ g } CO_2} = \textbf{0.341 mol } CO_2$$

$$V = \frac{nRT}{P} = \frac{0.341 \text{ mol } (0.08206 \text{ L·atm/K·mol}) \, (295 \text{ K})}{1.04 \text{ atm}} = \textbf{7.94 L}$$

Using dimensional analysis as a check:

$$\text{L } CO_2 \,\Big|\!\Big|\; \frac{0.08206 \text{ L·atm}}{\underset{\uparrow}{\text{K·mol}}} \times 295 \text{ K} \times \frac{1}{1.04 \text{ atm}} \times 0.341 \text{ mol} = \textbf{7.94 L}$$

get the proper
units on top

---

## Example 5.4 C - Practice with the Ideal Gas Law

0.500 L of $H_2(g)$ are reacted with 0.600 L of $O_2(g)$ **at STP** according to the equation

$$2H_2(g) + O_2(g) \longrightarrow 2H_2O(g)$$

What volume will the $H_2O$ occupy at 1.00 atm and 350°C?

## Strategy

Because you have two reactants, this may be a **limiting reactant** problem. Just as with any such problem, you must find out how many moles of each reactant you have so that you can determine the limiting reactant. Once you have done those calculations, you can determine the moles and then volume that the product occupies.

ideal gas law                    limiting reactant

reactant volume ---------------> moles of reactants -------------------> moles of product

ideal gas law

volume of product <--------------------

Use the fact that the reactants are initially at STP to help you solve the problem.

## Solution

$$\text{moles } H_2 \text{ at STP} \left\| \frac{1 \text{ mol}}{22.4 \text{ L}} \times 0.500 \text{ L} = \textbf{0.0223 moles } H_2 \right.$$

$$\text{moles } O_2 \text{ at STP} \left\| \frac{1 \text{ mol}}{22.4 \text{ L}} \times 0.600 \text{ L} = \textbf{0.0268 moles } O_2 \right.$$

To determine the limiting reactant,

$$\text{moles } H_2O \text{ from } H_2 \left\| 0.0223 \text{ mol } H_2 \times \frac{2 \text{ mol } H_2O}{2 \text{ mol } H_2} = 0.0223 \text{ mol } H_2O \right.$$

$$\text{moles } H_2O \text{ from } O_2 \left\| 0.0268 \text{ mol } O_2 \times \frac{2 \text{ mol } H_2O}{1 \text{ mol } O_2} = 0.0536 \text{ mol } H_2O \right.$$

**The limiting reactant is $H_2$, and 0.0223 mol $H_2O$ will be formed.**

$$\text{Volume } H_2O = \frac{nRT}{P} = \frac{0.0223 \text{ mol } (0.08206 \text{ L·atm/K·mol}) \ (623 \text{ K})}{1.00 \text{ atm}} = \textbf{1.14 L}$$

---

## Example 5.4 D - Density and Molar Mass

A gas at 34°C and 1.75 atm has a density of 3.4 g/L. Calculate the molar mass (M.M.) of the gas.

## Solution

$$\text{M.M.} = \frac{dRT}{P}$$

$$\text{M.M.} = \frac{3.4 \text{ g/L} \ (0.08206 \text{ L·atm/K·mol})(307 \text{ K})}{1.75 \text{ atm}}$$

As a check use dimensional analysis:

$$\frac{\text{grams}}{\text{mol}} \left\| \frac{3.4 \text{ g}}{\underset{\uparrow}{\text{L}}} \times \frac{0.08206 \text{ L·atm}}{\underset{\uparrow}{\text{K·mol}}} \times \frac{1}{1.75 \text{ atm}} \times 307 \text{ K} = \textbf{48.9 g/mol} \right.$$

density             R

Problems 22-28 at the end of this study chapter cover material from this section.

## 5.5 - Dalton's Law of Partial Pressures

At the end of this section you will be able to solve a variety of problems relating to **partial pressure, mole fraction** and **total pressure**.

Dalton's Law of Partial Pressures states that for a mixture of gases in a container, the **total pressure** is the **sum** of the pressures that each gas would exert if it were alone. Your textbook shows that because the ideal gas law holds,

$$P_{Total} = P_1 + P_2 + \ldots + P_n = (n_1 + n_2 + \ldots + n_n) \left[ \frac{RT}{V} \right],$$

because $\frac{RT}{V}$ will be the same for each of the different gases in the same container.

The **KEY PROBLEM SOLVING STRATEGY** with regard to partial pressure problems is to **USE THE IDEAL GAS LAW TO INTERCONVERT BETWEEN PRESSURE AND MOLES OF EACH GAS.** Let's try the following example.

## Example 5.5 A - Partial Pressures

A volume of 2.0 L of He at 46°C, and 1.2 atm pressure, was added to a vessel that contained 4.5 L of $N_2$ at STP. What is the **total pressure** and **partial pressure of each gas at STP** after the He is added?

### Solution

Find the **number of moles** of He at the original conditions. This will ultimately lead to finding the partial pressure of He at STP.

$$n = \frac{PV}{RT}$$

$$n_{He} = \frac{(1.2 \text{ atm}) (2.0 \text{ L})}{(0.08206 \text{ L·atm/K·mol}) (319 \text{ K})} = 0.091_7 \text{ mol He}$$

When the gases are combined under STP conditions, the partial pressure of He will change. That of $N_2$ (already at STP) will remain the same.

$$P_{He} = \frac{nRT}{V} = \frac{(0.091_7 \text{ mol}) (0.08206 \text{ L·atm/K·mol}) (273 \text{ K})}{4.5 \text{ L}} = 0.45_7 \text{ atm}$$

$$P_{N_2} = 1.00 \text{ atm}$$

$$P_{Total} = 1.00 \text{ atm} + 0.46 \text{ atm} = 1.46 \text{ atm}$$

The equations for **MOLE FRACTION** are derived in your textbook. Recalling,

$$\chi_i = \frac{n_i}{n_{total}} = \frac{P_i}{P_{total}}$$

The key idea here is that once you have either the number of moles **OR** the pressure of each component of your system, you can calculate the mole fraction. The next example has several parts.

---

## Example 5.5 B - Mole Fraction and Partial Pressure

    a.  Calculate the number of moles of $N_2$ present in the previous example (5.5 A).
    b.  Calculate the mole fractions of $N_2$ and He given the following data from example 5.5 A.

        i.  mole data
       ii.  pressure data

## Solution

    a.  In calculating moles of $N_2$ that occupy 4.5 L (as given in 5.5 A), take advantage of STP conditions.

$$\text{True at STP:} \quad \frac{1 \text{ mol } N_2}{22.4 \text{ L}} = \frac{X \text{ mol } N_2}{4.5 \text{ L}}$$

$$X = 0.20_1 \text{ mol } N_2$$

    b.  The total number of moles = mol $N_2$ + mol He = $0.20_1 + 0.091_7 = 0.293$ mol

$$\chi_{N_2} = \frac{0.20_1 \text{ mol}}{0.293 \text{ mol}} = \textbf{0.69} \qquad \text{dimensionless, because the units cancel on top and bottom}$$

$$\chi_{He} = \frac{0.091_7 \text{ mol}}{0.293 \text{ mol}} = \textbf{0.31}$$

Using partial pressure data, the total pressure was 1.46 ($1.45_7$) atm.

$$\chi_{N_2} = \frac{1.00 \text{ atm}}{1.45_7 \text{ atm}} = \textbf{0.69}$$

$$\chi_{He} = \frac{0.45_7 \text{ atm}}{1.45_7 \text{ atm}} = \textbf{0.31}$$

---

## Example 5.5 C - Vapor Pressure

The vapor pressure of water in air at 28°C is 28.3 torr. Calculate the mole fraction of water in a sample of air at 28°C and 1.03 atm pressure.

## Solution

$$\chi_{H_2O} = \frac{P_{H_2O}}{P_{air}} = \frac{28.3 \text{ torr}}{783 \text{ torr}} = \textbf{0.036}$$

---

Problems 29-33 at the end of this study chapter cover material from this section.

---

## 5.6 - The Kinetic Molecular Theory of Gases

When you finish this section you will be able to use the basic assumptions of the kinetic molecular theory to:

- define temperature, and
- calculate root mean square velocity.

At the beginning of this section, your textbook makes clear that the **KINETIC MOLECULAR THEORY** is simply a **model** that attempts to explain the properties of an ideal gas. The postulates of the kinetic molecular theory are listed and discussed in the <u>blue box in this section of your textbook</u>. The postulates are:

1.  The volume of the individual particles of a gas can be assumed to be negligible.

2.  The particles are in constant motion. The collisions of the particles with the walls of the container are the cause of the pressure exerted by the gas.

3.  The particles are assumed to exert no forces on each other.

4.  The average kinetic energy of a collection of gas particles is assumed to be directly proportional to the Kelvin temperature of the gas.

Your textbook uses these postulates to derive the ideal gas law.

The idea that **TEMPERATURE IS A MEASURE OF THE AVERAGE KINETIC ENERGY OF A GAS** is of critical importance.

$$(KE)_{average} = \frac{3}{2}RT$$

Because $KE = 1/2\ mv^2$ (where m = mass, and v = velocity), $1/2\ mv^2 = 3/2\ RT$, and velocity (as indicated by the random motions of the particles of a gas) increases with higher temperature. TEMPERATURE IS A MEASURE OF KINETIC ENERGY.

*Root Mean Square Velocity*

The expression dealing with the average velocity of gas particles is called the **root mean square velocity** and is derived in your textbook.

$$\mu_{rms} = \sqrt{\frac{3RT}{M}}$$

where $\mathbf{R} = 8.3145\ J/K{\cdot}mol = \dfrac{8.3145\ kg{\cdot}m^2/s^2}{K{\cdot}mol}$   ( because $1\ J = 1\ kg{\cdot}m^2/s^2$)

   $\mathbf{T}$ = temperature in Kelvins
   $\mathbf{M}$ = mass of a mole of the gas in **kilograms** ("kg/mol")

This is certainly a situation where the proper use of units will help you make sure that you have the correct answer.

## Example 5.6 - Root Mean Square Velocity

Calculate the root mean square velocity for the atoms in a sample of oxygen gas at

a.  0°C
b.  300°C

## Prediction

As the temperature is increased, we would expect the average kinetic energy of our particles to increase. The mass of the particles is constant, therefore, we would expect the rms velocity to increase when the temperature is raised from 0°C to 300°C.

## Solution

a.    at 0°C:   $R = 8.3145$ kg·m$^2$/s$^2$ K·mol
$T = 273$ K

$$M = \frac{kg}{mol} \left|\left| \frac{32.0 \text{ g}}{1 \text{ mol}} \times \frac{1 \text{ kg}}{1000 \text{ g}} \right.\right. = 0.0320 \text{ kg/mol}$$

$$\mu_{rms} = \left[ \frac{3 \left[ \frac{8.3145 \text{ kg·m}^2/\text{s}^2}{K·mol} \right] (273 \text{ K})}{0.0320 \text{ kg/mol}} \right]^{1/2} = \left[ \frac{2.128 \times 10^5 \text{ m}^2}{\text{s}^2} \right]^{1/2}$$

$$= 461 \text{ m/s}$$

b.    at 300°C:   $R$ = as above
$M$ = as above
$T$ = 573 K

$$\mu_{rms} = \left[ \frac{3 \left[ \frac{8.3145 \text{ kg·m}^2/\text{s}^2}{K·mol} \right] (573 \text{ K})}{0.0320 \text{ kg/mol}} \right]^{1/2} = \left[ \frac{4.467 \times 10^5 \text{ m}^2}{\text{s}^2} \right]^{1/2}$$

$$= 668 \text{ m/s}$$

The increase in rms velocity agreed with our prediction.

---

Finally, your textbook points out two very important ideas. The first is that particles collide with one another and exchange energy after traveling a very short distance. This **distance between collisions** is called the **mean free path** and is typically about 10$^{-8}$ m. Because the exchange of energy happens at different times, particles are speeding up and slowing down. They have an average rms velocity, but rarely have precisely that velocity. **Therefore, particles have a large range of velocities, the average of which is the rms velocity.**

---

Problems 34-38 at the end of this study chapter cover material from this section.

---

## 5.7 - Effusion and Diffusion

When you finish this section you will be able to calculate relative rates of effusion from molar masses and vice-versa.

Your textbook points out that **diffusion and effusion** are two different processes.

- **Diffusion** relates to the mixing of gases.
- **Effusion** relates to the passage of a gas through an orifice into an evacuated chamber.

*Graham's Law of Effusion*

If the kinetic energies of two gases, 1 and 2, in a system are the same at a given temperature, then for a mole of the particles

$$KE_{avg} = N_A \left(1/2\ m_1 \bar{u}_1^{\,2}\right) = N_A \left(1/2\ m_2 \bar{u}_2^{\,2}\right)$$

or, if M, molar mass $\dfrac{m}{N_A}$,*

$$1/2\ M_1 \bar{u}_1^{\,2} = 1/2\ M_2 \bar{u}_2^{\,2}$$

by crossmultiplying and cancelling, this becomes

$$\frac{M_1}{M_2} = \frac{u_2^{\,2}}{u_1^{\,2}}, \quad \text{or} \quad \sqrt{\frac{M_1}{M_2}} = \frac{u_{rms2}}{u_{rms2}} = \frac{\text{rate of effusion}_2}{\text{rate of effusion}_1}$$

This is Graham's Law of Effusion.  The **heavier** the molecule, the **slower** the rate of effusion through a small orifice.

---

# Example 5.7 - Graham's Law of Effusion

How many times faster than He would $NO_2$ gas effuse?

## Solution

$$M_{NO_2} = 46.01\ \text{g/mol}$$

$$M_{He} = 4.003\ \text{g/mol}$$

$$\sqrt{\frac{M_{NO2}}{M_{He}}} = \frac{\text{rate}_{He}}{\text{rate}_{NO_2}}$$

$$\sqrt{\frac{46.01}{4.003}} = \frac{\text{rate}_{He}}{\text{rate}_{NO_2}}$$

$$3.39 = \frac{\text{rate}_{He}}{\text{rate}_{NO_2}}$$

**He would effuse 3.39 times as fast as $NO_2$.**

Note:  you could have solved the problem as

$$\sqrt{\frac{M_{He}}{M_{NO_2}}} = \frac{\text{rate}_{NO_2}}{\text{rate}_{He}}$$

in this case, $0.295 = \dfrac{\text{rate}_{NO_2}}{\text{rate}_{He}}$.  The conclusion is the same.

---

*Note that "M" has changed meanings.  In Chapter 4 it meant "molarity (mol/L)."  Here, it means "molar mass."  Such shifts in meaning are rare, but they do happen.*

## Does the Answer Make Sense?

$NO_2$ is considerably heavier.  We would therefore expect it to effuse more slowly.

---

With regard to **diffusion**, the important idea is that even though gases travel very rapidly (hundreds of meters per second), their motions are in all directions, so mixing is relatively slow.  In the case of diffusion, the basic structure of Graham's Law holds.

$$\frac{\text{Distance traveled}_2}{\text{Distance traveled}_1} = \sqrt{\frac{M_1}{M_2}}$$

---

## Problems 39-41 at the end of this study chapter cover material from this section.

---

## 5.8 - Real Gases

When you finish this section you will be able to:

- describe why and how gases deviate from ideal behavior, and
- solve problems relating to the extent of the deviation.

We know that no gas behaves in a truly ideal fashion.  This section is devoted to **determining the deviation of a real gas from ideality**.  There are two major corrections that your textbook deals with:

1.  *Volume Correction*:  Because gas molecules **do** take up space, the free volume of the container is not as large as it would be if it were empty.

$$\underset{\uparrow}{\text{Volume}_{\text{available}}} = V_{\text{container}} - \overset{\overset{\text{correction factor}}{\downarrow}}{\underset{\underset{\text{\# moles of gas}}{\uparrow}}{nb}}$$

Therefore, pressure can be expressed

$$P = \frac{nRT}{(V - nb)}$$

We are half-way home.

2.  *Pressure Correction*:  Because gas molecules can interact with each other, they do not <u>collide with the walls of the container</u> ("exert pressure") to as great an extent as when there is no intramolecular interaction (as with an ideal gas).  Therefore,

$$P_{\text{observed}} = P_{\text{ideal}} - \text{a correction factor}$$

As discussed in your textbook, the correction factor that accounts for the decrease in pressure due to intramolecular interactions equals

$$P_{\text{obs}} = P_{\text{ideal}} - a\left[\frac{n}{V}\right]^2$$

where a = a constant which depends upon the gas, and
n = the number of moles of gas.

Combining $P_{obs}$ (we will call this "P") and $V_{corrected}$ into the ideal gas equation, the **CORRECTED** equation is called **van der Waals's equation** and is given by

$$\left[P + a\left[\frac{n}{V}\right]^2\right]\ (V - nb) = nRT$$

correction        correction
for pressure     for volume

The constants a and b have been tabulated for different gases, and are given in your textbook. Though the numbers get a bit messy, all you are really doing is determining corrected values for P and V.

Interactions between molecules are greatest at **low temperatures (low rms velocities), high pressures and low volumes**. Thus, deviations from ideality are expected to be greatest under these conditions.

---

## Example 5.8 - van der Waal's Equation

Calculate the pressure exerted by 0.3000 mol of He in a 0.2000 L container at -25°C

    a.   using the ideal gas law, and
    b.   using van der Waals's equation.

### Strategy

    a.   PV = nRT for this ideal situation.

    b.   $P = \frac{nRT}{V}$. But, correcting for non-ideality,

$$\left[P + a\left[\frac{n}{V}\right]^2\right]\ (V - nb) = nRT$$

$$\text{or}\quad P + a\left[\frac{n}{V}\right]^2 = \frac{nRT}{(V - nb)}$$

$$\text{or}\quad P = \frac{nRT}{(V - nb)} - a\left[\frac{n}{V}\right]^2$$

This is one of the few equations that you will be working with in general chemistry that you really cannot derive. You just have to know it.

From Table 5.3 in your textbook

$$a = 0.034\ atm \cdot L^2/mol^2 \qquad b = 0.0237\ L \cdot mol$$

### Solution

    a.  $P = \frac{nRT}{V} = \frac{0.3000\ mol\ (0.08206\ L \cdot atm/K \cdot mol)\ (248.2\ K)}{0.2000\ L}$

    $P_{ideal} = \textbf{30.6 atm}$

    b.  $P = \frac{0.3000\ mol\ (0.08206\ L \cdot atm/K \cdot mol)\ (248.2\ K)}{0.2000\ L - 0.3000\ mol\ (0.0237\ L/mol)} - \frac{0.034\ atm\ L^2}{mol^2}\left[\frac{0.3000\ mol}{0.2000\ L}\right]^2$

    $P_{nonideal} = 31.67\ atm - 0.077\ atm = \textbf{31.6 atm}$

There exists a pressure difference of 1 atmosphere between the ideal equation and the van der Waals's equation in this case. The error is about 3 percent.

In rereading this section in the text, you need to keep in mind the **conditions** under which gases deviate from ideality. Be able to discuss **why** certain conditions lead to non-ideal behavior.

---

Problems 42-47 at the end of this study chapter cover material from this section.

---

## 5.9 - Chemistry in the Atmosphere

When you finish this section you will be able to list and describe those reactions that are important in the study of the atmosphere.

### Composition of Our Atmosphere

In the first part of this section, your textbook makes some key points with regard to the **composition of our atmosphere.**

1.  The atmosphere is composed of 78% $N_2$, 21% $O_2$, 0.9% Ar and 0.03% $CO_2$ along with trace gases.

2.  The composition of the atmosphere varies as a function of distance from the earth's surface. Heavier molecules tend to be near the surface due to gravity.

3.  Upper atmospheric chemistry is largely affected by ultraviolet, x-ray, and cosmic radiation emanating from space. The ozone layer is especially reactive to ultraviolet radiation.

4.  Manufacturing and other processes of our modern society affect the chemistry of our atmosphere. Air pollution is a direct result of such processes.

### Air Pollution

The reactions that cause air pollution are extremely complex and only partially understood. Your textbook gives background reactions on **photochemical smog** and **acid rain.** This serves as but a tiny introduction to the myriad of pollution related problems and mechanisms.

The general reactions causing smog are

1.  $N_2(g) + O_2(g) \xrightarrow{\text{high temperatures}} 2NO(g)$

2.  $2NO(g) + O_2(g) \longrightarrow 2NO_2(g)$

3.  $NO_2(g) \xrightarrow{\text{radiant energy}} NO(g) + O(g)$

4.  $O(g) + O_2(g) \longrightarrow O_3(g)$
    $\qquad\qquad\qquad$ ozone

Higher ozone levels that are characteristic of smog cause lung and eye irritation and can be very dangerous for people with asthma, emphysema, and other respiratory conditions. Ozone can also react with other pollutants as discussed in the text, to form the hydroxyl radical (OH), a very reactive oxidizing agent which can react with nitrogen oxides and hydrocarbons to further increase levels of air pollutants.

## Acid Rain

Your textbook points out a number of gas precursors of acid rain. One example is $SO_3(g)$, from the reaction of sulfur (from coal) and oxygen. Read the <u>Chemical Impact in this chapter of your textbook</u>, then do the following example problem.

---

## Example 5.9 - Acid Rain

List the reactions that lead to acidification of rainwater.

### Solution

As discussed in the Chemical Impact section, there are three main acids produced: $HNO_2(aq)$, $HNO_3(aq)$, and $H_2SO_4(aq)$.

The reactions are:

1.  $2NO_2(g) + H_2O(l) \longrightarrow HNO_2(aq) + HNO_3(aq)$

2a. $2SO_2(g) + O_2(g) \longrightarrow 2SO_3(g)$

2b. $SO_3(g) + H_2O(l) \longrightarrow H_2SO_4(aq)$

---

**END OF CHAPTER FIVE**

# SELF-TEST QUESTIONS

1. In a science demonstration, 4 to 6 plastic garbage bags are arranged under a 2 m × 2 m piece of plywood. A volunteer stands on the plywood and others blow into the bags to "levitate" the volunteer. If four bags are used and a 140 lb person is on the plywood, what is the pressure that must be supplied by each of the four blowers (in Pa)? What if 6 bags and blowers were used? (Assume the plywood has no mass).

2. How much force is required to inflate a high pressure bicycle tire to 95 pounds per square inch (655 kPa) with a hand pump that has a plunger with area of 5.0 $cm^2$?

3. An object exerts a force of 500 N and sits on an area 4.5 m × 1.5 m. Calculate the amount of pressure exerted by the object in torr.

4. Calculate the density of mercury. (This can be done using the fact that 760 mm Hg = 101, 325 Pa). F = ma where a = 9.81 $m/s^2$. Hint: Consider a column 76.0 cm high with cross section of 1 $cm^2$.

5. During your travels through deep space, you discover a new solar system. You land on the outermost planet and determine that the acceleration due to gravity is about 2.7$m/s^2$. If your mass back on earth is 72 kg, what force would you exert on a scale in pounds while standing on the planet's surface?

6. As you proceed on to the next planet, some of your unbreakable equipment breaks, including that top-of-the-line machine which determines acceleration due to gravity.

   a. How do you determine the acceleration due to gravity of this planet?
   b. Calculate the acceleration due to gravity if your 72 kg mass exerts a force of 18 pounds on the planet's surface.

7. A diver at a depth of 100 ft (pressure is approximately 3 atm) exhales a small bubble of air with a volume equal to 100 mL. What will the volume of the bubble be (assume the same amount of air) at the surface?

8. What would the volume of gas contained in an expandable 1.0 L cylinder at 15 MPa (1 MPa = $10^6$ Pa) be at 1 atm (assume constant temperature)?

9. A sample tube containing 103.6 mL of CO gas at 20.6 torr is connected to an evacuated 1.13 liter flask (the new volume is the sum of those of the tube and flask). What will be the pressure when the CO is allowed into the flask?

10. A gas has a pressure of 3.2 atm and occupies a volume of 45 L. What will the pressure be if the volume is compressed to 27 L at a constant temperature?

11. The volume of a gas (held at constant pressure) is to be used "as a thermometer". If the volume at 0°C is 75.0 $cm^3$ what is the temperature when the measured volume is 56.7 $cm^3$?

12. If a 16.6 L sample of a gas contains 9.2 moles of $F_2$, how many moles of gas would there be in 750 mL sample at the same temperature and pressure?

13. A 11.2 L sample of a gas is determined to contain 0.50 moles of $N_2$. At the same temperature and pressure how many moles of gas would there be in a 20 L sample?

14. Consider a 3.57 L sample of an unknown gas at a pressure of $4.3 \times 10^3$ Pa. If the pressure is changed to $2.1 \times 10^4$ Pa at a constant temperature, what will the new volume of the gas be?

15. A weather balloon is filled with 0.295 $m^3$ of helium on the ground at 18°C and 756 torr. What will the volume of the balloon be at an altitude of 10 km where the temperature is -48°C and the pressure is 0.14 atm?

16. A sample of gas occupies 3.8 L at 15°C and 1 atm. What does the temperature need to be for the gas to occupy 8.3 L at 1 atm?

17. Calculate the volume of $O_2$ present in a sample containing 0.89 moles of $O_2$ at a temperature of 40°C and a pressure of 1 atm.

18. Water is decomposed to $H_2(g)$ and $O_2(g)$ by electrolysis. By measuring the current it was determined that 0.365 moles of water were decomposed. After the gases are dried and collected at 24.5°C and 757 torr what are the volumes of each?

19. What pressure would be exerted by 50 g of He at 25°C in a volume of 350 L?

20. A vacuum line used in a research lab has a volume of 1.013 L. The temperature in the lab is 23.7°C and the vacuum line is evacuated to a pressure of $1 \times 10^{-6}$ torr. How many gas molecules (atoms) remain?

21. A 10.5 g sample of $CO_2$ gas occupies a volume of 7 L at a pressure of 1.5 atm. What must the temperature of the gas be?

22. The density of liquid nitrogen is 0.808 g/mL at -196°C. What volume of nitrogen gas at STP must be liquified to make 10.0 L of liquid nitrogen?

23. Calculate the volume occupied by 2.5 mol of an ideal gas at STP.

24. A hydrocarbon (compound contain only hydrogen and carbon) was analyzed and found to be 85.7 mass percent carbon and 14.3 mass percent hydrogen. At 26°C and 745 torr pressure a sample with a volume of 1.13 L had a mass of 1.904 g. Determine the molecular formula. (You may wish to review Chapter 3).

25. An unknown gas has a density of 7.06 g/L at a pressure of 1.5 atm and 280 K. Calculate the molar mass of the gas.

26. $HCl(g)$ can be prepared by reaction of NaCl with $H_2SO_4$. What mass of NaCl is required to prepare enough HCl to fill a 340 mL cylinder to a pressure of 151 atm at 20°C?

27. You are not sure whether to fill a balloon with He or hot air. To what temperature would the air have to be heated for a balloon to rise to the same height as a balloon filled with He at 25°C?

28. A 27.7 mL sample of $CO_2(g)$ was collected over water at 25°C and 1 atm. What is the pressure due to $CO_2(g)$ (the vapor pressure of water at 25°C is 23.8 torr)? What will the volume of $CO_2(g)$ be at the same temperature and pressure after removing the water vapor?

29. A gas-tight vessel has a volume of 1000 m³. It is filled with air at 27°C and 1 atm. Assuming air to be 79% $N_2$ and 21% $O_2$ (by volume), calculate the following:

    a. partial pressure of $N_2$
    b. partial pressure of $O_2$

30. A gaseous mixture of $O_2$, $H_2$, and $N_2$ has a total pressure of 1.5 atm and contains 8.2 g of each gas. Find the partial pressure of each gas in the mixture.

31. The mole fraction of argon in dry air is 0.00934. How many liters of air at STP will contain enough argon to fill a 35.4 L cylinder to a pressure of 150 atm at 20°C?

32. Assume that the mole fraction of nitrogen in the air is 0.8902. Calculate the partial pressure of $N_2$ in the air when the atmospheric pressure is 820 torr.

33. The lower limit of flammability for $H_2(g)$ in air at room temperature and 1 atm is a partial pressure of 0.040 atm. Assuming a 4:1 ratio of $N_2$ to $O_2$ in air, what is the ratio of $H_2$ to $O_2$ at its flammability limit?

34. Calculate the temperature of a mole of oxygen molecules if the internal energy is $1.16 \times 10^4$ J. Assume ideal gas behavior.

35. Calculate the root mean square speed of $O_2$ gas moleucles at 300 K.

36. What happens to the average kinetic energy of a mole of an ideal gas if:

    a.   the volume is doubled resulting in a decrease in pressure at constant temperature?
    b.   the temperature is increased at a constant pressure?
    c.   absolute zero is obtained?

37. Ammonia ($NH_3(g)$) and HCl(g) react to form a solid precipitate, $NH_4Cl$.  Two cotton swabs, one moistened with ammonia the other with hydrochloric acid, are inserted into opposite ends of a 1 meter long glass tube. How far from the hydrochloric acid end of the tube would you expect to see the white $NH_4Cl$ precipitate?

38. Calculate the rate of effusion of $PH_3$ molecules through a small opening if $NH_3$ molecules pass through the same opening at a rate of 8.02 $cm^3$/s.  Assume the same temperature and equal partial pressures of the two gases.

39. What are the relative rates of diffusion for methane, $CH_4$ and oxygen, $O_2$?  If $O_2(g)$ travels 1.0 m in a certain amount of time, how far will methane be able to travel under the same conditions?

40. Which gas would effuse faster:  Ne or $CO_2$?  How much faster?

41. Arrange the following according to expected values for b (volume correction value) in the van der Waal's equation:

    He, $CO_2$, $H_2O$, HF, $SF_6$.

42. Put the following gases in order from smallest to largest according to van der Waals constant "a":

    $H_2$, $N_2$, $CH_4$, Ne, $H_2O$.

43. Put the following gases in order from smallest to largest according to van der Waals constant "b":

    Kr, $Cl_2$, $NH_3$, $O_2$, He.

44. Calculate the pressure exerted by 1 mole of Xe(g) using the ideal gas law and van der Waal's equation.

    a.   in a 100.0 L container at 23°C, and
    b.   in a 1.000 L container at 23°C

45. Calculate the pressure exerted by 100 moles of $Cl_2$ gas in a 20 L container at 25 C using the van der Waals equation and constants in Table 5.3 of your textbook.

46. Calculate the density of a mixture of 21% $O_2$ and 79% $N_2$ as well as for He at 760 torr and 0°C.  Will the ratio of densities be different at 100 atm and 6°C?

# ANSWERS TO SELF-TEST QUESTIONS

1.  40 Pa for each of 4 blowers; 27 Pa for each of 6.

2.  328 N = 74 lbs

3.  0.56 torr

4.  13.6 g/cm$^3$ (Steps:  1.  Force = P × A; 2.  mass = F/a; 3.  density = mass/volume, so ultimately, density = P × area/(volume × acceleration))

5.  43 pounds

6.  a.   Find out what force you exert on the surface and work backwards.
    b.   1.1 m/s$^2$

7.  300 mL

8.  150 L

9.  1.73 torr

10. 5.3 atm

11. 206 K, -67°C

12. 0.42 moles

13. 0.89 moles

14. 0.73 L

15. 1.6 m$^3$

16. 356°C

17. 22.9 L

18. 8.95 L H$_2$, 4.47 L O$_2$

19. 873 atm

20. 3 × 10$^{13}$

21. 536 K or 263°C

22. 6.4 × 10$^3$ L

23. 56 L

24. C$_3$H$_6$

25. 108 g/mol

26. 125 g

27. 2140 K or 1867°C

28. 736 torr, 26.8 mL

29. a. 0.79 atm
    b. 0.21 atm

30. $O_2$ = 0.0831 atm; $H_2$ = 1.322 atm; $N_2$ = 0.0952 atm

31. $5.27 \times 10^5$ L of air

32. 730 torr

33. 0.2 $H_2$:1 $O_2$; 0.04 $H_2$:0.2 $O_2$

34. 5450 K

35. 484 m/s

36. a. no change
    b. increase
    c. goes to zero

37. 40 cm

38. 5.67 $cm^3$/s

39. 1.41 $CH_4$:1 $O_2$; 1.41 m

40. Ne would effuse 1.48 times as fast as $CO_2$

41. (smallest) He, HF, $H_2O$, $CO_2$, $SF_6$ (largest)

42. Ne, $H_2$, $N_2$, $CH_4$, $H_2O$

43. He, $O_2$, $NH_3$, Kr $Cl_2$

44. a.  ideal = 0.243 atm, van der Waal's = 0.243 atm
    b.  ideal = 24.3 atm, van der Waal's = 21.41 atm

45. 7.84 atm

46. air = 1.29 g/L; He = 0.179 g/L; no

# CHAPTER 6

# Thermochemistry

|  | | Page in this |
|---|---|---|
| Section | Contents | Study Guide |
| 6.1 | The Nature of Energy | 135 |
| 6.2 | Enthalpy and Calorimetry | 138 |
| 6.3 | Hess's Law | 142 |
| 6.4 | Standard Enthalpies of Formation | 144 |
| 6.5 | Present Energy Sources | 146 |
| 6.6 | New Energy Sources | 146 |

This chapter serves as an introduction to the chemistry of energy production and exchange. Additional, more advanced, material will be presented in Chapter 16.

## 6.1 - The Nature of Energy

When you finish this section you will be able to solve problems relating to:

- work, in terms of pressure and volume, and
- the First Law of Thermodynamics.

### Basic Terms

Your textbook describes a number of new terms for you. The terms, along with brief definitions, are given below:

- **Thermodynamics:** The study of energy and its interconversions.
- **Energy:** The capacity to produce work or heat.
- **Kinetic Energy:** The energy of motion. Kinetic Energy = $1/2$ mass $\times$ (velocity)$^2$
- **Potential Energy:** Energy that can be converted to useful work.
- **Heat:** Involves transfer of energy between two objects.
- **Work:** Force $\times$ distance.

- **State Function:** A property that is independent of pathway. That is, it does not matter how you get there, the difference in the value is the same. For example, you can drive from New York to Los Angeles via many different routes. No matter which one you take, you are still going from New York to Los Angeles. The actual distance between the two cities is the same. **Energy** is a state function, **work** and **heat** are not.

Four more definitions will set the stage for thermodynamics. The **universe** is composed into the **system** and the **surroundings**.

- **System:** That which we are focusing on.
- **Surroundings:** Everything else in the universe.
- **Exothermic:** Energy (as heat) flows **out of the system**.
- **Endothermic:** Energy (as heat) flows **into the system**.

## Work

We will take a different approach to this topic than your textbook does. Let's look at work from the point of view of **units**. We learned in the review chapter on <u>gas laws</u> that

$$\text{Force} = \text{mass} \times \text{acceleration} = \text{kg} \times \text{m/s}^{-2}.$$

$$\text{Work} = \text{force} \times \text{distance} = \text{kg m/s}^{-2} \times \text{m} = \text{kg m}^2/\text{s}^{-2}.$$

$$\textbf{1 Joule} = \textbf{1 kg m}^2/\textbf{s}^{-2}.$$

---

## Example 6.1 A - The Units of Work

If pressure = force/area, what are the units of pressure × volume?

## Solution

$$P = F/A = \text{kg m s}^{-2}/\text{m}^2 = \text{kg m}^{-1}/\text{s}^{-2} \ (= 1 \text{ Pascal})$$

$$P \times V = \text{kg m}^{-1} \text{ s}^{-2} \times \text{m}^3 = \text{kg m}^2/\text{s}^{-2} \ (= 1 \text{ Joule})$$

Therefore, $P \times V$ has the same units as force × distance (work), and both are measures of **energy**.

## Conclusion

For an ideal gas,

$$\textbf{WORK} = \textbf{P}\Delta\textbf{V}$$

This equation holds at constant pressure.

---

The **sign conventions** for work are as follows.

- When the **system expands**, it is doing positive work on the surroundings, therefore it is doing **negative work on the system**.

- When the **system contracts**, the surroundings have done work on the system, therefore there is **positive work done on the system**.

From the point of view of the system, then,

$$w = -P\Delta V$$

---

# Example 6.1 B - Work

Calculate the work (with the proper sign!) associated with the contraction of a gas from 75 L to 30 L (work is done "on the system") at a constant **external** pressure of 6.0 atm in,

   a.  L • atm, and
   b.  Joules (1 L • atm = 101.3 J).

## Helpful Hint

Keep in mind that **system compression is positive work** and **system expansion is negative work.**

## Solution

   a.  The work is positive (compression).

   $\Delta V$ = change in volume = $V_{final}$ - $V_{initial}$ = 30 L - 75 L = **-45 L**

   $w = -P\Delta V$ = -6.0 atm (-45 L) = $^{+}$**270 L • atm**

   b.  w (in Joules) = $^{+}270 \text{ L} \bullet \text{atm} \times \dfrac{101.3 \text{ J}}{1 \text{ L} \bullet \text{atm}}$ = $^{+}$**2.7 × 10⁴  J**

---

## The First Law

**The Law of Conservation of Energy,** also called **The First Law of Thermodynamics** is described in your textbook.  It states that **energy can be converted from one form to another, but can neither be created nor destroyed.**  Another way of stating the First Law is that

### THE ENERGY OF THE UNIVERSE IS CONSTANT

We know that energy can be changed through **work.**  As chemical bonds are made and broken, energy is converted between the potential energy (stored in chemical bonds) and  thermal energy (kinetic energy) as heat.

The change in the internal energy of the system, which is equal in size but opposite in sign to that of the surroundings, is equal to the **sum of the heat and work.**

$$\Delta E = q + w$$

Your textbook points out that the **SIGN** of the energy change must be viewed from the point of view of the **SYSTEM.**

$\Delta E$ = - **means the system loses energy.**
$\Delta E$ = + **means the system gains energy.**

## Example 6.1 C - The First Law

Calculate the change in energy of the system if 38.9 J of work is done *by* the system with an associated heat loss of 16.2 J.

### Strategy

The most important part of problem solving in thermodynamics is getting the signs correct.

q = "-," because heat is **lost**.
w = "-," because work is done **by** the system.

### Solution

$\Delta E = q + w = -16.2\ J + (-38.9\ J) = $ **-55.1 J**

The system has **lost** 55.1 J of energy.

## Example 6.1 D - Practice with Heat and Work

A piston is compressed from a volume of 8.3 L to 2.8 L against a constant pressure of 1.9 atm. In the process, there is a heat gain by the system of 350 J. Calculate the change in energy of the system.

### Solution

$$w = -P\Delta V = -1.9\ atm\ (-5.5\ L) = +10.45\ L \cdot atm$$

$$10.45\ L \cdot atm \times \frac{103.1\ J}{L \cdot atm} = +1059\ J$$

$$q = +350\ J$$

$$\Delta E = q + w = +1059 + 350 = 1409\ J = \textbf{1400 J}\ \text{(to 2 significant figures)}$$

Problems 1-5 at the end of this study chapter cover material from this section.

## 6.2 - Enthalpy and Calorimetry

When you finish this section you will be able to solve problems relating to both **enthalpy** and **calorimetry**.

### Enthalpy

Your textbook derives and defines a term called **enthalpy** (H). It is a **state function**, so that the change in H is independent of pathway. That is,

$$\Delta H = H_{products} - H_{reactants}$$

The change in enthalpy ($\Delta H$) of the system is equal to the energy flow as heat *AT CONSTANT PRESSURE*.

$$\Delta H = q_p$$

If $\Delta H => 0$, the reaction is **endothermic**. (Heat is absorbed by the system.)
If $\Delta H <= 0$, the reaction is **exothermic**. (Heat is given off by the system.)

---

# Example 6.2 A - Enthalpy

Upon adding solid potassium hydroxide pellets to water, the following reaction takes place:

$$NaOH(s) ----> NaOH(aq)$$

For this reaction at constant pressure, $\Delta H = -43$ kJ/mol. Answer the following questions regarding the addition of 14 g of NaOH to water:

  a.   Does the beaker get warmer or colder?
  b.   Is the reaction exo- or endothermic?
  c.   What is the enthalpy change for the dissolution?

## Solution

  a.   If $\Delta H < 0$, then heat is given off by the system. The beaker therefore gets *warmer*.

  b.   If heat is given off by the system, the reaction is *exothermic*.

  c.   $kJ \left\|\right. \dfrac{-43 \text{ kJ}}{\text{mol}} \times \dfrac{1 \text{ mol NaOH}}{40.0 \text{ g NaOH}} \times 14 \text{ g NaOH} = \textbf{-15 J} = \Delta H$

## Calorimetry

**Calorimetry** is the experimental technique used to determine the heat exchange ($q$) associated with a reaction.

$$\text{At constant pressure, } q = \Delta H.$$
$$\text{At constant volume, } q = \Delta E.$$

In both cases, however, **heat gain or loss** is being determined. The amount of heat exchanged in a reaction depends upon:

  1.   **The net temperature change** during the reaction.
  2.   **The amount of substance.** The more you have, the more heat can be exchanged.
  3.   The **heat capacity (C)** of a substance.

$$C = \frac{\text{heat absorbed}}{\text{increase in temperature}} = J/°C$$

Some substances can absorb more heat than others for a given temperature change.

There are three ways of expressing heat capacity:

  1.   **Heat capacity** (as above) = J/°C.
  2.   **Specific heat capacity** = heat capacity per gram of substance (J/°C g or J/Kg)
  3.   **Molar heat capacity** = heat capacity per mole of substance (J/°C mol or J/K • mol)

You can solve calorimetry problems very well using dimensional analysis.

Before we solve numerical problems let's do a problem involving interpretation of specific heat capacities.

---

# Example 6.2 B - Specific Heat Capacity

Look at Table 6.1 in your textbook. Based on the values for specific heat capacity, which conducts heat better, water or aluminum? Why is this important in cooking?

## Solution

The heat capacity of water is **4.18 J/°C g**. This means that it takes 4.18 J of energy to raise the temperature of one gram of water 1°C.

The heat capacity of aluminum is **0.89 J/°C g**. This means that it takes 0.89J of energy to raise the temperature of one gram of aluminum 1°C.

In other words, it takes almost 5 times as much energy (4.18/0.89) to raise the temperature of an equivalent amount of water by 1°C. Therefore, **Aluminum conducts heat better** because less heat causes an equal rise in temperature. This is important in cooking because pots made of aluminum transfer heat very well to food. Note that iron pots conduct heat even more readily than aluminum (but they are more difficult to take care of).

Your textbook discusses doing **constant pressure** calorimetry using a "coffee cup calorimeter." In this case, **ΔH** = **$q_p$** in units of Joules. Remember that you may use dimensional analysis to solve calorimetry problems.

---

# Example 6.2 C - Constant Pressure Calorimetry

Recall from Example 6.2 A that the enthalpy change (ΔH) per mole of NaOH is  -43 kJ/mol when

$$NaOH(s) ----> NaOH(aq)$$

If 10.0 g of solid NaOH is added to 1.00 L of water (specific heat capacity = 4.18 J/°C g) at 25.0°C in a constant pressure calorimeter, what will be the final temperature of the solution? (Assume the density of the final solution is 1.05 g/mL.)

## Setting Up the Problem

We need to know three things:

1. **Mass of the solution** = 1.00 L × 1050 g = 1050 g.
2. **Heat capacity of the solution** = 4.18 J/°C g.
3. **The enthalpy of the reaction** = -43 kJ = -43,000 J.

We want to know the **change in temperature**, ΔT. We can solve using dimensional analysis. (Keep in mind the temperature will **rise** because heat is evolved.)

## Solution

$$°C \left\| \frac{1°C\ g}{4.18\ J} \times \frac{1}{1050\ g} \times 43,000\ J = 9.8°C \right.$$

$$\underset{\substack{\uparrow \\ \text{1/specific} \\ \text{heat capacity}}}{} \quad \underset{\substack{\uparrow \\ \text{1/soln} \\ \text{mass}}}{} \quad \underset{\substack{\uparrow \\ \Delta H}}{}$$

The **final temperature** will equal 25.0°C + 0.98°C = **26.0°C**.

---

Constant volume calorimetry is discussed in your textbook. The **bomb calorimeter** is used for this application. In this case, because $\Delta V = 0$, no work is done, and $\Delta E = q_v$ in units of joules. Here as well, dimensional analysis works well.

Each bomb calorimeter is different. The **heat capacity (J/°C)** of the bomb and its parts must be determined using a known substance before the energy (or heat) of combustion can be determined.

---

# Example 6.2 D - Constant Volume Calorimetry

The heat of combustion of glucose, $C_6H_{12}O_6$, is 2800 kJ/mol. A sample of glucose weighing 5.00 g was burned with excess oxygen in a bomb calorimeter. The temperature of the bomb rose 2.4°C. **What is the heat capacity of the calorimeter?**

A 4.40 g sample of propane ($C_3H_8$) was then burned with excess oxygen in the same bomb calorimeter. The temperature of the bomb increased 6.85°C. Calculate $\Delta E_{combustion}$ of propane.

## Strategy

There are two parts to this problem. First, we must calculate the heat capacity of the bomb calorimeter using the data for glucose. Second, we can use this heat capacity to determine the energy (heat) of combustion of propane, $C_3H_8$. Remember that the energy of combustion is expressed in **kJ/mol**. A useful beginning is to convert grams of substance to moles of substance.

## Solution

A. *Heat Capacity of the Calorimeter*

$$\text{moles } C_6H_{12}O_6 = 5.00 \text{ g} \times \frac{1 \text{ mol}}{180.0 \text{ g}} = 2.78 \times 10^{-2} \text{ moles}$$

$$\textbf{heat capacity} = \frac{kJ}{°C} \ \bigg|\bigg| \ \frac{2800 \text{ kJ}}{mol} \times 2.78 \times 10^{-2} \text{ mol} \times \frac{1}{2.4°C} = \textbf{32.4 kJ/°C}$$

B. *Energy of Combustion of Propane*

$$\text{moles } C_3H_8 = 4.40 \text{ g} \times \frac{1 \text{ mol}}{44.0 \text{ g}} = 0.100 \text{ moles}$$

$$\Delta E_{combustion} = \frac{kJ}{mol} = \frac{32.4 \text{ kJ}}{°C} \times 6.85°C \times \frac{1}{0.100 \text{ mol}} = \textbf{-2220 kJ/mol}$$

(Note: We add the *negative* sign because heat is *evolved*.)

---

Problems 6-12 at the end of this study chapter cover material from this section.

---

## 6.3 - Hess's Law

When you finish this section you will be able to use Hess's Law to calculate enthalpy changes for a variety of reactions.

The critical point that is made in this section is that **enthalpy changes are state functions.** The implication is that **it does not matter if ΔH for a reaction is calculated in one step or a series of steps. This idea is called Hess's Law.**

By using values of ΔH of known reactions, we can use Hess's Law to solve for enthalpies of reactions whose values we do not know.

---

## Example 6.3 A - Hess's Law

Given the following reactions and ΔH values,

a.   $2N_2O(g) \longrightarrow O_2(g) + 2N_2(g)$              $\Delta H_a$ = -164 kJ
b.   $2NH_3(g) + 3N_2O(g) \longrightarrow 4N_2(g) + 3H_2O(l)$        $\Delta H_b$ = -1012 kJ

Calculate ΔH for

$$4NH_3(g) + 3O_2(g) \longrightarrow 2N_2(g) + 6H_2O(l).$$

**Let's Work the Problem Together**

The idea is to manipulate equations "a" and "b" so that they add up to the desired equation.  <u>There are three ways that we can manipulate equations</u>:

1.   **We can reverse the entire equation**.  By doing this, the products become reactants and vice versa.
2.   **We can multiply the entire equation by a factor** such as 3, 2, 1/2, or 1/3.
3.   **We can do both #1 and #2.**

The most important thing to keep in mind is that **WHEN YOU MANIPULATE AN EQUATION, YOU MUST MANIPULATE THE ΔH VALUE IN EXACTLY THE SAME WAY!**

If you multiply an equation by 2, you must multiply ΔH by 2.  If you reverse the equation, you must multiply ΔH by -1.  (An exothermic reaction becomes endothermic and vice versa.)

**An Effective Route To Solving Hess's Law Problems**

In my experience, the best way to solve Hess's Law problems is to **find a substance that only appears once in the reactants**.  Modify that reaction so the substance appears where it should be, and in the correct amount, as in the final reaction.  The entire substance equation must therefore be correct.  In our example, $NH_3$ appears only once in the reactants ($N_2O$ appears in **both** equations ... STAY AWAY FROM $N_2O$!!).

We have $2NH_3$ on the left hand side.  We want $4NH_3$ on that side.  Therefore, we must multiply equation "b" and $\Delta H_b$ by +2, which gives

$$4NH_3(g) + 6N_2O(g) \longrightarrow 8N_2(g) + 6H_2O(l)    \Delta H_b = -2024 \text{ kJ}$$

Oxygen appears only once in the reactants.  Therefore, if we modify equation "a" to get the correct amount of $O_2$

in the proper place, we should be done.  (We have modified both equations.)  We need to **reverse** equation "a" and **multiply** it by 3 to get $3O_2$ on the left side.  This will agree with the desired reaction.  Remember to multiply $\Delta H_a$ by -3 as well!  This gives,

$$3O_2(g) + 6N_2(g) \longrightarrow 6N_2O(g) \quad \Delta Ha = +492 \text{ kJ}$$

Let's get the final $\Delta H$ by adding our new "a" and "b,"

$$4NH_3(g) + 6N_2O(g) \longrightarrow 8N_2(g) + 6H_2O(l) \quad \Delta H_b = -2024 \text{ kJ}$$
$$3O_2(g) + 6N_2(g) \longrightarrow 6N_2O(g) \qquad\qquad \Delta H_a = +492 \text{ kJ}$$

$$3O_2(g) + 6N_2(g) + 4NH_3(g) + 6N_2O(g) \longrightarrow 6N_2O(g) + 8N_2(g) + 6H_2O(l)$$

and cancelling common terms,

$$3O_2(g) + 4NH_3(g) \longrightarrow 2N_2(g) + 6H_2O(l) \quad \Delta H = -1532 \text{ kJ}$$

Getting the correct final reaction serves as your check of correctness.

---

# Example 6.3 B - Practice with Hess's Law

Given the following reactions and $\Delta H$ values,

a.  $B_2O_3(s) + 3H_2O(g) \longrightarrow B_2H_6(g) + 3O_2(g)$    $\Delta H = +2035 \text{ kJ}$
b.  $2H_2O(l) \longrightarrow 2H_2O(g)$    $\Delta H = +88 \text{ kJ}$
c.  $H_2(g) + 1/2\ O_2(g) \longrightarrow H_2O(l)$    $\Delta H = -286 \text{ kJ}$
d.  $2B(s) + 3H_2(g) \longrightarrow B_2H_6(g)$    $\Delta H = +36 \text{ kJ}$

Calculate $\Delta H$ for

$$2B(s) + 3/2\ O_2(g) \longrightarrow B_2O_3(s)$$

## Solution

Keep the strategy in mind!  Boron and $B_2O_3$ only appear once.  Work with them first.

1.  Working with B(s), we see that **equation "d" is correct as is.**

2.  Working with $B_2O_3(s)$, we must **multiply equation "a" by -1** (thus reversing the equation).

Now we must work with things that are present more than once.

3.  Working with $O_2(g)$, we see that we must **multiply equation "c" by -3** to guarantee a total of **3/2 $O_2$ (g)** on the left side when all equations are added together (remember, we already have 3 $O_2$ (g) on the left side from equation "a".)

Three of our equations are now set.  Equation "b" involves $H_2O(l)$ and $H_2O(g)$.  Because we multiplied equation "c" by -3, we have $3H_2O$ (l) on the left-hand side to cancel.  Therefore, we must **multiply equation "b" by -3/2.**

### Checking the Equations and Determining $\Delta H$

a.  $B_2H_6(g) + 3O_2(g) \longrightarrow B_2O_3(s) + 3H_2O(g)$    $\Delta H = -2035 \text{ kJ}$

b.  $3H_2O(g) \longrightarrow 3H_2O(l)$    $\Delta H = -132 \text{ kJ}$

c.  $3H_2O(l) \longrightarrow 3H_2(g) + 3/2 \, O_2(g)$            $\Delta H = +858 \, kJ$

d.  $2B(s) + 3H_2(g) \longrightarrow B_2H_6(g)$            $\Delta H = +36 \, kJ$

$$B_2H_6(g) + 3O_2(g) + 3H_2O(g) + 3H_2O(l) + 2B(s) + 3H_2(g) \longrightarrow$$

$$B_2O_3(s) + 3H_2O(g) + 3H_2O(l) + 3H_2(g) + 3/2 \, O_2(g) + B_2H_6(g)$$

Cancelling we get

$$2B(s) + 3/2 \, O_2(g) \longrightarrow B_2O_3(s) \quad \Delta H = -1273 \, kJ$$

Look at Sample Exercise 6.8 in your textbook. Some of the equations here were turned around and multiplied by an integer value, but you notice that the $\Delta H$ values ultimately agree with those in your textbook. Hess's Law illustrates that enthalpy is a state function.

---

# Problems 13-15 at the end of this study chapter cover material from this section.

---

## 6.4 - Standard Enthalpies of Formation

When you finish this section you will be able to use your knowledge of standard states and standard enthalpies of formation to calculate $\Delta H$ for a variety of reactions.

Look at your textbook, where the STANDARD ENTHALPY OF FORMATION ($\Delta H°_f$) of a compound is defined. There are some important points that are worth going over.

1.  $\Delta H°_f$ is always given **per mole** of compound formed.
2.  $\Delta H°_f$ involves formation of a **compound** from its **elements** with the substances in their **standard states**.
3.  Your textbook lists the following standard state conditions:

    **For an element:**
    •it is the form which the element exists in at 25°C and 1 atmosphere.

    **For a compound:**
    •For a gas it is a pressure of exactly 1 atmosphere.
    •For a substance in solution, it is a concentration of exactly 1 M.
    •For a pure solid or liquid, it is the pure solid or liquid.

4.  $\Delta H°f$ for an element in its standard state (such as $Ba(s)$ or $N_2(g)$) equals 0.

---

## Example 6.4 A - Standard Enthalpies of Formation

By consulting Appendix Four of your textbook, and from your knowledge of standard states, list the standard enthalpy of formation for each of the following substances.

a.  $Al_2O_3(s)$            d.  $SO_4{}^{2-}(aq)$
b.  $Ti(s)$            e.  $F_2(g)$
c.  $P_4(g)$

## Solution

    a.   -1676 kJ/mol

    b.   0 kJ/mol (The solid is the standard state of titanium.)

    c.   59 kJ/mol

    d.   -909 kJ/mol

    e.   0 kJ/mol (The gaseous diatom is the standard state of fluorine.)

---

The key to calculating standard enthalpy changes in reactions is to remember "products minus reactants." More correctly,

$$\Delta H^{\circ}_{reaction} = \Sigma n_p \Delta H^{\circ}_{f(products)} - \Sigma n_r \Delta H^{\circ}_{f(reactants)}.$$

This reads "the sum of the $\Delta H^{\circ}_f$ for n moles of each of the products minus the sum of the $\Delta H^{\circ}_f$ for n moles of each of the reactants."

**Keep in mind that just as in Hess's Law problems, when you multiply the substance by an integer coefficient in a balanced equation, you must multiply the $\Delta H^{\circ}_f$ value by that integer as well!

Let's work the following example together.

---

# Example 6.4 B - Calculating Standard Enthalpies of Formation

Using the data in <u>Appendix Four of your textbook</u>, calculate $\Delta H^{\circ}_f$ for the following reaction:

$$2C_3H_6(g) + 9O_2(g) \longrightarrow 6CO_2(g) + 6H_2O(l)$$

## Solution

The $\Delta H^{\circ}_f$ for the reaction $= \Sigma n_p \Delta H^{\circ}_{f(products)} - \Sigma n_r \Delta H^{\circ}_{f(reactants)}.$

$$\Sigma n_p \Delta H^{\circ}_{f(products)} \quad = 6 \times \Delta H^{\circ}_{f_{H_2O(l)}} + 6 \times \Delta H^{\circ}_{f_{CO_2(g)}}$$

$$= 6 \text{ mol} \times -286 \text{ kJ/mol} + 6 \text{ mol} \times -393.5 \text{ kJ/mol}$$

$$= \textbf{-4077 kJ}$$

$$\Sigma n_r \Delta H^{\circ}_{f(reactants)} \quad = 2 \times \Delta H^{\circ}_{f_{C_3H_6(g)}} + 9 \times \Delta H^{\circ}_{f_{O_2(g)}}$$

$$= 2 \text{ mol} \times 20.9 \text{ kJ/mol} + 9 \text{ mol} \times 0 \text{ kJ/mol}$$

$$= \textbf{+41.8 kJ}$$

Finally,

$\Delta H^{\circ}_f = (-4077 \text{ kJ}) - (+41.8 \text{ kJ}) = \textbf{-4119 kJ}$

## Example 6.4 C - Practice with Standard Enthalpies

Sample Exercise 6.10 in your textbook describes the "thermite" reaction in which molten iron is made from the reaction of aluminum powder and iron oxide. A variation on that reaction was described in the October 1984 Journal of Chemical Education. The reaction is,

$$2Al(s) + Cr_2O_3(s) ----> Al_2O_3(s) + 2Cr(s)$$

 a.   Calculate $\Delta H°_f$ for this reaction.
 b.   Which reaction yields more energy **per gram of metal formed**, thermite or this one?

## Solution

 a.   $\Delta H°_f = [-1676\ kJ + 2 \times 0] - [2 \times 0 + (-1128\ kJ)]$
          ↑             ↑          ↑              ↑
       $Al_2O_3(s)$   $Cr(s)$   $Al(s)$      $Cr_2O_3(s)$

   $\Delta H°_f$ = **-548 kJ**

 b.   $\Delta H°_f$(thermite) = **-850 kJ**

In the thermite reaction, 2 moles, or 111.7 g, of Fe are formed. Therefore, the heat evolved per gram of Fe formed equals:

$$\frac{-850\ kJ}{111.7\ g\ Fe} = \textbf{-7.61 kJ/g Fe formed}$$

$\Delta H°_f$(chromium formation) = -548 kJ. There are 2 moles, or 104.0 g of chromium formed. The heat evolved per gram of chromium formed equals:

$$\frac{548\ kJ}{104.0\ g\ Cr} = \textbf{-5.27 kJ/g Cr formed}$$

Therefore, the **thermite reaction yields more energy per gram of metal formed.**.

Problems 16-18 at the end of this study chapter cover material from this section.

## 6.5 and 6.6 - Present and New Energy Sources

Carefully read Sections 6.5 and 6.6 in your textbook. This material deals with the application of thermodynamics to the "real world," and as such is perhaps the most interesting in the chapter! When you are done, use the following review questions to test your understanding of the material.

## Section 6.5

 1.   What is **petroleum**?
 2.   What was the most likely way in which petroleum was formed?
 3.   What gases make up **natural gas**?
 4.   How are hydrogens separated from one another?

5.  What are the major uses of the various petroleum fractions?
6.  What is **pyrolitic cracking**?
7.  There are two reasons that lead additives are a problem in gasoline.  Discuss them.
8.  What are the four stages of **coal**?  How do they differ?
9.  What are some of the problems with the use of coal as an energy source?
10. What is the greenhouse effect?  Why will increased amounts of $CO_2$ exacerbate this effect?

## Section 6.6

11. What is meant by **coal gasification**?
12. Outline the process of coal gasification.
13. Define **synthetic gas**.
14. Why is hydrogen a theoretically good choice for use as a fuel?
15. Why is hydrogen a poor **practical** choice for use as a fuel?  (List reactions and thermodynamic values.)
16. What is the most common method of producing ethanol?
17. What are some of the advantages of alcohols in fuels?

---

**END OF CHAPTER SIX**

## SELF-TEST QUESTIONS

1. A system does 3 J of work on the surroundings and 12 J of work are added to the system.

   a. What is the energy change of the system?
   b. Of the surroundings?

2. One hundred joules of work are required to compress a gas. At the same time, the gas gives off 23 J of heat to the surroundings. What is the energy change of the system?

3. A gas expands from 10 L to 20 L against a constant pressure of 5 atm. During this time it absorbs 2 kJ of heat. Calculate the work done in kJ.

4. A piston expands against 1 atm of pressure, from 11.2 L to 29.1 L. This is done without any transfer of heat.

   a. Calculate the change in energy of the system.
   b. Calculate the change in energy for the above change if, in addition, the system absorbs 1,037 J of heat from the surroundings.

5. If the internal energy of a thermodynamic system is decreased by 300 J when 75 J of work is done *on* the system, how much heat was transferred, and in which direction, to or from the system?

6. A gas is compressed against a constant pressure of 3.4 atm from 27.9 L to 16.3 L. During this process, there is a heat gain by the system of 122 J. Calculate the change in energy of the system.

7. If 596 J of heat are added to 29.6 g of water at 22.9°C in a coffee cup calorimeter, what will be the final temperature of the water?

8. A 5.037 g piece of iron heated to 100°C is placed in a coffee cup calorimeter that initially contains 27.3 g of water at 21.2°C. If the final temperature is 22.7°C, what is the specific heat capacity of the iron (J/g°C)?

9. One liter of an ideal gas at 0°C and 10 atm was allowed to expand to 1.89 L against a constant external pressure of 1 atm, at a constant temperature. The enthalpy change ($\Delta H$) for this process is -901 J. Calculate q, w, and $\Delta E$.

10. The heat capacity of a bomb calorimeter was determined by burning 6.79 g of methane (heat of combustion = -802 kJ/mol) in the bomb. The temperature changed by 10.8°C.

    a. What is the heat capacity of the bomb?
    b. A 12.6 g sample of acetylene, $C_2H_2$, produced a temperature increase of 16.9°C in the same calorimeter. What is the heat of combustion of acetylene (kJ/mol)?

11. A sample of $C_6H_5COOH$ (benzoic acid) weighing 1.221 g was placed in a bomb calorimeter and ignited in a pure $O_2$ atmosphere. A temperature rise from 25.24°C to 31.67°C was noted. The heat capacity of the calorimeter was 5.020 kJ/°C and the combustion products were $CO_2$ and $H_2O$. Calculate the $\Delta H$ in kJ/mol for the reaction.

12. When 1.5 L of 1 M $Na_2SO_4$ solution at 30°C is added to 1.5 L of 1 M $Ba(NO_3)_2$ solution at 30°C in a calorimeter, a white solid ($BaSO_4$) forms. The temperature of the mixture increases to 42°C. Assuming that the specific heat capacity of the solution is 6.37 J/°C g and that the density of the final solution is 2 g/mL, calculate the enthalpy change per mole of $BaSO_4$ formed.

13. Calculate $\Delta H$ for

$$N_2(g) + O_2(g) \longrightarrow 2NO(g)$$

    given:

    a. $N_2(g) + 2O_2(g) \longrightarrow 2NO_2(g)$    $\Delta H_{298} = 66.4$ kJ/mol
    b. $2NO(g) + O_2(g) \longrightarrow 2NO_2(g)$    $\Delta H_{298} = -114.1$ kJ/mol

14. Given

    a.   $2H_2(g) + C(s) \longrightarrow CH_4(g)$       $\Delta H_{298} = -74.81$ kJ/mol

    b.   $2H_2(g) + O_2(g) \longrightarrow 2H_2O(l)$     $\Delta H_{298} = -571.66$ kJ/mol

    c.   $C(s) + O_2(g) \longrightarrow CO_2(g)$       $\Delta H_{298} = -393.52$ kJ/mol

    Calculate $\Delta H$ for

$$CH_4(g) + 2O_2(g) \longrightarrow CO_2(g) + 2H_2O(l)$$

15. Given the following thermochemical data, calculate the $\Delta H°_f$ for:

$$Ca(s) + 2H_2O(l) \longrightarrow Ca(OH)_2(s) + H_2$$

    a.   $H_2(g) + 1/2O_2(g) \longrightarrow H_2O(l)$    $\Delta H° = -285$ kJ

    b.   $CaO(s) + H_2O(l) \longrightarrow Ca(OH)_2(s)$    $\Delta H° = -64$ kJ

    c.   $Ca(s) + 1/2O_2(g) \longrightarrow CaO(s)$    $\Delta H° = -635$ kJ

16. Using standard heats of formation (Appendix 4 in your textbook) calculate $\Delta H$ for the following reactions.

    a.   $2H_2O_2(l) \longrightarrow 2H_2O(l) + O_2(g)$    $\Delta H°_f\ H_2O_2(l) = {}^-187.8$ kJ/mol

    b.   $HCl(g) \longrightarrow H^+(aq) + Cl^-(aq)$

    c.   $2NO_2(g) \longrightarrow N_2O_4(g)$

    d.   $C_2H_2(g) + H_2(g) \longrightarrow C_2H_4(g)$

    e.   $2NaOH(s) + CO_2(g) \longrightarrow Na_2CO_3(s) + H_2O(g)$

17. Calculate the standard change in enthalpy for the following thermite reaction by using enthalpies of formation:

$$2Al(s) + Fe_2O_3(s) \longrightarrow Al_2O_3(s) + 2Fe(s)$$

NOTE: This reaction occurs when a mixture of powdered aluminum and iron (III) oxide are ignited with a magnesium fuse.

## ANSWERS TO SELF-TEST QUESTIONS

1.  a.  9 J
    b.  -9 J

2.  77 J

3.  w = -5 kJ

4.  a.  -1.81 kJ
    b.  -776 J

5.  -375 J was transferred *from* the system

6.  +4117 J

7.  27.7°C

8.  0.441 J/g°C

9.  q = 901 J;          w = -901 J;       $\Delta E = 0$

10. a.  31.5 kJ/°C
    b.  1100 kJ/mol

11. -3228 kJ/mol

12. $\Delta H = -459$ kJ/mol

13. 180.5 kJ/mol

14. -890.36 kJ

15. -414 kJ

16. a.  -196 kJ     c.  -58 kJ     e.   -125
    b.  -75 kJ      d.  -175 kJ

17. -850 kJ

# CHAPTER 7

# Atomic Structure and Periodicity

| Section | Contents | Page in this Study Guide |
|---|---|---|
| 7.1 | Electromagnetic Radiation | 151 |
| 7.2 | The Nature of Matter | 153 |
| 7.3 | The Atomic Spectrum of Hydrogen | 155 |
| 7.4 | The Bohr Model | 155 |
| 7.5 | The Quantum Mechanical Model of the Atom | 157 |
| 7.6 | Quantum Numbers | 158 |
| 7.7 | Orbital Shapes and Energies | 160 |
| 7.8 | Electron Spin and the Pauli Principle | 160 |
| 7.9 | Polyelectronic Atoms | 160 |
| 7.10 | History of the Periodic Table | 160 |
| 7.11 | The Aufbau Principle and the Periodic Table | 162 |
| 7.12 | Further Development of the Polyelectronic Model | 165 |
| 7.13 | Periodic Trends in Atomic Properties | 167 |
| 7.14 | The Properties of a Group:  The Alkali Metals | 168 |

I find the material in this chapter to be among the most interesting and useful in all of chemistry.  The coverage of the electromagnetic spectrum leads into the electronic structure of atoms. From this information, we can rationalize, and predict, such properties of atoms as size, ionization energy, and the way in which they will form bonds (a discussion which continues in chapters 8 and 9).

## 7.1 - Electromagnetic Radiation

When you finish this section you will be able to:

- List the regions of the electromagnetic spectrum.
- State the relationship between the wavelength and frequency of a given region.
- Convert between wavelength and frequency.

This topic is important for many reasons, as is pointed out in your textbook.  From my point of view, the most fascinating aspect has to do with outer space.  Electromagnetic radiation is the **only** source of information from celestial objects other than the Moon, Mars and Venus.  We can't directly touch the stars, but we can receive their radiation and learn from this.

The definitions for **electromagnetic radiation**, **wavelength**, and **frequency** are given in Section 7.1 of your textbook. Be able to define these terms. You should also know the different regions of the electromagnetic spectrum and the wavelengths covered by each region. The relationship between frequency and wavelength is

$$\nu = c/\lambda$$
$$\uparrow \quad \uparrow$$
$$\text{nu} \quad \text{lambda}$$

**wavelength** ($\lambda$) is in meters.
**frequency** ($\nu$) is in sec$^{-1}$, or Hz.
**speed of light** (c) is in meters/sec.

In my class, we often greet each other with "What's nu (new)?" "Why, c/$\lambda$, of course! What's nu with you?" This way you will remember that **frequency is inversely proportional to wavelength**.

---

## Example 7.1 A - True-False on the Electromagnetic Spectrum

Answer the following questions as "True" or "False."

1.  Blue light has a shorter wavelength than red light.
2.  X-rays have lower frequencies than radio waves.
3.  Microwaves have higher frequencies than gamma rays.
4.  Visible radiation composes the major portion of the electromagnetic spectrum.

## Solution

1.  True       2.  False       3.  False       4.  False

---

## Example 7.1 B - Wavelength - Frequency Conversion

The Chemical Impact at the end of this section of your textbook discusses the polar bear's ability to convert ultraviolet light from the sun to energy. If U. V. radiation with a frequency of $1.76 \times 10^{15}$ s$^{-1}$ is converted by the bear, what wavelength does this correspond to?

## Solution

$$\nu = c/\lambda, \text{ so } \lambda = c/\nu$$

$$\lambda = \frac{2.9979 \times 10^8 \text{ m/s}}{1.76 \times 10^{15} \text{ s}^{-1}} = 1.70 \times 10^{-7} \text{ m} = \mathbf{170 \text{ nm}}$$

## Does the Answer Make Sense?

As discussed in your textbook, the visible region goes from about 400 nm to 700 nm. The ultraviolet region has a **lower wavelength** (higher frequency), so the answer does make sense.

---

Problems 1-4 at the end of this study chapter cover material from this section.

---

## 7.2 - The Nature of Matter

When you finish this section you will be able to solve problems relating to

- the interconversion between energy, wavelength, and frequency, and
- the de Broglie equation.

### Relating Energy, Wavelength and Frequency

There are two critically important equations introduced at the beginning of Section 7.2 of your textbook. The first one is:

$$\Delta E = h\nu$$

where $\Delta E$ is the change in energy for a system (in Joules **PER PHOTON**),
      **h** is Planck's constant ($6.626 \times 10^{-34}$ J s), and
      **$\nu$** is the frequency of the wave (in $s^{-1}$).

*5.23 × 10¹³* 

$5.23 \times 10^{13}$

Only certain specific amounts of energy can be gained or lost in a substance. These **quanta** have magnitudes that depend on the substance.

Given the observed frequency change, the change in energy can be calculated for the absorption or emission of "**photons**" (the term used if the quanta of energy are viewed as **particles**).

From section 7.1, you know that $\nu$ and $\lambda$ are related by $\nu = c/\lambda$. Therefore,

$$\Delta E = h\nu = \frac{hc}{\lambda}$$

This is the second equation you should know. These two equations are among the most important in all of chemistry, as you shall see in succeeding chapters.

Let's have some practice at interconversion between energy, frequency, and wavelength.

$\nu = \dfrac{c}{\lambda}$

$\lambda = \dfrac{c}{\nu}$

## Example 7.2 A - Interconverting Between Energy, Frequency and Wavelength

Sodium atoms have a characteristic yellow color when excited in a flame. The color comes from the emission of light of 589.0 nm.

    a.  What is the frequency of this radiation?
    b.  What is the energy of this radiation per photon? Per mole of photons?

$m = \dfrac{h}{\lambda \nu}$

$\nu = \dfrac{h}{m\lambda}$

### Solution

$\Delta E = h\nu = \dfrac{hc}{\lambda}$

    a.  $\nu = \dfrac{c}{\lambda} = \dfrac{2.9979 \times 10^8 \ \text{m/s}}{5.890 \times 10^{-7} \ \text{m}} = \mathbf{5.090 \times 10^{14} \ s^{-1}}$

    b.  $\Delta E = h\nu = 6.626 \times 10^{-34} \ \text{J s} \times 5.090 \times 10^{14} \ s^{-1} = \mathbf{3.372 \times 10^{-19} \ J}$

$\dfrac{6.626 \times 10^{-32} \times 2.2297 \times 10^8}{5.89 \times 10^{-7}}$

This value is *per photon*.  There are $\dfrac{6.022 \times 10^{23} \text{ photons}}{\text{mole photons}}$  (Avogadro's number).  To convert $\Delta E$ per photon to to $\Delta E$ per mole,

$$\frac{J}{\text{mole}} = \frac{3.372 \times 10^{-19} \text{ J}}{\text{photon}} \times \frac{6.022 \times 10^{23} \text{ photons}}{\text{mole photons}} = 2.031 \times 10^5 \text{ J/mole}$$

$$\Delta E = 203.1 \text{ kJ/mole}$$

---

## Example  7.2 B - Practice with Interconversions

It takes 382 kJ of energy to remove one mole of electrons from gaseous cesium.  What is the wavelength associated with this energy?

### Strategy

The conversion between energy and wavelength requires the value of energy to be **per photon**.  This is our first conversion.  We may then convert directly to wavelength.

### Solution

$$\frac{kJ}{\text{photon}} = \frac{3.82 \times 10^5 \text{ J}}{\text{mol}} \times \frac{1 \text{ mol}}{6.022 \times 10^{23} \text{ photons}} = 6.343 \times 10^{-19} \text{ J (per photon)}$$

$$\Delta E = \frac{hc}{\lambda} \text{ , so } \lambda = \frac{hc}{\Delta E}$$

$$\lambda = \frac{6.626 \times 10^{-34} \text{ J s} (2.9979 \times 10^8 \text{ m/s})}{6.343 \times 10^{-19} \text{ J}} = 3.13 \times 10^{-7} \text{ m} = \textbf{313 nm}$$

This lies in the ultraviolet region.

---

### de Broglie's Equation

Let's review your textbook's derivation of de Broglie's equation.  The goal is to relate wavelength ($\lambda$) to mass (m) of a particle.

$$E = mc^2, \text{ and } E = \frac{hc}{\lambda}$$

This gives us $m = \dfrac{h}{\lambda c}$, or for a particle which is not moving at the speed of light (c), but rather at some velocity

(v), **de Broglie's equation** is

$$m = \frac{h}{\lambda v} \text{ or } \lambda = \frac{h}{mv}$$

The important point is that **at least on an atomic scale**, electromagnetic radiation has mass, and particles with mass exhibit a characteristic wavelength.  This is called the **dual nature of light**.

---

## Example 7.2 C - de Broglie's Equation

What is the wavelength of an electron (mass = $9.11 \times 10^{-31}$ kg) traveling at $5.31 \times 10^6$ m/s?

### Solution

Before we "plug and chug," recall that $1 \text{ J} = 1 \text{ kg m}^2 \text{ s}^{-2}$.

$$\lambda = \frac{h}{mv} = \frac{6.626 \times 10^{-34} \text{ J s}}{9.11 \times 10^{-31} \text{ kg } (5.31 \times 10^6 \text{ m/s})} = \frac{6.626 \times 10^{-34} \text{ kg m}^2 \text{ s}^{-2} \text{ s}}{9.11 \times 10^{-31} \text{kg } (5.31 \times 10^6 \text{ m s}^{-1})} = 1.37 \times 10^{-10} \text{ m} = \mathbf{0.137 \text{ nm}}$$

Keep in mind that your answer is probably reasonable if the dimensions work out!!

---

The importance of the **dual nature of light** is outlined in the paragraphs on **diffraction** at the end of <u>Section 7.2 in your textbook.</u>

---

Problems 5-11 at the end of this study chapter cover material from this section.

---

## 7.3 and 7.4 - The Atomic Spectrum of Hydrogen and the Bohr Atom

When you finish this section you will be able to calculate the energies or associated wavelengths corresponding to electron transitions in the hydrogen atom.

The key idea for Section 7.3 is the difference between a **continuous spectrum** and a **discrete or line spectrum**.

- A **continuous spectrum** contains all the wavelengths over which the spectrum is continuous.
- A **line spectrum** contains certain specific wavelengths which are characteristic of the substance emitting those wavelengths.

It is concluded in your textbook that the fact that hydrogen has a line spectrum shows that only certain energy transfers are allowed in hydrogen. **There are specific energy levels among which the hydrogen electron can shift**. These energy levels are said to be **quantized**.

At the beginning of Section 7.4, your textbook discusses Neils Bohr's reasoning for relating the **energy levels** to the **observed wavelengths** emitted by the hydrogen atom. Although Bohr's ultimate conclusions have since been enhanced, his work represented a great leap forward in 1913. The equation that is descriptive of Bohr's model is

$$E = -2.178 \times 10^{-18} \text{ J} \left[ \frac{z^2}{n^2} \right]$$

where E is the energy (in Joules)
z is the nuclear charge (1 for hydrogen's one proton)
n is an integer related to orbital position. The farther out from the nucleus, the higher the value of n. If an electron is given enough energy, it goes away from the nucleus. We say it is **ionized**, and $\mathbf{n = \infty}$.

**The lowest energy state is n = 1, the ground state.**
**The highest energy state is n = ∞ , where an electron is ionized.**

## Example 7.4 A - Energy Levels in the Bohr Atom

Calculate the energy corresponding to the n = 3 electronic state in the Bohr hydrogen atom.

### Solution

In the hydrogen atom, z = 1. In this problem, n = 3.

$$E_3 = -2.178 \times 10^{-18} \text{ J} \left[ \frac{z^2}{n^2} \right] = -2.178 \times 10^{-18} \text{J} \left[ \frac{1^2}{2^2} \right] = -2.42 \times 10^{-19} \text{ J}$$

This is fairly straightforward. A more challenging but **chemically meaningful** calculation is presented in the next problem.

## Example 7.4 B - Changes in Energy in the Bohr Atom

Calculate the energy **change** corresponding to the exciting of an electron from the n = 1 to n = 3 electronic state in the hydrogen atom.

### Solution

In this problem, we need the change in energy.

$$\Delta E = E_{final} - E_{initial} = E_3 - E_1$$

There are two possible approaches to the calculation. The first is to calculate each energy separately and take the difference. The second is to subtract the equations, then factor and calculate the change using one equation.

Approach 1:

$$\Delta E = -2.178 \times 10^{-18} \text{ J} \left[ \frac{1^2}{3^2} \right] = -2.178 \times 10^{-18} \text{ J} \left[ \frac{1^2}{1^2} \right]$$

Approach 2:

$$\Delta E = -2.178 \times 10^{-18} \text{ J} \left[ \frac{1}{3^2} - \frac{1}{1^2} \right] = -2.178 \times 10^{-18} \left[ \frac{-8}{9} \right]$$

Both approaches will yield

$$\Delta E = +1.94 \times 10^{-18} \text{ J}$$

The "+" sign is very important! It means energy was **absorbed** to excite the electron. In other words, the system has **gained energy**.

## Example 7.4 C - Wavelength from Energy in the Bohr Atom

What wavelength of electromagnetic radiation is associated with the energy change in promoting an electron from the n = 1 to n = 3 level in the hydrogen atom?  (Use the value of $\Delta E$ from the previous example.)

**Solution**

$$\lambda = \frac{hc}{\Delta E}$$

$$\Delta E = \frac{hc}{\lambda} \text{ , so } \lambda = \frac{hc}{\Delta E}$$

$$\lambda = \frac{6.626 \times 10^{-34} \text{ J s } (2.9979 \times 10^8 \text{ m/s})}{1.94 \times 10^{-18} \text{ J}} = 1.02 \times 10^{-7} \text{ m} = \textbf{102 nm}$$

This energy corresponds to radiation in the ultraviolet region of the spectrum.

(Note:  Even if $\Delta E$ = "-" (as it will when energy is released), you **must** assign a "+" sign to $\lambda$ because <u>you can't have a negative wavelength!</u>)

---

Problems 12-17 at the end of this study chapter cover material from this section.

---

## 7.5 - The Wave Mechanical Model of the Atom

When you finish this section you will be able to:

- Describe the basic terms and ideas in the wave mechanical model of the atom.
- Solve problems relating to the Heisenberg Uncertainty Principle.

As is pointed out in <u>Section 7.5 of your textbook</u>, there is a key assumption of the wave mechanical model regarding electron motion around the hydrogen atom.  This assumption is that **the electron is assumed to behave as a standing wave.**  Only certain orbits are shaped such that the "wave" (electron) can fit.

The **wave function** of an electron represents the allowed coordinates where the electron *may* reside in the atom. Each wave function is called an **orbital**.

Read the discussion on the **Heisenberg Uncertainty Principle** in your textbook.  The principle says that "there is a limit to just how precisely we can know both the position and momentum of a particle at a given time."  It turns out that when the radiation used to locate a particle hits that particle, it changes its momentum.  Therefore, the position and momentum cannot both be measured exactly.  As one is measured more precisely, the other is known less precisely.  The relation is given by

$$\Delta x \cdot \Delta(mv) \geq \frac{h}{4\pi}$$

where $\Delta \mathbf{x}$ is the uncertainty in the particle's position
$\Delta \mathbf{mv}$ is the uncertainty in the particle's momentum
$\mathbf{h}$ is Planck's constant

The smallest possible uncertainty is $\mathbf{h/4\pi}$.

The Heisenberg Uncertainty Principle tells us that we cannot know exactly where an electron is around the atom at any given time.  However, we can know the **probability** of it being within a certain region.

The **square of the wave function** is called the **probability distribution** and it represents the probability (statistical likelihood) of finding the electron at a particular area around the nucleus.

This section in your textbook ends by noting that there is a probability (exceedingly small, but it does exist) of finding the electron anywhere in the universe. Therefore, the (arbitrary) limit of the size of an orbital is "the radius of the sphere that encloses 90% of the total electron probability." In other words, at any time, t, there is a 90% chance of finding the electron in that orbital. It is also helpful to think of an orbital as a "3-dimensional electron density map" in which there is a finite probability of finding the electron.

---

Problems 18-19 at the end of this study chapter cover material from this section.

---

## 7.6 - Quantum Numbers

When you finish this section you will be able to:

- Describe the principle quantum numbers n, l and $m_l$.
- Decide which values for these numbers are valid for a variety of problems.

This section in your textbook begins by defining the term "**quantum numbers**" as being **a series of numbers which describe several properties of an orbital.**

### *Quantum Numbers*

| Name | Designation | Property of the Orbital | Possible Range of Values |
|------|-------------|-------------------------|--------------------------|
| principle quantum number | n | related to **size and energy** of the orbital | integers > 0 (1,2,3,...) |
| angular momentum quantum number | l | related to the **shape** of the orbital | integers from 0 to n-1* |
| magnetic quantum number | $m_l$ | related to the **position of the orbital in space relative to other orbitals** | integers from -1 to +1 and 0 |

Table 7.2 in your textbook gives the allowed values for the quantum number for the first four levels of the hydrogen atom. In examining the table, the **most important** thing that you can do is **derive** the values of l and ml that go with each value of n. For example, if n=2, and if l can go from 0 to n-1, then l = 0 or l = 1.

$$n = 2 \begin{cases} \text{l} = 0 \\ \text{l} = 1 \end{cases}$$

---

*l = 0 is called "s"; l = 1 is "p"; l = 2 is "d"; l = 3 is "f"; l = 4 is "g"; l = 5 is "h" and so on through the alphabet. As your textbook outlines, this labeling system arose historically from early spectral line studies: s = "sharp," p = "principle," d = "diffuse," and f = "fundamental."*

For l = 0 ("s"), ml can only equal 0 (-l to +l).
For l = 1 ("p"), ml can equal -1, 0 or +1.

Therefore, there are 4 energy levels associated with n = 2, **one for 2s and 3 for 2p**.

---

## Example 7.6 A - Quantum Numbers

If for a given atom l = 11 (this is not physically realistic; rather, it is an exercise in understanding), how many energy levels are possible?  What is the subshell designation for this orbital?

### Solution

If l = 11, $m_l$ can range from -11 to +11 and can include 0.  Therefore there are **23 possible energy levels**.  If l = 3 is called "f" and l = 4 is a "g," **l = 11 would be an "n" orbital**.  (Remember that this example is hypothetical.)

---

## Example 7.6 B - Practice with Quantum Numbers

Which of the following sets of quantum numbers are not allowed in the hydrogen atom?  For each incorrect set, state why it is incorrect.

   a.   $n = 1, l = 0, m_l = 1$
   b.   $n = 2, l = 2, m_l = 1$
   c.   $n = 5, l = 3, m_l = 2$
   d.   $n = 6, l = -2, m_l = 2$
   e.   $n = 6, l = 2, m_l = -2$

### Solution

   a.   Not allowed.  ($m_l$ can't be greater than l.)
   b.   Not allowed.  (l can't be greater than or equal to n.)
   c.   Allowed.
   d.   Not allowed.  (l can't be negative.)
   e.   Allowed.

---

Problems 20-21 at the end of this study chapter cover material from this section.

---

## 7.7 - 7.10 - Orbital Shapes and Energies (7.7)
### Electron Spin and thePauli Principle (7.8)
### Polyelectronic Atoms (7.9)
### History of the Periodic Table (7.10)

Sections 7.7 - 7.9 contain information regarding electronic orbitals and how they are arranged around the nucleus. Section 7.10 deals with the history of the periodic table.

These sections contain descriptive, not mathematical, material. As such, I will help you review the high points by having you answer nonmathematical questions about the material in these sections.

### Section 7.7

1. What are **nodes**?
2. How many nodes are in a 5s orbital?
3. Draw a 2p orbital.
4. How do the three 2p orbitals differ from one another?
5. In what ways do d orbitals differ from p orbitals?
6. Why doesn't a 2d orbital exist?
7. What is the chemical meaning of "**degenerate**?"
8. What is meant by the "**ground state**?" What is the ground state orbital for hydrogen?
9. How can a 1s electron be excited to a higher energy level?

### Section 7.8

10. What property of an electron does the quantum number given by ms represent?
11. What is the **Pauli Exclusion Principle**? What is its consequence regarding the number of electrons that an orbital can hold?

## Example 7.8 A - Practice with the Four Quantum Numbers

Which of the following sets of quantum numbers are not allowed? For each incorrect set state why it is incorrect.

  a.  $n = 3, l = 3, m_l = 0, m_s = -1/2$
  b.  $n = 4, l = 3, m_l = 2, m_s = -1/2$
  c.  $n = 4, l = 1, m_l = 1, m_s = +1/2$
  d.  $n = 2, l = 1, m_l = -1, m_s = -1$
  e.  $n = 5, l = -4, m_l = 2, m_s = +1/2$
  f.  $n = 3, l = 1, m_l = -2, m_s = -1/2$
  g.  $n = 3, l = 2, m_l = -1, m_s = 1$

## Solution

  a.  Not allowed. (l cannot be equal to n.)
  b.  Allowed.
  c.  Allowed.
  d.  Not allowed. (ms must be either +1/2 or -1/2.)
  e.  Not allowed. (l must be a **positive** integer.)
  f.  Not allowed. (ml must be between -l and +l.)
  g.  Not allowed. (ms must be either +1/2 or -1/2.)

Remember the key message from section 7.8 - **an orbital can hold a maximum of two electrons**, and they must have opposite spins.

# Example 7.8 B - Electons in Orbital

If each **orbital** can hold a maximum of two electrons (of opposite spin), how many electrons can each of the following hold?

    a. 2s   b. 5p   c. 4f   d. 3d   e. 4d

## Strategy

The key here is to figure out **how many orbitals** each contains. This is determined by the azimuthal quantum number, NOT by the principle quantum number. For example, a p orbital ($l = 1$) can have $m_l = +1, 0$ or $-1$. *Each $m_l$ can have 2 electrons*. Therefore, p can have a total of $3 \times 2$, or 6 electrons. It doesn't matter if it is a 5p, 3p or 2p. Each p can have up to 6 electrons.

## Solution

    a.   $2s \Rightarrow l = 0$, $m_l = 0$ (1 value) $\times$ 2 electrons = **2 electrons**
    b.   $5p \Rightarrow l = 1$, $m_l = +1, 0, -1$ (3 values) $\times$ 2 electrons = **6 electrons**
    c.   $4f \Rightarrow l = 3$, $m_l = +3, +2, +1, 0, -1, -2, -3$ (7 values) $\times$ 2 electrons = **14 electrons**
    d.   $3d \Rightarrow l = 2$, $m_l = +2, +1, 0, -1, -2$ (5 values) $\times$ 2 electrons = **10 electrons**
    e.   $4d \Rightarrow$ same as 3d

## Section 7.9

12. What are the three energy contributions that must be considered when describing the helium atom?
13. What does your textbook mean by the **electron correlation problem**? How do we deal with the problem?
14. Why does it take more energy to remove an electron from $Al^+$ than from $Al$?
15. What is meant by **effective nuclear charge**?
16. Why is $z_{eff}$ greater than $z_{actual}$?

## Section 7.10

17. What was the original basis of the construction of the periodic table?
18. What are triads?
19. List the properties that Mendeleev used to predict ekasilicon's position in the periodic table.
20. Several atoms of element 116 have been observed. Based on <u>Table 7.4 in your textbook</u>, what properties would you predict this element would have?
21. What is the important difference between Mendeleev's periodic table and the modern one?

## 7.11 - The Aufbau Principle and the Periodic Table

When you finish this section you will be able to:

- Write electron configurations using both the "conventional" (long-hand) form and the core element and valence electron (shorthand) form.
- Determine electron configurations based on position in the periodic table.

Read the statement of the <u>Aufbau Principle in your textbook</u>.  Your textbook presents **orbital diagrams** for the first ten elements, **hydrogen through neon**.  When constructing orbital diagrams and electron configurations, please keep the following in mind:

1. Electrons fill in order from lowest to highest energy.

2. The Pauli exclusion principle holds.  An orbital can only hold two electrons.

3. Two electrons in the same orbital must have opposite spins.

4. You must know how many electrons can be held by each azimuthal quantum number (ie; s can hold 2, 6 for p, 10 for d, 14 for f).

5. **Hund's rule** applies.  The lowest energy configuration for an atom is the one having the maximum number of **unpaired** electrons for a set of **degenerate** orbitals.  By convention, all unpaired electrons are represented as having parallel spins with the spin "up."

One excellent approach to address the order that electrons fill as the atomic number increases is shown in <u>Figure 7.27</u> in your textbook.  Practice writing the following triangle (use the **penetration effect**), discussed in <u>Section 7.12</u>, to justify the order of filling presented in the triangle.

$$1s^2$$
$$2s^2\ 2p^6$$
$$3s^2\ 3p^6\ 3d^{10}$$
$$4s^2\ 4p^6\ 4d^{10}\ 4f^{14}$$
$$5s^2\ 5p^6\ 5d^{10}\ 5f^{14}$$
$$6s^2\ 6p^6\ 6d^{10}\ 6f^{14}$$
$$7s^2\ 7p^6\ 7d^{10}\ 7f^{14}$$

To determine the order of filling, draw arrows from the upper right to the lower left.

$$1s^2 \quad {}^1$$
$$1\quad 2s^2\ 2p^6 \quad {}^3$$
$$2\quad 3s^2\ 3p^6\ 3d^{10} \quad {}^5$$
$$3\quad 4s^2\ 4p^6\ 4d^{10}\ 4f^{14} \quad {}^7$$
$$4\quad 5s^2\ 5p^6\ 5d^{10}\ 5f^{14}$$
$$5\quad 6s^2\ 6p^6\ 6d^{10}\ 6f^{14}$$
$$6\quad 7s^2\ 7p^6\ 7d^{10}\ 7f^{14}$$
$$7$$

Get the final filling pattern by following the arrows in order:

$$1s^2\ 2s^2\ 2p^6\ 3s^2\ 3p^6\ 4s^2\ 3d^{10}\ 4p^6\ 5s^2\ 4d^{10}\ 5p^6\ 6s^2\ 4f^{14}\ 5d^{10}\ 6p^6\ 7s^2 \ldots$$

$$\quad 1 \quad 2 \qquad 3 \qquad\quad 4 \qquad\qquad 5 \qquad\qquad\quad 6 \qquad\qquad\qquad 7$$

Let's apply this strategy toward determining the electron configuration for oxygen (8 electrons).

- The 1s and 2s levels hold a total of 4 electrons ($1s^2\ 2s^2$).
- The 2p level can hold up to 6 electrons. However, we only have 4 remaining, which means that we will have a $2p^4$.

$$O:\ 1s^2\ 2s^2\ 2p^4$$

Let's translate this to an orbital diagram.

$$\underset{1s}{\uparrow\downarrow}\quad \underset{2s}{\uparrow\downarrow}\quad \underset{2p}{\uparrow\downarrow\ \uparrow\ \uparrow}$$

The first three 2p electrons occupy their own **degenerate** 2p orbitals. The fourth electron shares a degenerate orbital, but does so with **opposite spin**.

---

# Example 7.11 A - Electron Configurations

Write electron configurations for each of the following neutral atoms (**use a configuration triangle**):

    a.  boron      b.  sulfur     c.  vanadium   d.  iodine

## Solution

    a.  B (5 electrons):  $1s^2\ 2s^2\ 2p^1$
    b.  S (16 electrons):  $1s^2\ 2s^2\ 2p^6\ 3s^2\ 3p^4$
    c.  V (23 electrons):  $1s^2\ 2s^2\ 2p^6\ 3s^2\ 3p^6\ 4s^2\ 3d^3$
    d.  I (53 electrons):  $1s^2\ 2s^2\ 2p^6\ 3s^2\ 3p^6\ 4s^2\ 3d^{10}\ 4p^6\ 5s^2\ 4d^{10}\ 5p^5$

---

# Example 7.11 B - Orbital Diagrams

Draw orbital diagrams for the following:

    a.  sodium     b.  phosphorus    c.  chlorine

## Strategy

Write the electron configuration for the atom. The only sticking points will be how unpaired electrons fill degenerate orbitals (singly, if possible). The inner electrons will all be paired. The outer electrons may be unpaired. You must deal with each atom separately.

## Solution

    a.  Na (11 electrons):  $1s^2\ 2s^2\ 2p^6\ 3s^1$

$$\underset{1s}{\uparrow\downarrow}\quad \underset{2s}{\uparrow\downarrow}\quad \underset{2p}{\uparrow\downarrow\ \uparrow\downarrow\ \uparrow\downarrow}\quad \underset{3s}{\uparrow}$$

    b.  P (15 electrons):  $1s^2\ 2s^2\ 2p^6\ 3s^2\ 3p^3$

$$\underset{1s}{\uparrow\downarrow}\quad \underset{2s}{\uparrow\downarrow}\quad \underset{2p}{\uparrow\downarrow\ \uparrow\downarrow\ \uparrow\downarrow}\quad \underset{3s}{\uparrow\downarrow}\quad \underset{3p}{\uparrow\ \uparrow\ \uparrow}$$

c.   Cl (17 electrons):  $1s^2\ 2s^2\ 2p^6\ 3s^2\ 3p^5$

$$\underset{1s}{\uparrow\downarrow}\quad \underset{2s}{\uparrow\downarrow}\quad \underset{2p}{\uparrow\downarrow\ \uparrow\downarrow\ \uparrow\downarrow}\quad \underset{3s}{\uparrow\downarrow}\quad \underset{3p}{\uparrow\downarrow\ \uparrow\downarrow\ \uparrow}$$

---

Look at the term "**core**" and "**valence**" <u>electrons in your textbook</u>.  Know how to define these terms.

Notice in the problem we just completed that phosphorous and chlorine have the same core electronic structure. This **core** has the **same electron configuration as neon**.  We can therefore write a shorthand version of electron configurations:

$$P = [Ne]\ 3s^2\ 3p^3 \quad Cl = [Ne]\ 3s^2\ 3p^5$$
$$\quad\ \ \uparrow\qquad\ \uparrow$$
$$\quad\ \ \text{core}\quad\text{valence}$$

Using the same strategy,

$$\text{yttrium (39 electrons)} = 1s^2\ 2s^2\ 2p^6\ 3s^2\ 3p^6\ 4s^2\ 3d^{10}\ 4p^6\ 5s^2\ 4d^1$$
$$\text{(longhand)}$$

$$= \mathbf{[Kr]\ 5s^2\ 4d^1}$$
$$\text{(shorthand)}$$

Neon and krypton are both **noble gases (group 8A)** and are atoms which have complete inner energy levels, and an outer energy level with complete s and p orbitals.

---

# Example 7.11 C - Shorthand Configurations

Write the shorthand configuration for the atoms in example 7.11 A, and state for parts a, b, and d how many **valence electrons** the element has.

## Solution

a.   B:  $[He]\ 2s^2\ 2p^1$  (3 valence electrons)
b.   S:  $[Ne]\ 3s^2\ 3p^4$  (6 valence electrons)
c.   V:  $[Ar]\ 4s^2\ 3d^3$  (transition metal, 2 valence electrons)
d.   I:  $[Kr]\ 5s^2\ 4d^{10}\ 5p^5$  (7 valence electrons . . . n = 4 level is complete)

Notice that boron is in group 3A
                sulfur is in group 6A
                vanadium is a transition metal
                iodine is in group 7A

The group number indicates the number of valence electrons for nontransition metals.  This should help you in determining electron configurations.

Also, sulfur is in period 3.  It is filling n = 3 electronic orbitals.  Iodine is filling n = 5 electronic orbitals.  It is in group 5.  Use the group and period locations to help do electron configurations quickly and correctly.  Two final ideas:

• With transition metals, if it is possible to have a $3d^5$ (half-filled) or $3d^{10}$ (completely filled) electronic configuration at the expense of a filled 4s, that will happen.  Thus, Cu is $[Ar]\ 4s^1\ 3d^{10}$, NOT $[Ar]\ 4s^2\ 3d^9$.  The same holds true for 4s/5d filling, lanthanides and actinides.

- Know what is meant by representative, d-transition, f-transition and noble gas elements.

---

# Example 7.11 D - Practice with Electron Configurations

Write shorthand electron configurations for the following real and hypothetical atoms:

a.  Sr          b.  Mo          c.  Ge          d.  Q (hypothetical, 111 electrons)

## Solution

a.  **Sr** is in Group 2A, period 5.  **[Kr] 5s$^2$**
b.  **Mo** is the fourth transition metal in period 5.  [Kr] 5s$^2$ 4d$^4$ <u>BUT</u>, it can have a half-filled 4d if the configuration is 5s$^1$ 4d$^5$.  Therefore, Mo = **[Kr] 5s$^1$ 4d$^5$**.
c.  **Ge** is in group 4A, period 4.  **[Ar] 4s$^2$ 3d$^{10}$ 4p$^2$**.
d.  **Q** would be the ninth transition metal in period 7, [Rn] 7s$^2$ 6d$^9$ <u>BUT</u>, it can have a completed 6d if the configuration is 7s$^1$ 6d$^{10}$.  Therefore, the likely configuration is **[Rn] 7s$^1$ 5f$^{14}$ 6d$^{10}$**.

---

Problems 22-25 at the end of this study chapter cover material from this section.

---

# 7.12 - Further Development of the Polyelectronic Model

When you finish this section you will be able to calculate the **ionization energy** and the **effective nuclear charge** of a species.

Your textbook points out that we make an important simplification in dealing with **polyelectronic atoms**.  Rather than take electron-proton and electron-electron interaction into account, we put all interactions under the umbrella of **one electron at a time interacting with the nucleus**.  This is what leads to the idea of an **effective nuclear charge ($z_{eff}$)**.

In the first part of this section, your textbook deals with the calculation of **ionization energies** in species where there is only **one electron**.  In such cases (H, He$^+$, Li$^{2+}$, As$^{32+}$, ... )

$$z = z_{eff}$$

Therefore, the equation we used in section 7.4 holds:

$$E = -2.178 \times 10^{-18} \left[ \frac{z^2}{n^2} \right] \text{ (per photon)}$$

We often measure ionization energy on a molar scale, so

$$E = -1312 \text{ kJ/mol} \left[ \frac{z^2}{n^2} \right]$$

where z = the nuclear charge (# of protons in the nucleus)
       n = the principle energy level where the electron resides

## Example 7.12 A - Ionization Energy of a Monoelectronic Species

Calculate the ionization energy for a mole of $N^{6+}$ ions with their electrons in the ground state (n = 1).

### Strategy

Recall that ionization energy (I.E.) means the energy required to excite the electron to the n = ∞ state. As a result,

$$I.E. = E_\infty - E_n$$

E will equal 0 (n = ∞), therefore, I.E. = $-E_n$

Also, for nitrogen, z = 7.

### Solution

$$I.E. = -E_1 = +1312 \text{ kJ/mol} \left[ \frac{7^2}{1^2} \right] = 6.429 \times 10^4 \text{ kJ/mol}$$

This very large number makes sense because there is a large positive charge acting on just one electron.

---

Your textbook deals next with the case where more than one electron is present. For such situations, **we know I.E.** The goal is to **calculate $z_{eff}$**. Once we calculate this value for a variety of electrons in an atom, we need to interpret what the results mean in terms of atomic structure.

---

## Example 7.12 B - Calculating $z_{eff}$

The ionization energy for a 3p electron in phosphorus is 1060 kJ/mol (see Table 7.5 in your textbook). What is the effective nuclear charge on the 3p electron?

### Solution

$$E = 1312 \text{ kJ/mol} \left[ \frac{z_{eff}^2}{n^2} \right]$$

$$\text{Rearranging, } \frac{E \times n^2}{1312 \text{ kJ/mol}} = z_{eff}^2; \frac{1060 \text{ kJ/mol } (9)}{1312 \text{ kJ/mol}} = z_{eff}^2$$

$$z_{eff}^2 = 7.27; z_{eff} = 2.70$$

This says that the 3p electron on phosphorous is **shielded** by the inner electrons. It is not shielded completely because, if it had been, $z_{eff}$ would have been equal to 1.

---

The **penetration effect** is used to explain the order in which electrons fill energy levels. For example, 4s fills before 3d because an electron in a 4s orbital has a greater likelihood of being near the nucleus ("penetrating") than a 3d.

## 7.13 - Periodic Trends in Atomic Properties

When you finish this section you will be able to use your understanding of electronic configuration to predict **trends in ionization energy, electron affinity and atomic radius**.

**Ionization Energy** increases as successive electrons are removed from an atom because:

1.  The value for $z_{eff}$ increases because there are fewer electron-electron repulsions and a higher positive to negative charge ratio than before.
2.  Upon going from f to d to p to s electrons, there is a higher penetration effect. For example, the removal of a 3p will require more energy than the removal of a 3d from the same atom.
3.  When you remove all of the electrons from an energy level, you begin removing **core electrons**, which are more tightly bound to the nucleus than valence electrons.

## Example 7.13 A - Ionization Energy

Examine Table 7.5 in your textbook. Justify the large increases in ionization energy at $I_5$ and $I_7$ for **sulfur**.

## Solution

$I_5$ represents removal of the last 3p electron. The remaining electrons are 3s and core electrons. The 3s electron has a much greater penetration effect than the 3p. Therefore, $I_5$ is much larger than $I_4$.

$I_7$ represents ionization of a core (2p) electron. This electron is much closer to the nucleus, thus requiring more energy to ionize it.

Note that:

1.  First ionization energy **increases** as we go **across a period**.
2.  First ionization energy **decreases** as we go **down a group**.
3.  There are anomalies that exist, such as the decrease from P to S. (Can you explain why?)

**Electron Affinity** is the **change in energy** associated with the **addition of an electron** to a **gaseous atom**. In keeping with thermodynamic convention, if the addition is **exothermic**, the energy change will be **negative**. Although electron affinity *generally* increases from left to right across a period, there are several exceptions. For example, the electron affinity of phosphorous is lower than sulfur. That is because P is $3p^3$ (half filled) while S is $3p^4$. If you put and extra electron on phosphorous, it must share an orbital, thus forcing electron-electron repulsion. These repulsions already exist in the 3p orbitals of sulfur. The trend in electron affinity is less predictable.

The **atomic radius** of an atom **decreases** from left to right **across a period**. This is because the $z_{eff}$ increases. Atomic radius **increases** going **down a group**. This is because of increased orbital size. (See Figure 7.36 in your textbook.) You must consider **ionic radius** in terms of the following questions: What will adding an electron do to electron-electron repulsions? What will subtracting an electron do to the effective nuclear charge?

## Example 7.13 B - Trends in Atomic Radius

Order the atoms or ions in the following groups from smallest to largest radius.

   a.  Cs, Si, F, Ca, Ga          b.  $Ca^{2+}$, $I^-$, I, Li

## Solution

   a.  F, Si, Ga, Ca, Cs
       ↑                ↑
       smallest      largest

       Cs is in period 6, group 1
       F is in period 2,  group 7

   b.  I is large.  Adding an electron forces extreme electron-electron repulsion making it larger.  Calcium is large, but taking its 4s electrons away markedly increases the $z_{eff}$.

       $Ca^{2+}$, Li, I, $I^-$
       ↑         ↑
       smallest  largest

Problems 26-31 at the end of this study chapter cover material from this section.

## 7.14 - The Properties of a Group:  The Alkali Metals

The goal of your textbook in this section is to show you how understanding electron configurations and the structure of the periodic table can allow you to predict the properties of this group.

This section begins with a review of the information contained in the periodic table.

   1.  Elements in a group exhibit similar properties.
   2.  It is primarily the number of valence electrons that determine an atom's chemistry.
   3.  Electron configurations can be gleaned from the periodic table.
   4.  Learn the names of the different groups (halogens, active metals, lanthanides, etc.).  See Figure 7.37 in your textbook.
   5.  Metals tend to lose electrons (have low ionization energies).  Non metals tend to gain electrons.  Metalloids (semimetals) have properties of both.

The focus of the chapter is on **Group I metals**.

**Properties as We go Down the Group**

   A.  The first ionization energy decreases.
   B.  The atomic radius increases.
   C.  The density increases.
   D.  The reactivity increases (they lose electrons readily).
   E.  The melting and boiling points decrease.

## Example 7.14 - The Alkali Metals

Explain the trends that the alkali metals follow in properties "A" through "D" above based on your knowledge of electronic configurations and atomic structure.

## Solution

A. **First ionization energy:** As we go down the group, the valence electron falls in a higher energy level. The shielding is less, making it easier to ionize the electron.

B. **Atomic radius:** Again, the valence electron occupies a higher energy level, thereby being farther from the nucleus.

C. **Density:** Atomic mass increases faster than atomic size.

D. **Reactivity:** The electrons are easier to ionize, therefore reactions which require less energy are possible.

Problems 32-33 at the end of this study chapter cover material from this section.

**END OF CHAPTER SEVEN**

# SELF-TEST QUESTIONS

1.  The visible region of the spectrum goes from 400 nm to 700 nm. What is the frequency range of the visible spectrum?

2.  List the regions of the electromagnetic spectrum and the wavelengths of radiation associated with each region.

3.  Calculate the frequency of blue light of wavelength $4.5 \times 10^2$ nm.

4.  Calculate the wavelength of green light of frequency $5.7 \times 10^{14}$ Hz.

5.  Red light with a wavelength of 6708 Å is emitted when lithium is heated in a flame.

    a.  What if the frequency of this radiation?
    b.  What is the energy of this radiation per photon? Per mole of photons?

6.  It takes $6.72 \times 10^{-18}$ J of energy to remove an electron from an unknown atom. What is the maximum wave length of light that can do this?

7.  A carbon-oxygen double bond in certain organic molecule absorbs radiation that has a frequency of $6.0 \times 10^{13}$ s$^{-1}$.

    a.  To what region of the spectrum does this radiation belong?
    b.  What is the wavelength of this radiation?
    c.  What is the energy of this radiation per photon? Per mole of photons?
    d.  A carbon-oxygen bond in a different molecule absorbs radiation with frequency equal to $5.4 \times 10^{13}$ s$^{-1}$. Does this radiation give more or less energy?

8.  Calculate the energy of a photon that is emitted at a wavelength of $5.69 \times 10^3$ nm.

9.  Many spectroscopists prefer using frequencies to wavelengths when describing electromagnetic radiation. Can you think of an advantage to the use of frequencies? (A main concern of spectroscopists is the energy of radiation that is either emitted of absorbed.)

10. Calculate the wavelength of a thoroughbred racehorse, which weighs 600 pounds, moving with a speed of 40 mi/hr.

11. What are the wavelengths associated with the following:

    a.  An alpha particle (mass = $6.64 \times 10^{-27}$ kg) traveling at $3.0 \times 10^8$ m/s?
    b.  A 1000 kg automobile traveling at 100 km/hr?

12. Make a plot of energy vs. n for the Bohr hydrogen atom for n = 1 to n = 50.

    a.  What is the energy of the Bohr hydrogen atom when n = ∞?
    b.  What is the ionization energy for the Bohr hydrogen atom (i.e. the energy required to move an    electron from n = 1 to n = ∞)?

13. Calculate the wavelength of light that must be absorbed by a hydrogen atom in its ground state to reach the excited state of $\Delta E = +2.914 \times 10^{-18}$ J.

14. How much energy is required to ionize a mole of hydrogen atoms?

15. Calculate the wavelength of light emitted in the spectral transition of n = 4 to n = 2 in the hydrogen atom.

16. What region of the spectrum would you look in to find the radiation associated with the n = 4 to n = 1 transition of the Bohr hydrogen atom?

17. What region of the spectrum would you look in to find the radiation associated with the spectral transition of n = 3 to n = 2 in the hydrogen atom?

18. Use the wave mechanical model to explain the quantized nature of the orbits of a hydrogen atom.

19. A chemistry book lists the radius of the hydrogen orbital as 1 Å. Will the electron ever be further than 1 Å from the nucleus?

20. Which of the following sets of quantum numbers are allowed?

   a.   $n = 7, l = 7, m_l = 0$ ↲
   b.   $n = 7, l = 0, m_l = 1$ ↳
   c.   $n = 7, l = 5, m_l = -3$ ↴
   d.   $n = 3, l = -1, m_l = 0$ ↳
   e.   $n = 0, l = 0, m_l = 0$ ↴

21. What is the maximum number of electrons allowed in the n states corresponding to the M, N, and O shells?

22. Write n, l, $m_l$, and $m_s$, quantum numbers for the 4 electrons of a boron atom.

23. What is the electron configuration for calcium?

24. How many half filled orbitals do each of the following have in the ground state?

   a.   O           d.   Mn         f.   Cf
   b.   B            e.   K           g.   Zn
   c.   Ar

25. Indicate the higher of the two energy states in each of the following pairs:

   a.   3d or 4s       b.   4p or 5s       c.   4s or 4p

26. Arrange the following atoms in order of increasing $z_{eff}$ for the highest-energy electron: Te, In, Mg, Ga, Xe, Ca.

27. The first and second ionization energies for argon are 1525 and 2665 kJ per mole. Calculate $z_{eff}$ for the first and second electrons removed.

28. Calculate the energy required to form one mole of sodium ions from a gas of sodium atoms if $z_{eff} = 3.40$.

29. In which orbital would an electron have a greater likelihood of being near the nucleus: 4f or 6s?

30. Order the following groups from smallest to largest radius.

   a.   Ar, Cl⁻, K⁺, S²⁻
   b.   C, Al, F, Si
   c.   Na, Mg, Ar, P
   d.   I⁻, Ba²⁺, Cs⁺, Xe

31. Which of the following will have the most exothermic electron affinity? the least?

   a.   Ge, Si, C       b.   Cl, Cl⁻, Cl⁺.

32. Properties of the alkali metals are discussed in Section 7.14. List some properties that you would expect for the alkaline earths, Be, Mg, Ca, Sr, and Ba.

33. Which elements are metalloids and why are they called metalloids?

## ANSWERS TO SELF-TEST QUESTIONS

1.  $7.5 \times 10^{14} \text{ s}^{-1}$ to $4.3 \times 10^{14} \text{ s}^{-1}$

2.  See Figure 7.2 in your textbook.

3.  $6.7 \times 10^{14}$ Hz

4.  $5.3 \times 10^{-7}$ m

5.  a.   $4.472 \times 10^{14} \text{ s}^{-1}$      b.   $2.963 \times 10^{-19}$ J, 178.5 kJ

6.  29.6 nm

7.  a.   infrared          c.   $4.0 \times 10^{-20}$ J, 24 kJ
    b.   $5 \times 10^{-6}$ m         d.   less

8.  $3.49 \times 10^{-20}$ J

9.  direct porportionality between frequency and energy

10. $1.4 \times 10^{-37}$ m

11. a.   $3.3 \times 10^{-16}$ m           b.   $2.4 \times 10^{-38}$ m

12. a.   0           b.   $2.18 \times 10^{-18}$ J

13. $6.817 \times 10^{-8}$ m

14. $1.31 \times 10^6$ J

15. 664 nm

16. ultraviolet

17. infrared

18. The upshot is, the Schroedinger equation shows that electrons can reside in certain mathematically allowed locations.

19. yes

20. b, c

21. 18, 32, and 50

22. 1, 0, 0, -1/2;  1, 0, 0, +1/2;  2, 0, 0, +1/2;  2, 1, 0, -1/2

23. Ca: $1s^2 \, 2s^2 \, 2p^6 \, 3s^2 \, 3p^6 \, 4s^2$

24. a.   2           e.   1
    b.   1           f.   4
    c.   2           g.   0
    d.   5

25. a.   3d               b.   5s               c.   4p

26. Mg < Ca < Ga < In < Te < Xe

27. 3.23 and 4.28

28. 496 kJ

29. 6s

30. a.  $K^+$, Ar, $Cl^-$, $S^{2-}$    c.  Ar, P, Mg, Na
    b.  F, C, Si, Al    d.  $Ba^{2+}$, $Cs^+$, Xe, $I^-$

31. a.  C, Ge    b.  $Cl^+$, $Cl^-$

32. Examples are reaction with water to give bases, and metallic character.

33. Si, Ge, As, Sb, Te, Po, and At are called metalloids because they exhibit both metallic and nonmetallic properties under certain circumstances.

# CHAPTER 8

# Bonding - General Concepts

| Section | Contents | Page in this Study Guide |
|---|---|---|
| 8.1 | Types of Chemical Bonds | 174 |
| 8.2 | Electronegativity | 175 |
| 8.3 | Bond Polarity and Dipole Moments | 177 |
| 8.4 | Ions:  Electron Configurations and Sizes | 178 |
| 8.5 | Formation of Binary Ionic Compounds | 180 |
| 8.6 | Partial Ionic Character of Covalent Bonds | 181 |
| 8.7 | The Covalent Chemical Bond - A Model | 181 |
| 8.8 | Covalent Bond Energies and Chemical Reactions | 182 |
| 8.9 | The Localized Electron Bonding Model | 183 |
| 8.10 | Lewis Structures | 183 |
| 8.11 | Exceptions to the Octet Rule | 186 |
| 8.12 | Resonance | 188 |
| 8.13 | Molecular Structure:  The VSEPR Model | 190 |

In this chapter, you will use many of the concepts you learned in chapter 7, especially electronic configurations.  You will learn why different types of bonds form, the nature of those bonds and a model to predict the three-dimensional structure of molecules formed from covalent bonds.

## 8.1 - Types of Chemical Bonding

When you finish this section you will be able to list and define **three types of bonding** and solve for **the energy of interaction of an ionic solid.**

Your textbook says that bonds are *"forces that hold groups of atoms together and make them function as a unit."* Bonds form because the energy of the system is lower than if bonds did not form.

**Ionic bonding** is due to electrostatic attraction.  It results from the loss of an electron from an alkaline or alkaline earth metal and its gain by a nonmetal.

Your textbook gives a formula for the energy of an ion pair called the **energy of interaction.**  Let's use that formula to calculate the stability gain when $Mg^{2+}$ and $O^{2-}$ interact.

# Example 8.1 - Coulomb's Law

Calculate the energy of interaction (in kJ/mol) between $Mg^{2+}$ and $O^{2-}$ if the distance between the centers of $Mg^{2+}$ and $O^{2-}$ is 0.205 nm (2.05 Å).

## Solution

$$E = 2.31 \times 10^{-19} \text{ J nm} \left[\frac{Q_1 Q_2}{r}\right] = 2.31 \times 10^{-19} \text{ J nm} \left[\frac{(+2)(-2)}{0.205 \text{ nm}}\right]$$

$$= -4.51 \times 10^{-18} \text{ J/ion pair} = \mathbf{-4.51 \times 10^{-21} \text{ kJ/ion pair}}$$

For a mole of ion pairs,

$$E = \frac{-4.51 \times 10^{-21} \text{ kJ}}{\text{ion pair}} \times \frac{6.022 \times 10^{23} \text{ ion pairs}}{\text{mole}} = \mathbf{-2710 \text{ kJ/mol}}$$

## Commentary

Note how much higher this value is than for the NaCl example in your textbook!  Part of the reason is because the **radius is smaller** (owing to the higher charge on Mg and O).  The more important contributer is the **higher charge** on each ion.  That allows for more powerful electrostatic interaction which gives a more stable bond.

---

**Covalent Bonding** occurs when bonds form between similar kinds of atoms for the same reason as between disimilar atoms:  the energy of the system is lowered as a result of the bond formation.

Look at Figure 8.1 in your textbook.  There is a particular distance apart where the combination of repulsive and attractive forces allow the system to have a minimum energy.  In this case, we have a **covalent bond** in which **electrons are shared by both nuclei approximately equally.**  Examples are $S_8$, graphite and diamond.

In the case where there is **unequal sharing of electrons**, a **polar covalent** bond exists.  Charges are not distributed equally in such a molecule.  Positive and negative poles exist.  Examples of polar covalent bonds are C-Cl, H-Cl, and O-H.

In summary, the nature of the bond will depend upon the ability of each atom in the bond to attract electrons to itself.  This is called the **electronegativity**.

---

Problems 1-3 at the end of this study chapter cover material from this section.

---

# 8.2 - Electronegativity

When you finish this section you will be able to use the positions of atoms in the periodic table and their electronegativities to predict relative bond polarities.

Recall that unequal sharing of electrons in a bond results in a polar covalent bond.  Ionic bonds result from the transfer of electrons between atoms.

Your textbook says that we can get a measure of the degree of ionic character of a bond by **measuring bond energies**. The more the electrostatic interaction that occurs between two atoms the greater will be the difference in bond energies when compared to perfectly covalent bonds involving the atoms. This <u>difference in bond energy is called</u> $\Delta$ and is the relative **electronegativity** differences between the bonding atoms.

For the representative elements, **electronegativity decreases going down a group** and **increases going across a period**. Thus, francium has the lowest electronegativity and fluorine has the highest.

The greater the $\Delta$, the more ionic character the bond has. If $\Delta = 0$, the bond is perfectly covalent. There are no precise cut-offs. All bonds have some ionic and some covalent character.

---

## Example 8.2 A - Electronegativities of Atoms

Look at the periodic table at the front of your textbook. Based only on their positions in the table, place the following atoms in order of increasing electronegativity.

Sr, Cs, Se, O, Ba

### Solution

Remember the trend: electronegativity increases from the lower left to the upper right of the periodic table. Based on their positions, the order is:

**Cs < Ba < Sr < Se < O**
least electronegative            most electronegative

---

Try the next example, keeping in mind that the determinant of bond polarity is the value of $\Delta$.

---

## Example 8.2 B - Bond Polarity

Using <u>Figure 8.3 in your textbook</u>, calculate $\Delta$ for each of the following bonds and order the set from most covalent to most ionic character.

a.  Na-Cl        b.  Li-H        c.  H-C        d.  H-F        e.  Rb-O

### Solution

a.  Na = 0.9, Cl = 3.0        $\Delta = 3.0 - 0.9 = \mathbf{2.1}$
b.  Li = 1.0, H = 2.1        $\Delta = 2.1 - 1.0 = \mathbf{1.1}$
c.  H = 2.1, C = 2.5        $\Delta = 2.5 - 2.1 = \mathbf{0.4}$
d.  H = 2.1, F = 4.0        $\Delta = 4.0 - 2.1 = \mathbf{1.9}$
e.  Rb = 0.8, O = 3.5        $\Delta = 3.5 - 0.8 = \mathbf{2.7}$

In order of increasing ionic character:

**H-C, Li-H, H-F, Na-Cl, Rb-O**
most covalent                    most ionic

---

The information gained from electronegativities will be used in the next section to figure out the polarities of molecules.

Problems 4-9 at the end of this study chapter cover material from this section.

## 8.3 - Bond Polarity and Dipole Moments

When you finish this section you will be able to determine whether or not simple molecules have a dipole moment.

Recall from Section 8.2 that it is possible to determine the **polarity of a bond** by the **size of** $\Delta$. If a molecule is diatomic (two atoms), there is often only one bond and that will determine whether the molecule is polar.

For instance, in our example 8.2, we determined that H-F was polar, with fluorine being the more electronegative atom. A **partial negative charge** ($\delta$-) resides on the fluorine atom and a **partial positive charge** ($\delta$+) resides on the hydrogen atom.

```
-+---->
H-----F
δ+    δ-
```

The **arrow** points to the **center of negative charge** while the **tail** is at the **center of positive charge**. A **dipole moment** means that the molecule has **two poles**.

The situation is clear-cut with HF. It becomes more difficult with 3 or more atoms in a molecule because the **individual dipoles can cancel each other out**. Look at Table 8.2 in your textbook. This shows how individual bond polarities can cancel each other out to yield a molecule with no dipole moment. Although you will be able to derive the geometries in Table 8.2 later on, you should memorize them for now.

## Example 8.3 A - Dipole Moment

Does $CHCl_3$ (a tetrahedral molecule with carbon at the center) have a dipole moment? If so, show the orientation of the dipole moment.

### Strategy

Perform the following steps:

1. Look up the electronegativity of each atom.
2. Draw the molecule in 3-dimensional space.
3. Determine the polarity of each bond and the net polarity on each atom.
4. Draw the dipoles and determine the direction (if any) of the molecular dipole moment.

### Solution

C = 2.5, H = 2.1, Cl = 3.0

## Example 8.3 B - Practice with Dipole Moments

For each of the following, determine the orientation of the dipole moment (if any).

    a.  HI        b.  $N_2$        c.  $CCl_2F_2$ (carbon is the central atom)

## Solution

    a.  H-----I  $\Rightarrow$  H-----I  $\Rightarrow$  H-----I
        2.1   2.5         $\delta$+  $\delta$-      -+---->

    b.  N-----N  $\Rightarrow$  No dipole moment ($\Delta = 0$)
        3.0   3.0

    c.  C = 2.5, Cl = 3.0, F = 4.0  $\Rightarrow$

(Fluorine is more electronegative than chlorine, therefore the dipole moment on this molecule is tipped slightly toward the fluorines.)

Problems 10-12 at the end of this study chapter cover material from this section.

## 8.4 - Ions:  Electron Configurations and Sizes

When you finish this section you will be able to predict

- the formulas of simple ionic compounds and
- the relative sizes of ions.

Your textbook deals only with **nonmetals, representative metals** and ionic bonds in this discussion.  It has been observed that atoms that form bonds in stable compounds have a **noble gas electronic configuration.**  (They are **isoelectronic** with a noble gas.)

## Example 8.4 A - Ionic Electron Configuration

Write the electronic configuration and determine the charge on each of the following atoms when it forms its most stable ion (noble gas electronic configuration).

    a.  Mg        b.  P        c.  Br        d.  Rb

## Strategy and Solution

Determine how many electrons the atom must gain or lose (metals will lose, nonmetals will gain) to obtain the electronic structure of the **nearest noble gas**.

a. Magnesium is $1s^2\ 2s^2\ 2p^6\ 3s^2$. It will **lose** the two 3s electrons to have the electronic configuration of neon. Magnesium will therefore ionize to $Mg^{2+}$.

$$Mg^{2+} = 1s^2\ 2s^2\ 2p^6 \text{ (isoelectronic with neon)}$$

b. P (nonmetal) will gain 3 electrons to be isoelectronic with argon.

$$P^{3-} = 1s^2\ 2s^2\ 2p^6\ 3s^2\ 3p^6$$

c. Br (nonmetal) will gain one electron to be isoelectronic with krypton.

$$Br^- = 1s^2\ 2s^2\ 2p^6\ 3s^2\ 3p^6\ 4s^2\ 3d^{10}\ 4p^6$$

d. Rb (metal) will lose one electron to be isoelectronic with krypton.

$$Rb^+ = 1s^2\ 2s^2\ 2p^6\ 3s^2\ 3p^6\ 4s^2\ 3d^{10}\ 4p^6$$

Note that $Br^-$, $Rb^+$, and Kr are all **isoelectronic**.

---

To determine the formula of binary ionic compounds remember that **chemical compounds are electrically neutral** (the sum of the cation charges must equal the anion charges).

For example, the formula of the ionic compound formed from combining magnesium and chlorine can be determined by:

1. assessing the charge on the ions, and
2. determining how many of each ion is required to combine to make the compound electrically neutral.

The ions are $Mg^{2+}$ and $Cl^-$. In order to maintain electronic neutrality,

$$Mg^{2+} + 2Cl^- \longrightarrow MgCl_2.$$

---

## Example 8.4 B - Formulas of Binary Ionic Compounds

Determine the formula for each of the following sets of atoms when they combine to form a binary ionic compound.

a. K and Br    b. Sr and F    c. Al and Se    d. Ba and O

## Solution

a. $K^+ + Br^- \longrightarrow$ **KBr**
b. $Sr^{2+} + 2F^- \longrightarrow$ **SrF$_2$**
c. $2Al^{3+} + 3Se^{2-} \longrightarrow$ **Al$_2$Se$_3$**
d. $Ba^{2+} + O^{2-} \longrightarrow$ **BaO**

We have maintained electronic neutrality throughout.

We discussed **ion size** in our review of chapter 7. As we go down a group, ion size increases (higher energy levels have a larger average distance). Because the ratio of protons to electrons becomes greater, **cations are smaller than their neutral atoms.** Because of electron-electron repulsion and less effective shielding **anions are always larger than their neutral atoms.** *The larger the charge, the more severe the effect.*

Thus, $S^{2-}$ is larger than S.
  $Ca^{2+}$ is smaller than Ca.
  $S^{2-}$ is **isoelectronic** with $Ca^{2+}$. Because $Ca^{2+}$ has 4 more protons than $S^{2-}$, it is much smaller. Therefore, we can conclude that for an isoelectronic series, **the higher the nuclear charge (z), the smaller the ion.**

---

## Example 8.4 C - Ion Size

Order the following ions from smallest to largest.

  a.   $O^{2-}$, $Na^+$, $Mg^{2+}$, $F^-$
  b.   $Se^{2-}$, $Te^{2-}$, $Rb^+$, $Mg^{2+}$

## Solution

  a.   These ions are all **isoelectronic with neon.** Therefore, the smallest will have the highest positive charge. The ions will get larger until we have the highest negative charge.

$$Mg^{2+} < Na^+ < F^- < O^{2-}$$
  smallest                    largest

  b.   Because $Se^{2-}$ and $Te^{2-}$ are anions, they will be larger than $Rb^+$ and $Mg^{2+}$. Because $Te^{2-}$ is farther down the periodic table than $Se^{2-}$ it will be larger. Magnesium is higher on the periodic table and has a greater charge than rubidium, therefore it will have the smallest ion.

$$Mg^{2+} < Rb^+ < Se^{2-} < Te^{2-}$$
  smallest                    largest

---

Problems 13-18 at the end of this study chapter cover material from this section.

---

## 8.5 - Binary Ionic Compounds

When you finish this section you will be able to explain the different steps involved when elements form ionic solids. We will pay special attention to the idea of lattice energy.

The theme of this section is that there are **many separate processes** which go into **forming an ionic solid.** As we have said before, *the ionic compound forms because its energy is lower than if its elements remained separated.* However, not every part of the process is energetically favorable. Your textbook points out that it is the **lattice energy** (the energy released when an ionic solid is formed from its gaseous ions) that is the most favorable, and more than makes up for some parts of the process that are energetically unfavorable. Let's examine the formation of the **KCl ionic solid** ("salt") from **K(s) and $Cl_2(g)$.**

**Processes That Must Occur**

1. K(s) must form K(g); (**Energy of Sublimation = +64 kJ**).
2. K(g) must form $K^+$(g); (**First Ionization Energy = +419 kJ**).
3. $1/2 \, Cl_2$(g) must form Cl(g); (**Bond Energy × 1/2 = +120 kJ**).
4. Cl(g) must form $Cl^-$(g); (**Electron Affinity = -349 kJ**).
5. $K^+$(g) must combine with $Cl^-$(g) to form KCl(s); (**Lattice Energy = -690 kJ**).

The net energy of formation ($\Delta H°_f$) equals the sum of the energy changes, **-436 kJ**. So you see that the value for any of the processes that make up salt formation can be obtained if you **understand the processes involved**, and are given suitable data.

---

# Example 8.5 - Calculation of Lattice Energy

Given the following data, determine $\Delta H°_f$ for LiBr.

$$Li(s) + 1/2 \, Br_2(g) \longrightarrow LiBr(s)$$

| | |
|---|---|
| ionization energy for Li = | + 520 kJ/mol |
| electron affinity for Br = | - 324 kJ/mol |
| energy of sublimation for Li = | + 161 kJ/mol |
| lattice energy = | - 787 kJ/mol |
| bond energy of $Br_2$ = | + 193 kJ/mol |

## Solution

While $\Delta H°_f$ will be the sum of all the energy changes, be careful to **multiply the bond energy by 1/2**, because we have 1/2 of a mole of $Br_2$(g).

$$520 + (-324) + (+161) + (-787) + (+96) = \text{-334 kJ} = \Delta H°_f$$

Note that here, as in the previous example, the lattice energy gave the biggest push to salt formation.

---

The remainder of the section makes the point that the **higher the charge** on each ion, the greater the lattice energy will be. This value counteracts the higher endothermic ionization energies, thus resulting in a **more energetically stable crystal**.

---

# 8.6 and 8.7 - Partial Ionic Character of Covalent Bonds and the Covalent Chemical Bond - a Model

Read Sections 8.6 and 8.7 in your textbook. If you can answer the questions posed in this review section, then you understand the material that your textbook is trying to get across.

# 8.6

1. Why do we say that there are **no totally ionic** bonds?
2. How do we determine the **percent ionic character** of a bond?
3. Why is it ambiguous to say that $Na_2CO_3$ or $KHSO_3$ are ionic compounds?
4. How do we define (as an operating definition) an ionic compound (salt)?

## 8.7

5. (Review) What is a chemical bond?
6. (Review) Why do chemical bonds occur?
7. Why is it useful to think of each bond in a molecule as being **environment independent**?
8. Knowing the energy of stabilization of $CCl_4$, discuss how we would determine the bond energy of a C-H bond in $CHCl_3$.
9. What is **a model**?
10. Why do we develop models?
11. List the fundamental properties of a model.
12. What are the **limitations** of models?
13. Is a wrong model useless? Why or why not?

---

## 8.8 - Covalent Bond Energies and Chemical Reactions

When you finish this section you will be able to calculate heats of reaction from bond energies.

Recall from the last section that your textbook calculated the average bond energy for a C-H bond for methane ($CH_4$). The assumptions made were:

1. that the energy needed to break each C-H bond was the same (413 kJ), and
2. that each of the bonds was not sensitive to its environment.

Assumption #2 is not really correct, as can be shown by the inset in which hydrogens and ethane ($C_2H_6$) are replaced by bromine, chlorine, and fluorine (see your textbook). In spite of this, average bond energies can help to give us fair estimates of heats of reaction, as such they are useful tools.

Look at Table 8.4 in your textbook. This table gives **average bond energies** for many covalent bonds. Notice that **multiple bonds** (bonds that involve **sharing more than 2 electrons** between atoms) require more energy to break than single bonds.

Your textbook uses the example of the combination of hydrogen and fluorine gas to make hydrogen fluoride. It states that

$$\Delta H = \Sigma \text{ D (bonds broken) - } \Sigma \text{ D (bonds formed)}$$

where $\Delta H$ = heat of reaction and D = bond energy per mole of bonds.

The key to calculating the heat of reaction from the average bond energy is to **carefully list all bonds broken and all bonds formed**. Also, recognize that *it takes energy to break bonds* (endothermic, $\Delta H = +$) while *energy is released when bonds are formed* (exothermic, $\Delta H = -$).

---

## Example 8.8 - Heat of Reaction from Bond Energy

Using data from Table 8.4 in your textbook, calculate $\Delta H$ for the following reaction. Compare this to $\Delta H°_f$ calculated from the thermodynamic values in the appendix. (See chapter 6 if you need a review of this.) ($\Delta H°_f(CF_4)$ = -680 kJ/mol)

$$CH_4(g) + 4F_2(g) \text{ ----> } CF_4(g) + 4HF(g)$$

## Strategy and Solution

Let's make a list of bonds broken and bonds formed.

| Bonds Broken | Energy per Bond (kJ) | | Total Energy (kJ) |
|---|---|---|---|
| 4 C-H | $413 \times 4$ bonds | = | 1652 |
| 4 F-F | 154 | = | 616 ($154 \times 4$ moles) |
| **Total Energy to Break Bonds** | | = | **2268 kJ** |

| Bonds Formed | Energy per Bond (kJ) | | Total Energy (kJ) |
|---|---|---|---|
| 4 C-F | $485 \times 4$ bonds | = | 1940 |
| 4 H-F | 565 | = | 2260 ($565 \times 4$ moles) |
| **Total Energy to Form Bonds** | | = | **4200 kJ** |

$\Delta H$ = (Energy to break bonds) - (Energy to form bonds)

= 2268 kJ - 4200 kJ

**= -1932 kJ**

This reaction is exothermic.

$$
\begin{aligned}
\text{The value for } \Delta H^\circ_f &= [\Delta H^\circ_f(CF_4) + 4\Delta H^\circ_f(HF)] - [\Delta H^\circ_f(CH_4) + 4\Delta H^\circ_f(F_2)] \\
&= [-680 + 4\,(-271)] - [-75 + 4\,(0)] \\
&= \textbf{-1689 kJ}
\end{aligned}
$$

---

Problems 19-22 at the end of this study chapter cover material from this section.

---

## 8.9 - Covalent Bonding Models

If you can answer the following review questions, you understand the important ideas in this brief section.

1. What are the two covalent bonding models?
2. What is the difference in the **basic assumptions** about electron position in these models?
3. Why is the MO model somewhat unsuitable for use with larger molecules?
4. What information can the MO model give us that the LE model doesn't?
5. What are **lone pairs** and **bonding pairs**?
6. What are the parts of the LE model?

---

## 8.10 - Lewis Structures

When you finish this section you will be able to draw **Lewis structures** for a variety of **covalent molecules**.

**Lewis structures** are often used to depict **bonding pairs** and **lone pairs** in molecules. We are concerned only with **valence** electrons because these are the ones (for period 1 and 2 atoms) that are involved in bond making and breaking.

Individual atoms are represented with Lewis structrues by putting valence electrons (as dots or circles) around the atomic symbol. For example, magnesium ([Ne] $3s^2$) would be represented as

· **Mg** ·

$Mg^{2+}$ ([Ne]) would be Mg.  It no longer has its valence electrons.  Chlorine  ([Ne] $3s^2 3p^5$) would be  $:\overset{\cdot\cdot}{\underset{\cdot\cdot}{Cl}}\cdot$

and $Cl^-$ ([Ne] $3s^2 3p^6$) would be  $:\overset{\cdot\cdot}{\underset{\cdot\cdot}{Cl}}:$

---

# Example 8.10 A - Lewis Dot Structures

Draw Lewis dot structures for the following atoms or ions.

   a.  N          b.  $N^{3-}$          c.  I          d.  Ba          e.  $Ba^{2+}$

## Solution

   a.  N = [He] $2s^2 2p^3$  (5 valence electrons)  $:\overset{\cdot}{N}\cdot$
       (I drew the $2s^2$ on the left and the three 2p electrons singly.  You can draw a maximum of 2 electrons on any side, unless there is a triple bond involved.)
   b.  $N^{3-}$ = [He] $2s^2 2p^6$  (8 valence electrons)  $:\overset{\cdot\cdot}{N}:$
   c.  I = [Kr] $5s^2 4d^{10} 5p^5$  (7 valence electrons)  $:\overset{\cdot}{I}\cdot$
   d.  Ba = [Xe] $6s^2$  (2 valence electrons)  **Ba:**
   e.  $Ba^{2+}$ = [Xe]  (0 valence electrons)  **Ba**


**EVERY PERIOD 1 AND 2 ELEMENT (with the exception of H, He, B and Be) CAN FORM COMPOUNDS OF LOWEST ENERGY IF THEIR HIGHEST ENERGY LEVELS ARE FILLED ($s^2 p^6$).  THIS IS CALLED THE OCTET RULE**.  If an ion or atom observes the octet rule, we will say that it is "happy."  Hydrogen is "happy" if its 1s orbital is filled.  We will discuss boron and beryllium later on.

Your textbook discusses a strategy for drawing Lewis structures.  I will propose a different strategy.  Use the one you feel more comfortable with.


### Kelter Strategy for writing Lewis Structures

1.  **Total number of valence electrons in the system**:  Sum the number of valence electrons on all of the atoms.  Add the total negative charge if you have an anion.  Subtract the charge if you have a cation.

    eg. $CO_3^{2-}$ ;  C has 4 valence electrons          = 4
                 O has 6 valence electrons × 3 atoms    = 18
                 <u>charge on the ion is -2 so add          =  2</u>
                 **electrons in the system             = 24**

2.  **Number of electrons if each atom is to be happy**:  Atoms in our examples will need 8 electrons (octet rule) or 2 electrons (hydrogen).  Using $CO_3^{2-}$ as an example:

    C needs 8 electrons          = 8
    O needs 8 electrons × 3 atoms     = 24
    **electrons for happiness          = 32**

    The -2 charge comes as a result of the electrons in the system.  The charge is **never** counted towards happiness.

3. **The number of bonds in the system:** Covalent bonds are made by sharing electrons. You need 32 electrons and you have 24. You are **8 electrons deficient**. If you make **4 bonds** (with 2 electrons per bond), you will make up the deficiency. Therefore,

$$\# \text{ bonds } = \frac{\text{"#2" - "#1"}}{2} = \frac{32 - 24}{2} = 4 \text{ bonds}$$

4. Draw the structure. The central atom is carbon. The oxygens surround it. Because there are **4 bonds**, there will be **two single bonds** and **one double bond**. Each bond accounts for two electrons. Then, complete the octets by putting electrons around each atom. Double check your results by counting total electrons in the system.

= 24 electrons

We will discuss resonance in this system later on.

---

# Example 8.10 B - More Lewis Structures

Using the steps outlined above (or in your textbook) write Lewis structures that obey the octet rule for the following:

a.  $Cl_2$     b.  $CH_2Cl_2$     c.  $H_2CO$     d.  $NH_3$

## Solution

a.  1.  Total of valence electrons = 7 per Cl × 2 = **14**
    2.  Total if happy = 8 per Cl × 2 = **16**
    3.  # bonds = (16 - 14)/2 = **1 bond**

= 14 electrons

b.  1.  Total of valence electrons =  4 for the C  =  4
                                       7 per Cl × 2 = 14
                                       1 per H × 2 =  2
                      valence electrons  = **20**

    2.  Total if happy =          8 for the C  =  8
                                   8 per Cl × 2 = 16
                                   2 per H × 2 =  4
                      electrons if happy = **28**

    3.  # bonds =          (28 - 20)/2  = **4 bonds**

= 20 electrons    (hydrogen does NOT get a complete octet!)

c.   1.   Total of valence electrons = 4 for the C  = 4

6 for the O  = 6

1 per H × 2 = 2

**valence electrons = 12**

2.   Total if happy =          8 for the C  = 8

8 for the O  = 8

2 per H × 2 = 4

**electrons if happy = 20**

3.   # bonds =             (20 - 12)/2   = **4 bonds**

= 12 electrons

(**\* Only C, O, N, S, and P can have double bonds.  Only C and N can have <u>stable</u> triple bonds.**)

d.   1.   Total of valence electrons = 5 for the N   = 5

1 per H × 3 = 3

**valence electrons  = 8**

2.   Total if happy = 8 for the N           =  8

2 per H × 3 =  6

**electrons if happy = 14**

3.   # bonds = (14 - 8)/2                 = **3 bonds**

= 8 electrons

---

Problems 23-25 at the end of this study chapter cover material from this section.

---

## 8.11 - Exceptions to the Octet Rule

When you finish this study section you will be able to draw Lewis structures for covalent molecules whose central atom does not obey the octet rule.

Although your textbook deals with boron and beryllium in this section, we will focus on the more general case of **central atoms that can exceed the octet rule**.  This can happen because there are **empty d orbitals** that shared electrons can occupy.  This can happen with **third period (row) elements**.

### *Writing Lewis Structures for Exceptions to the Octet Rule*

To determine if you have an exception to the rule, proceed as if your molecule **obeys** the octet rule.  Let's use $ICl_3$ as an example.

1.   Total of valence electrons = 7 for the I    =   7

7 per Cl × 3 = 21

**valence electrons = 28**

2.  Total if happy =    8 for the I    = 8
                        8 per Cl × 3  = 24
           **electrons if happy  = 32**

3.  # bonds =         (32 - 28)/2    = **2 bonds**

CAUTION!  We have 2 bonds for 3 chlorine atoms!  **There are not enough bonds to account for all the atoms!**
This is how you know that we have an exception to the octet rule.

To write the Lewis structure of exceptions, draw the structure with **one bond to each ligand, complete the octets**
and **add any extra electrons to the central atom.**  With $ICl_3$,

$$:\ddot{Cl}\diagdown \underset{|}{\underset{:\ddot{Cl}:}{I}} \diagup \ddot{Cl}:$$

= 28 valence electrons (I has <u>10 electrons</u>)

---

# Example 8.11 - Exceptions to the Octet Rule

Write Lewis structures for the following molecules:

a.  $IF_2^-$        b.  $CO_2$        c.  $XeF_4$

## <u>Solution</u>

a.  1.  Total of valence electrons = 7 for the I    =  7
                                     7 per F × 2   = 14
                                     <u>add 1 for "-" =   1</u>
                    **valence electrons  = 22**

    2.  Total if happy =            8 for the I    =  8
                                    8 per F × 2   = 16
                       **electrons if happy  = 24**

    3.  # bonds = (24 - 22)/2 = **1 bond**  <u>NOT ENOUGH!  EXCEPTION!</u>

$$\left[:\ddot{F}-\ddot{I}-\ddot{F}:\right]^-$$

b.  1.  Total valence electrons = **16**
    2.  Total if happy = **24**
    3.  # bonds = **4 bonds**  Not an exception!

$$\ddot{O}=C=\ddot{O}$$

c.  1.  Total of valence electrons = **36**  (Xe = 8!)
    2.  Total if happy = **40**
    3.  # bonds = **2 bonds**  <u>Exception!</u>

Problems 26-27 at the end of this study chapter cover material from this section.

## 8.12 - Resonance

When you finish this section, you will be able to draw reasonable resonance structures.

Recall the example of $CO_3^{2-}$ that we did at the beginning of section 8.10. We determined the Lewis structure to be

However, the double bond could have been placed on any of the three oxygens:

Measurements in bond lengths suggest that **all three C-O bond lengths are equivalent.** Electrons move around the entire molecule. Therefore, the actual structure is a **time-average** of all these structures. These structures are called **resonance** structures. The Lewis structure of the molecule can be drawn either of three ways. The double bond seems to resonate between the carbon and oxygen atoms. A molecule can have resonance structures if it has multiple bonds.

**Formal Charge and Resonance**

Let's review your textbook's discussion on formal charge and how it is used to draw resonance structures. Formal charge is, "the difference between the number of valence electrons on the free atom and the number of valence electrons assigned to the atom in a molecule." Formal charge is a **computational device** based on a **localized electron (LE) model**, and as such is not perfectly correct. To determine formal charge (a somewhat more realistic estimate of charge distribution in a molecule), we need to know:

1.   How many electrons an atom "owns."

> **Electrons Owned  =**    [# valence electrons around the atom +
> # of bonds (which equals 1/2 # shared electrons)]

2.   The **formal charge** on an atom.

> **Formal Charge  =**    # valence electrons on the neutral atom
> - # electrons owned by the atom based on the
> resonance structure you drew

Let's look at $CO_3^{2-}$ again.

- Carbon owns 4 electrons (4 bonds).  The formal charge on carbon = **4 valence electrons** on the neutral atom minus **4 electrons owned = 0.**

- Oxygen$_a$ has 6 valence electrons and 1 bond.  It owns **7 total valence electrons.**  The formal charge = 6 valence electrons on the neutral atom minus 7 electrons owned = **-1.**

- Oxygen$_b$ has the same formal charge as oxygen$_a$ = **-1.**

- Oxygen$_c$ has 4 valence electrons and 2 bonds.  It owns **6 total valence electrons.**  The formal charge = 6 valence electrons on the neutral atom minus 6 electrons owned = **0.**

**The sum of the formal charges** $(0 + (-1) + (-1) + 0 = -2)$, **must always equal the charge on the ion** (or molecule, if that's what you are dealing with).  Your textbook says that if you can write, **nonequivalent** Lewis structures (different numbers of single and double bonds) for a molecule or ion, those with formal charges closest to zero and with any negative formal charges on the most electronegative atoms are considered to best describe the bonding.

---

# Example 8.12 A - Formal Charges

Assign formal charges to each atom in the following resonance structures of $CO_2$.

1.  $\overset{..}{\underset{..}{O}} = C = \overset{..}{\underset{..}{O}}$
    
    a        b

2.  $:O \equiv \overset{.}{C} - \overset{..}{\underset{..}{O}}:$
    
    a        b

Which structure is more likely to be correct?

## Solution

Let's establish formal charges for each atom.

Structure 1:
> C owns 4 electrons.  Formal charge = 0
> $O_a$ owns 6 electrons.  Formal charge = 0
> $O_b$ owns 6 electrons.  Formal charge = 0

Structure 2:
> C owns 4 electrons.  Formal charge = 0
> $O_a$ owns 5 electrons.  Formal charge = +1
> $O_b$ owns 7 electrons.  Formal charge = -1

Structure 1 is more likely because all formal charges are zero.

**Note:**  Structure 2 can be represented with its formal charges:

$$:O \equiv C - \overset{..}{\underset{..}{O}}:$$
$$\quad + \qquad\qquad -$$

## Example 8.12 B - Resonance

Draw all resonance structures and select the most stable one for (SCN)⁻.

### Solution

First determine the number of bonds for the Lewis structures.

There are **16 valence electrons** in the system.
The total for happiness is **24 electrons**.
There are (24 - 16)/2 = **4 bonds** in this ion.

Now let's draw some possible resonance structures based on 4 bonds.

a.  $:\ddot{S}{-}C{\equiv}N:$ ]⁻     b.  $:S{\equiv}C{-}\ddot{N}:$ ]⁻     c.  $\ddot{S}{=}C{=}\ddot{N}$ ]⁻

Now evaluate the formal charges on each atom.

a.  $:\ddot{S}{-}C{\equiv}N:$ ]⁻
$\quad$ -1 $\quad$ 0 $\quad$ 0

b.  $:S{\equiv}C{-}\ddot{N}:$ ]⁻
$\quad$ 0 $\quad$ 0 $\quad$ -1

c.  $\ddot{S}{=}C{=}\ddot{N}$ ]⁻
$\quad$ 0 $\quad$ 0 $\quad$ -1

Nitrogen is more electronegative than sulfur. Therefore, structure "c" is the most likely.

---

Problems 28-30 at the end of this study chapter cover material from this section.

---

## 8.13 - Molecular Structure:  The VSEPR Model

When you finish this section you will be able to predict molecular geometries by using the VSEPR model.

The Valence Shell Electron Pair Repulsion (VSEPR) model assumes that atoms will orient themselves so as to minimize electron pair repulsions around the central atom.

Memorize the information in Table 8.6 in your textbook. Each lone pair or bond around the central atom occupies a position in space. The effect of **lone pairs** around the central atom is to squeeze bonded pairs closer together (see your textbook's discussion regarding bond angles in $CH_4$, $NH_3$ and $H_2O$). Multiple bonds are counted as "one bonding pair" in the VSEPR model, because the double bonds are constrained in space. Let's determine the VSEPR structure of formaldehyde, $H_2CO$, together:

Step 1:   Determine the Lewis structure.

Total valence electrons = **12**
Total if happy = **20**
# bonds = **4**

$$\underset{H}{\overset{H}{\diagdown}}C{=}\ddot{O}{:}$$

Step 2:   Count the number of bonds and lone pairs on the central atom.

$$2 \text{ C}-\text{H} + 1 \text{ C}=\text{O} = \textbf{3 bonds}$$

Step 3:   Determine the geometry based on Table 8.6 in your textbook.

$$3 \text{ "electron pairs"} = \textbf{trigonal planar}$$

Now let's try $IBr_2^-$.

Step 1:   Total valence electrons = 22
          Total if happy = 24
          # bonds = 1   **Exception!**

$$:\overset{\cdot\cdot}{\underset{\cdot\cdot}{Br}}-\overset{\cdot\,\cdot\,\cdot}{I}-\overset{\cdot\cdot}{\underset{\cdot\cdot}{Br}}:$$

Step 2:   There are **2 bonds** and **3 lone pairs** around the central atom.  The total is **5 electron pairs.**

Step 3:   The molecule will orient itself with a **trigonal bipyramid** geometry.  The electron pairs will orient in the equatorial positions first (see discussion in the text) and the bonding pairs will make up the remaining positions:

$$\begin{array}{c} Br \\ | \\ :-I\cdot \\ | \\ Br \end{array}$$

The molecule will be **linear.**

(Note that when we draw VSEPR sturctures, we are only concerned with the central atom.  We often omit valence electrons around the ligands.)

---

# Example 8.13 - VSEPR Structures

Determine the geometry of each of the following molecules or ions.

   a.   $CO_2$        b.   $SO_4^{2-}$        c.   $BrF_3$        d.   $XeO_4$        e.   $ICl_3$

## Solution

   a.   The Lewis structure for $CO_2$ (**16 valence electrons**) is

$$\overset{\cdot\cdot}{O}=C=\overset{\cdot\cdot}{\underset{\cdot\cdot}{O}}$$

The double bonds each count in the VSEPR model as 1 restricted bond, so $CO_2$ acts as if it has **2 electron pairs** around it.  Geometry = **linear.**

   b.   $SO_4^{2-}$ = **32 valence electrons**

$$\left[ \begin{array}{c} :\overset{\cdot\cdot}{O}: \\ | \\ :\overset{\cdot\cdot}{O}-S-\overset{\cdot\cdot}{O}: \\ | \\ :\overset{\cdot\cdot}{O}: \end{array} \right]^{2-} \longleftrightarrow \left[ \begin{array}{c} :\overset{\cdot\cdot}{O}: \\ \| \\ :\overset{\cdot\cdot}{O}-S-\overset{\cdot\cdot}{O}: \\ \| \\ :\overset{\cdot\cdot}{O}: \end{array} \right]^{2-}$$

The double-bonded resonance structure is better because of the lower formal charges on the oxygens. It turns out that for purposes of the VSEPR model, both structures will give the same **4 electron pairs** (2 double bonds "count" as 2 electron pairs). Geometry = **tetrahedron**.

c.  $BrF_3$ = **28 valence electrons**

There are **2 lone pairs** and **3 bonding pairs = 5 electron pairs**. It is a **trigonal pyramid** basis. The electrons take up two equatorial spots, leaving two F's axial and one equatorial. Geometry = **T-shaped**.

d.  $XeO_4$ = **32 valence electrons**

There are 4 bonds, thus **4 electron pairs**. Geometry = **tetrahedral**.

e.  $ICl_2^+$ = **20 valence electrons**

There are 2 bonding pairs and 2 lone pairs, thus 4 electron pairs. It is a **tetrahedral basis**. Geometry = **V-shaped**.

---

# Problems 31-33 at the end of this study chapter cover material from this section.

---

**END OF CHAPTER EIGHT**

# SELF-TEST QUESTIONS

1. Indicate whether the bonds between the following would be primarily covalent, polar covalent, or ionic:

   a. O-H          c. H-Cl
   b. Cs-Cl        d. Br-Br

2. Calculate the energy of interaction for KCl if the internuclear distance is 0.314 nm.

3. Calculate the energy of interaction between $Ag^+$ and $Br^-$ if the internuclear distance of AgBr is 0.120 nm (in kJ/mole).

4. Using a periodic table order the following from lowest to highest electronegativity.

   a. Fr, Mg, Rb          c. P, As, Ga, O
   b. B, Al, C, N          d. Cl, S, P, Ar

5. Using the periodic chart of elements, place the following elements in order from the lowest to the highest electronegativity:

   F, Nb, N, Si, Rb, Ca, Pt

6. Using the Figure 8.3 in your textbook, calculate the difference in bond energy (Δ) for each of the following bonds:

   a. Cl-Cl          c. Fe-O          e. S-H
   b. K-Br            d. H-O

7. Place the following in order of increasing polarity:

   $NaBr$, $I_2$, $H_2O$, $MnO_2$, $CN^-$

8. Which of the following molecules contain polar covalent bonds? List in order of increasing polarity. (Use fig. 8.3 in your text.)

   $O_3$, $P_8$, NO, $CO_2$, $CH_4$, $H_2S$

9. How will the charge be distributed on each of the following molecules: HF, NO, CO, and HCl?

10. Determine the orientation of the dipole of the following, if any.

    a. $AlCl_3$ (planar with aluminum atom at the center)
    b. $CH_3F$ (tetrahedral with carbon at the center)
    c. $N_2O$ (linear with N-N-O structure)
    d. $AgCl_4$ (planar molecule, silver atom at center, Ag-Cl bonds 90° apart)

11. Which of the molecules in problem 10 contain one or more polar bonds?

12. Which of the following molecules would you expect to have a dipole moment of zero? Describe the dipole orientation of the other 2 molecules.

    a. KI                    c. $H_2Se$ (bent structure)
    b. $CF_4$ (tetrahedral structure)

13. Determine the charge on each of the following atoms and indicate which element they would be isoelectronic with if they lost or gained electrons:

    a. O          c. I          e. Na
    b. Be         d. Te

14. List four ions that are isoelectronic with argon and have charges from -2 to +2. Arrange these in order of increasing ionic radius.

15. Determine the formula of the binary compound formed from the following sets of atoms.

    a.  Ca and O
    b.  K and Cl
    c.  Rb and S
    d.  Ba and P

16. Predict formulas for the following binary ionic compounds.

    a.  Mg and N          c.  Ca and S
    b.  Na and F          d.  Sr and Te

17. Using shorthand notation, list the electron configurations of the ions in the compounds in problem 16.

18. Place the following in an order of increasing ionic size. Use the shorthand notation to list the electron configuration for each of the ions.

    a.  $Ba^{2+}$, $Te^{2-}$, $Cs^+$, $I-$
    b.  $Cs^+$, $S^{2-}$, $O^{2-}$, $K^+$

19. Using the bond energy values listed in Table 8.4 of your text, calculate the $\Delta H$ for the following reactions:

    a.  $2H_2(g) + O_2(g) \longrightarrow 2H_2O$
    b.  $2C_2H_6(g) + 7O_2(g) \longrightarrow 4CO_2(g) + 6H_2O(g)$
    c.  $HCN(g) + 2H_2(g) \longrightarrow CH_3NH_2(g)$

20. Use bond energy values from table 8.4 in your textbook to calculate $\Delta H$ for the following reactions.

    a.  $H-C\equiv C-H(g) + H_2(g) \longrightarrow CH_2=CH_2(g)$
    b.  $2CH_4(g) + 3O_2(g) \longrightarrow 2CO(g) + 4H_2O(g)$
    c.  $N_2(g) + 2H_2(g) \longrightarrow N_2H_4(g)(NH_2-NH_2)$

21. Compare the values obtained in parts a and b of problem 20 to $\Delta H$ values calculated from $\Delta H°_f$ data in Appendix 4 in your textbook.

22. Calculate the enthalpy of reaction $\Delta H°_f$ for the following reaction. Use the enthalpies of formation found in appendix four of your textbook.

$$H_2(g) + C_2H_4(g) \longrightarrow C_2H_6(g)$$

23. Draw Lewis dot structures for the following atoms, ions or molecules:

    a.  Sr          d.  Ga          g.  $NaNH_2$
    b.  $Br^-$       e.  $GaCl_4^-$   h.  $CSe_2$
    c.  ICN         f.  $P^{3-}$

24. Draw Lewis structures for the following:

    a.  $H^+$        c.  P           e.  $Cl^-$
    b.  C           d.  $P^{5+}$

25. Draw Lewis structures for the following:

    a.  $AsF_3$      c.  $H_3O^+$     e.  $NH_4$
    b.  $O_3$        d.  $BH_4^-$     f.  $O_2$

26. Draw Lewis dot structures for the following:

    a. $BCl_3$              c. $BrO_3^-$

    b. $AsF_5$              d. $S_2F_{10}$ (contains a S-S bond)

27. Draw Lewis dot structures for the following:

    a. $SbCl_3$

    b. $AlF_6^{3-}$

    c. $PCl_5$

28. Assign formal charges to each of the labeled atoms.

        a      b               c

29. Draw the remaining resonance forms for $N_2O_4$.

30. Draw a Lewis structure and any resonance forms of benzene, $C_6H_6$ (benzene consists of a ring of six carbon atoms with one hydrogen bonded to each carbon).

31. Predict the structure of each of the following molecules or ions:

    a. $SeF_6$              c. $ClF_4^+$           e. $CF_3Cl$ (carbon is central atom).

    b. $N_2O$              d. $ClO^-$

32. Using the VSPER model, determine the molecular geometry for each of the following molecules:

    a. $SCl_4$              c. $IF_4^-$            e. $TlCl_2^+$

    b. $H_2Se$             d. $SnCl_5^-$

33. Which of the molecules of ions in problem 31 contain polar covalent bonds? Are polar? For the polar species, which direction is the net dipole?

## ANSWERS TO PRACTICE PROBLEMS

1.  a.  polar covalent        c.  polar covalent
    b.  ionic                  d.  covalent

2.  $-7.36 \times 10^{-19}$ J

3.  E = -1159 kJ/mole

4.  a.  Fr, Rb, Mg            c.  Ga, As, P, O
    b.  Al, B, C, N           d.  Ar, P, S, Cl

5.  Rb < Ca < Nb < Si < Pt < N < F
    0.8   1.0   1.6   1.8   2.2   3.0  4.0

6.  a.  0                     c.  1.7          e.  0.4
    b.  2.0                   d.  1.4

7.  $I_2$ < $CN^-$ < $H_2O$ < NaBr < $MnO_2$
    0      0.5     1.4     1.9      2.0

8.  $O_3$, $P_8$ < $H_2S$, $CH_4$ < NO < $CO_2$

9.  HF;  NO;  CO;  HCl
    +-   +-   +-   +-

10. a.  no dipole             c.  negative toward O
    b.  negative toward F     d.  no dipole

11. all

12. **b.**  The opposing bond polarities in a tetrahedral structure cancel out.  Thus, $CF_4$ has no dipole moment.
    KI is a binary ionic compound which has a negative dipole toward I.
    Selenium will have a partial negative charge as its electronegativity is greater than that of hydrogen.
    Thus, the resulting dipole moment would be orientated as shown:

13. a.  $O^{2-}$, isoelectronic with neon
    b.  $Be^{2+}$, isoelectronic with helium
    c.  $I^-$, isoelectronic with xenon
    d.  $Te^{2-}$, isoelectronic with xenon
    e.  $Na^+$, isoelectronic with neon

14. $Ca^{2+}$, $K^+$, $Cl^-$, $S^{2-}$

15. a.  CaO                   c.  $Rb_2Cl$
    b.  KCl                   d.  $Ba_3P_2$

16. a.  $Mg_3N_2$            c.  CaS
    b.  NaF                   d.  SnTe

17. a.  [Ne], [Ne]           c.  [Ar], [Ar]
    b.  [Ne], [Ne]           d.  [Kr], [Xe]

18. a.   $Ba^{2+} < Cs^+ < I^- < Te^2$  all can be written as [Xe]
    b.   $K^+ < Cs^+ < O^{2-} < S^{2-}$, $K^+$ = [Ar], $Cs^+$ = [Xe], $O^{2-}$ = [Ne], $S^{2-}$ = [Ar]

19. a.   76 kJ
    b.   -3803 kJ
    c.   -984 kJ

20. a.   -169 kJ        b.   -1091 kJ        c.   +81 kJ

21. a.   6 kJ difference
    b.   -1089 kJ difference

22. $\Delta H_f^\circ$ = -136.7 kJ/mole

23. a.   Sr:        d.   :Ga        g.   $\left[ \text{H—N̈—H} \right]^-$

    b.   :B̈r:        e.   $\left[ \begin{array}{c} :\ddot{C}l: \\ :Cl\text{—Ga—}Cl: \\ :\ddot{C}l: \end{array} \right]^{1-}$        h.   :S̈e=C=S̈e:

    c.   :Ï—C≡N:        f.   P

24. a.   H        c.   :Ṗ·        e.   :C̈l:

    b.   ·Ċ·        d.   P

25. a.   :F̈—Äs—F̈:        c.   $\left[ \text{H—Ö—H} \right]^+$        e.   $\left[ \begin{array}{c} \text{H} \\ \text{H—N—H} \\ \text{H} \end{array} \right]^+$
         :F̈:                              H

    b.   :Ö=Ö—Ö:        d.   $\left[ \begin{array}{c} \text{H} \\ \text{H—B—H} \\ \text{H} \end{array} \right]^-$        f.   :Ö=Ö:

26. a.   Cl—B—Cl        c.   $\left[ :\ddot{O}\text{—Br—}\ddot{O}: \right]^-$
              Cl                      :Ö:

    b.   [arsenic fluoride structure]        d.   [disulfur fluoride structure]

27. a.   :C̈l—Sb—C̈l:        b.   $\left[ \begin{array}{c} :F: :F: \\ :F\text{—Al—}F: \\ :F: \end{array} \right]^{3-}$        c.   [phosphorus chloride structure]
              :C̈l:

28.  a.  0              d.  +1
     b.  0              e.  +1
     c.  -1             f.  0

29.

30.

31.  a.  octahedral        c.  distorted tetrahedron        e.  tetrahedral
     b.  linear            d.  linear

32.  a.  see-saw
     b.  bent
     c.  tetrahedral
     d.  trigonal bipyramidal
     e.  linear

33.  All contain polar covalent bonds b, c, d and e are polar.  Negative is toward:  O in b;  axial fluorine in C; O in d; fluorine in e.

# CHAPTER 9

# Covalent Bonds - Orbitals

| Section | Contents | Page in this Study Guide |
|---|---|---|
| 9.1 | Hybridization and the Localized Electron Model | 199 |
| 9.2 | The Molecular Orbital Model: Bond Order | 202 |
| 9.3 | Bonding in Homonuclear Diatomic Molecules | 204 |
| 9.4 | Bonding in Heteronuclear Diatomic Molecules | 206 |
| 9.5 | Combining the Localized Electron and Molecular Orbital Models | 207 |

Chapter 8 dealt with ways of representing structures on paper and with predicting the three dimensional structure that covalent molecules should have. This chapter reviews the two most important models which attempt to explain covalent molecular and ionic structure and shape - the **Localized Electron (LE) Model** and the **Molecular Orbital (MO) Model**.

In the **LE model**, electron pairs are still localized around specific atoms, but the orbitals around the central atom are modified.

In the **MO model**, all of the electrons in the molecule are combined into a set of molecular orbitals which describe bonding in the entire molecule. As we shall review, each model has its special strengths.

## 9.1 Hybridization and the Localized Electron Model

When you finish this section you will be able to describe the bonding in a variety of covalent molecules and ions using the concept of hybridization.

Your textbook points out the difficulty with assuming that the 4 C-H bonds in methane ($CH_4$) are made by interacting the 2s and 2p orbitals of carbon (valence orbitals) with the 1s orbitals of hydrogen. It says that the three 2p-1s C-H interactions should be located at 90° to one another. This does not agree with the observed bond angles of 109.5°. Recall that these bond angles **minimize electron-electron repulsions** in methane.

To work around this difficulty, the concept of **hybrid orbitals** is suggested. By combining the **one 2s** and the **three 2p** orbitals of carbon, four **hybrid** $sp^3$ orbitals are formed. These orbitals are equivalent, and that is consistent with the observed presence of 4 equivalent C-H bonds. Figure 9.6 in your textbook displays the interaction between the $sp^3$ hybrid orbitals of carbon and the 1s orbitals of the hydrogens.

**The bottom line is this**: when the VSEPR model indicates that you have a **tetrahedron** as the basis of your structure (4 effective electron pairs), the atom is $sp^3$ **hybridized**.

## Example 9.1 A - sp$^3$ Hybridization

Describe the bonding in the water molecule using the localized electron model.

### Strategy

We must establish the VSEPR structure. This is done by drawing the Lewis structure and determining the number of effective electron pairs around the central atom.

### Solution

The Lewis structure of $H_2O$ is (see sections 8.10 and 8.13)

Four effective electron pairs around the oxygen atom indicate a **tetrahedral** basis leading to a V-shaped structure. Therefore, the oxygen is sp$^3$ hybridized. Two sp$^3$ orbitals are occupied with 1s electrons from hydrogen. The other two sp$^3$ orbitals are occupied by lone pairs.

In molecules or ions with a **trigonal planar** configuration (3 effective electron pairs around an atom), **sp$^2$ hybridization** occurs. With this hybridization, an s orbital and two p orbitals are used. That leaves an **unhybridized** (unchanged) **p orbital perpendicular** to the sp$^2$ plane.

Look at Figure 9.11 in your textbook.

- Bonds formed from the overlap of orbitals in the plane between two atoms are called **sigma ($\sigma$) bonds.**
- Bonds formed by the overlap of unhybridized p orbitals (above and below the center plane) are called **pi ($\pi$) bonds.**
- A **single bond** is a $\sigma$ **bond.**
- A **double bond** consists of **one $\sigma$ and one $\pi$ bond.**
- A **triple bond** consists of **one $\sigma$ and two $\pi$ bonds.**

---

## Example 9.1 B - Sigma and Pi Bonds

How many $\sigma$ bonds are there in the commercial insecticide, "Sevin," shown below? How many $\pi$ bonds?

### Solution

There are 26 $\sigma$ bonds (single bonds and one of the bonds in a double bond).

There are 6 $\pi$ bonds (the other bond in a double bond).

Your textbook goes over a variety of different hybridization schemes, all of which center on the idea that **the hybridization of the orbitals of an atom depends on the total number of effective electron pairs around it**. In order to determine the hybridization of an atom, it is essential that you can figure out its VSEPR structure.

The following table summarizes effective electron pairs around an atom with hybridization. Remember that for purposes of the VSEPR model, **double and triple bonds count as only one effective electron pair.**

| Effective Electron Pairs Around an Atom | Arrangement | Hybridization |
|---|---|---|
| 2 | linear | $sp$ |
| 3 | trigonal planar | $sp^2$ |
| 4 | tetrahedral | $sp^3$ |
| 5 | trigonal bipyramid | $dsp^3$ |
| 6 | octahedral | $d^2sp^3$ |

Keep in mind that a ligand can have a hybridization different than that of the central atom. Each atom in a molecule must be considered separately based on the Lewis and VSEPR structures of that molecule.

# Example 9.1 C - Practice with Hybrid Orbitals

Give the hybridization and predict the geometry of each of the atoms in the following molecules or ions.

  a.   $IF_2^+$
  b.   $OSF_4$ (sulfur is the central atom)
  c.   $SiF_6^{2-}$
  d.   HCCH

## Solution

  a.   Lewis structure:

$$:\ddot{F}-\ddot{I}-\ddot{F}:$$

All three atoms have 4 effective electron pairs. They are **all $sp^3$ hybridized**. The VSEPR structure has a tetrahedral basis. Because the central atom (iodine) has two bonding pairs, it will take on a **V-shape**, but the bond angle will be smaller than 109.5°

  b.   Lewis structure:

Sulfur has 5 effective electron pairs. The VSEPR structure is a trigonal bipyramid. **Sulfur is $dsp^3$ hybridized**. Each of the fluorines has 4 effective electrons pairs. **The fluorines are $sp^3$ hybridized.**

  c.   Lewis structure:

Silicon has 6 effective electron pairs. The VSEPR structure is octahedral. **Silicon is $d^2sp^3$ hybridized.** Each of the fluorines has 4 effective electron pairs. **The fluorines are $sp^3$ hybridized.**

d.   Lewis structure:  H-C≡C-H

Each carbon has 2 effective electron pairs.  The VSEPR structure is linear.  **Each carbon is sp hybridized**.  Hydrogen atoms bond using 1s orbitals.  The orbitals are unhybridized.

---

## Example 9.1 D - Summing It All Up

Answer the following questions regarding Aspartame (Nutra-sweet).

a.   How many σ bonds are in the molecule?
b.   How many π bonds?
c.   What is the hybridization on carbon "a"?  Carbon "b"?
d.   What is the hybridization on nitrogen "a"?  Oxygen "a"?
e.   What is the $C_b$-$O_b$-$H_a$ bond angle?

## Solution

You must remember to complete the octets in this "shorthand" Lewis structure.  Lone pairs are often "assumed," so always be on the lookout.

a.   39 σ bonds
b.   6 π bonds
c.   $sp^3$, $sp^2$ (3 effective electron pairs)
d.   $sp^3$ (remember the "assumed" lone pair to complete the octet),
     $sp^2$ (3 effective electron pairs, including the two to complete the octet)
e.   $O_b$ is $sp^3$ hybridized (complete the octet!), therefore, the angle is based on a tetrahedron, with two lone pairs compressing the C-O-H bond angle to 104.5°.

---

Problems 1-2 at the end of this study chapter cover material from this section.

---

## 9.2 - The Molecular Orbital Model

When you finish this section you will be able to

 • state the major ideas of the molecular orbital (MO) model, and
 • compute bond orders for homonuclear diatoms of hydrogen and helium.

The localized electron model does an excellent job of predicting and justifying molecular shapes.  It does not deal with molecules with **unpaired electrons**.  It also neglects **bond energies**.

In addition to these things, the molecular orbital (MO) model also gives a view of electrons in a molecule which allows us to get a clearer understanding of what we had called resonance.

## KEY IDEAS OF THE MO MODEL

1. All the valence electrons in a molecule exist in a set of **molecular orbitals** of a given energy. The valence electrons of each atom are not acting independently, but rather act with other valence electrons to form a set of MO's.
2. Figure 9.27 in your textbook illustrates that there are **bonding** and **antibonding** MO's. Bonding results in **lower energy** than if no interaction occurred. Antibonding results in higher energy.
3. A molecule will be stable if there is more bonding than antibonding interaction.
4. **Bond order** (BO) is a measure of net bonding interactions.

$$BO = \frac{\text{\# bonding electrons - \# antibonding electrons}}{2}$$

5. BO must be greater than 0 for a stable molecule to form.
6. The higher the BO, the stronger the bond.

Your text goes over the bonding in $H_2$, $H_2^-$ and $He_2$. Let's try two more.

---

# Example 9.2 - MO Theory

Use MO theory to describe the bonding and stability of

   a.  $H_2^{2-}$   b.  $H_2^+$

## Solution

We must fill in the molecular orbital energy-level diagram for each of the species.

   a.  $H_2^{2-}$ has **4 electrons**. They will fill pairwise, and with **opposite spins**, the $\sigma^*_{1s}$ and $\sigma^*_{1s}$ orbitals.

**Bond order** = $\dfrac{\text{2 bonding electrons - 2 antibonding electrons}}{2}$ **= 0**

We would not expect $H_2^{2-}$ to be a stable ion.

   b.  $H_2^+$ has 1 electron. We must again fill in our MO energy-level diagram.

$$\text{Bond order} = \frac{1 \text{ bonding electron - 0 antibonding electrons}}{2} = 1/2$$

We would expect $H_2^+$ to form, but it would probably not be much more stable than two separate hydrogens.

---

## Problem 3 at the end of this study chapter covers material from this section.

---

## 9.3 - Bonding in Homonuclear Diatomic Molecules

When you finish this section you will be able to describe the bonding in homonuclear diatomic molecules with regard to:

- stability
- bond order, relative bond energy and bond length
- paramagnetism or diamagnetism

In the last section we saw how to use the MO model to assess the **stability** of diatomic molecules with 1s orbitals. In this section we will expand the discussion to **diatoms containing 2s and 2p orbitals**.

Bonds can be formed when atomic orbitals overlap in space. This is not possible with 1s orbitals in period two elements because each of the orbitals is too close to its own nucleus. Only the 2s and 2p orbitals (containing the valence electrons) can participate in bond formation.

We know that $\sigma$ **bonds** form **between nuclear centers**, and these MO's can be formed using 2s orbitals or **$2p_x$ orbitals**. In addition, $\pi$ **orbitals** can be formed by the overlap of both **$2p_y$ or $2p_z$ orbitals**. Also $\sigma$ and $\pi$ antibonding orbitals exist. (See Figure 9.34 in your textbook.)

Your textbook uses **paramagnetism** and **diamagnetism** to prove the relative positions of MO energy levels. **Paramagnetism** indicates **unpaired electrons** in a substance. This causes the substance to be **attracted to** a magnetic field. **Diamagnetism** indicates **paired electrons** in a substance. This causes the substance to be **repelled from** the magnetic field.

Paramagnetism is a much stronger effect than diamagnetism. It will dominate if both effects are present. (The substance will be attracted to a magnetic field.)

Your textbook describes how para- and diamagnetism help explain the order of molecular orbital energy-levels. This order, for valence electrons, is

$$
\begin{array}{cl}
& \sigma_{2p}^* \quad \text{---} \\
& \pi_{2p}^* \quad \text{--- ---} \\
& \sigma_{2p} \quad \text{---} \\
E \quad & \pi_{2p} \quad \text{--- ---} \\
& \sigma_{2s}^* \quad \text{---} \\
& \sigma_{2s} \quad \text{---}
\end{array}
$$

Filling molecular orbitals from atoms containing 2s and 2p electronic orbitals works the same as when filling from 1s orbitals. Just fill from lowest to highest energy and remember to fill degenerate orbitals separately, then pairwise with the electrons having opposite spins. Let's try the following example together.

## Example 9.3 A - $\sigma$ and $\pi$ Molecular Orbitals

Determine the following regarding $F_2^-$:

- electron configuration
- bond order
- para- or diamagnetism
- the bond energy relative to $F_2$

## Solution

To get the **electron configuration** we must fill in the molecular orbital energy-level diagram. We know that $F_2^-$ has **15 valence electrons** (7 on F, 8 on $F^-$).

**Electron Configuration** (valence only) $F_2^- = (\sigma_{2s})^2(\sigma_{2s}^*)^2(\pi_{2p})^4(\sigma_{2p})^2(\pi_{2p}^*)^4(\sigma_{2p}^*)^1$

**Bond Order** = $\dfrac{\text{8 bonding electrons - 7 antibonding electrons}}{2}$ = **1/2**

$F_2^-$ has one unpaired electron. It would be expected to be paramagnetic. Molecular fluorine, $F_2$, has a bond order of 1 and is therefore more stable than $F_2^-$. Its bond energy would be expected to be higher.

## Example 9.3 B - Practice with $\sigma$ and $\pi$ Orbitals

Determine the electron configurations and bond orders for $S_2^{2-}$ and $Cl_2^{2+}$. If they can exist, discuss their magnetism.

## Solution

$S_2^{2-}$ has **14 valence electrons** (6 + 6 from the sulfur atoms and 2- charge).

$Cl_2^{2+}$ has **12 valence electrons** (7 + 7 from the chlorine atoms and 2+ charge).

**Molecular Orbital Energy-Level Diagrams**

$$S_2^{2-} \qquad Cl_2^{2+}$$

**Electron Configuration of $S_2^{2-}$** = $(\sigma_{3s})^2(\sigma_{3s}^*)^2(\pi_{3p})^4(\sigma_{3p})^2(\pi_{3p}^*)^4$

**Electron Configuration of $Cl_2^{2+}$** = $(\sigma_{3s})^2(\sigma_{3s}^*)^2(\pi_{3p})^4(\sigma_{3p})^2(\pi_{3p}^*)^2$

**Bond Order of $S_2^{2-}$** = (8 - 6)/2 = **1**

**Bond Order of $Cl_2^{2+}$** = (8 - 4)/2 = **2**

$S_2^{2-}$ would be **diamagnetic**.
$Cl_2^{2+}$ would be **paramagnetic**.

---

Problems 4-5 at the end of this study chapter cover material from this section.

---

## 9.4 - Bonding in Heteronuclear Atoms

When you finish this section you will be able to describe the bonding in some heteronuclear diatomic atoms.

The **MO model** for homonuclear molecules **works well** for describing the bonding in atoms **adjacent to one another** on the periodic table. The model **breaks down** with two **very different** atoms.

Let's apply the model to several heteronuclear species.

---

## Example 9.4 - MO Model for Heteronuclear Atoms

Using the MO model, describe the bonding, magnetism and relative bond energies in the following species

$$O_2, O_2^- \text{ and } O_2^{2-}$$

## Solution

**Valence Electrons**
   $O_2$ has 12
   $O_2^-$ has 13
   $O_2^{2-}$ has 14

**Molecular Orbital Energy-Level Diagrams**

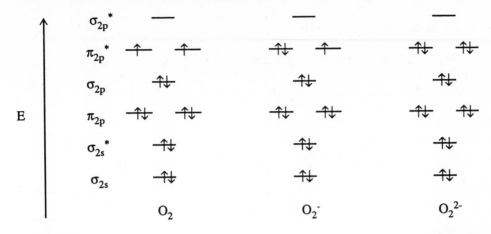

**Bond Orders**

$O_2 = (8 - 4)/2 = 2$

$O_2^- = (8 - 5)/2 = 1.5$

$O_2^{2-} = (8 - 6)/2 = 1$

Both $O_2$ and $O_2^-$ are expected to be paramagnetic. We expect $O_2^{2-}$ to be diamagnetic. The bond energy of $O_2$ is expected to be the highest, followed by $O_2^-$ with $O_2^{2-}$ having the lowest.

_____

In the MO description of HF your textbook notes that the MO model accounts for the polarity in that molecule. This is a great strength of the model.

_____

Problems 6-7 at the end of this study chapter cover material from this section.

_____

## 9.5 - Combining the Localized Electron and Molecular Orbital Model

When you finish this section you should be able to answer the following review questions:

- What is the primary problem with the localized electron (LE) model?
- Why is the use of resonance unsatisfactory for describing molecular bonding?
- What is the major disadvantage of the molecular orbital (MO) model?
- What bond is depicted as shifting in LE resonance structures? Why is that important in our combination model?
- In summary, what are the advantages and disadvantages of the LE model? The MO model?

_____

**END OF CHAPTER NINE**

# SELF-TEST QUESTIONS

1. Predict the geometries of the following compounds:

   a. $SF_2$         b. $SF_4$         c. $XeF_2$         d. $XeF_4$         e. $IF_5$         f. $ClF_3$

2. Predict the geometry about the indicated atom and identify the hybridization of each atom.

   a. The two carbon atoms and the nitrogen atom of glycine

   b. the carbon atom in $CF_2Cl_2$
   c. the phosphorous atom in $PCl_5$
   d. the nitrogen atom in $NH_2^-$

3. Determine the number of sigma and pi bonds in each of the following:

4. The structure of urea is

   a. How many $\sigma$ bonds are there?
   b. How many $\pi$ bonds?
   c. What is the hybridization at the carbon?
   d. How are the nitrogen atoms hybridized?
   e. What is the N-C-N bond angle expected to be?
   f. How many lone pairs of electrons are there?

5. Draw an energy level diagram for $He^+$. What is the bond order?

6. What is the bond order for $H_2^-$?

7. The elements N, O, and F exist as diatomic molecules. Use MO theory to explain why Ne, the next element on the periodic table, does not exist as diatomic molecules.

8. Which of the following are paramagnetic species: $Li_2$, $N_2$, $He_2^+$, $H_2$, $O_2$, $Be_2$?

9. For $N_2$, $N_2^+$, $N_2^-$ and $N_2^{2-}$, use molecular orbitals to predict bond order and if the species are diamagnetic or paramagnetic.

10. If liquid nitrogen were poured between the poles of a powerful magnet, would it "stick" to the poles or simply pour through the space between the poles? Why?

11. What is the bond order and expected magnetism of carbon monoxide? For $CO^-$?

12. Summarize the relative bond order for the diatomic molecules of the second period from boron to fluorine.

13. What is the bond order and expected magnetism of IBr?

14. Which of the following are diamagnetic: $CN^-$, $CN$, $CN^+$?

# ANSWERS TO SELF-TEST QUESTIONS

1.  a.  angular (like water)       c.  linear           e.  square pyramidal
    b.  see-saw                    d.  square planar    f.  T-shaped

2.  a.  carbons-$sp^3$, tetrahedral; $sp^2$, trigonal planar; nitrogen-$sp^3$, based on tetrahedron-trigonal pyramid
    b.  $sp^3$-tetrahedral
    c.  $dsp^3$-trigonal bipyramid
    d.  $sp^3$-based on tetrahedron-bent.

3.  a.  8 sigma, 2 pi      c.  7 sigma, 3 pi
    b.  5 sigma, 2 pi      d.  9 sigma, 9 pi

4.  a.  7      c.  $sp^2$      e.  120°
    b.  1      d.  $sp^3$      f.  6

5.  1/2

6.  1/2

7.  equal number of bonding and antibonding electrons

8.  $He_2^+$ and $O_2$

9.  $N_2$   : bond order = 3, diamagnetic
    $N_2^+$ : bond order = 2.5, paramagnetic
    $N_2^-$ : bond order = 2.5, paramagnetic
    $N_2^{2-}$ : bond order = 2, paramagnetic

10. Liquid nitrogen would simply pour through the space between the poles because it is diamagnetic as opposed to oxygen, which would be held between the poles because it is paramagnetic.

11. CO: bond order = 3, diamagnetic; $CO^-$: bond order = 2.5, paramagnetic

12. $B_2 < C_2 < N_2 > O_2 > F_2$

13. IBr: bond order = 1, diamagnetic

14. $CN^-$ and $CN^+$

# CHAPTER 10

# Liquids and Solids

| Section | Contents | Page in this Study Guide |
|---------|----------|--------------------------|
| 10.1 | Intermolecular Forces | 210 |
| 10.2 | The Liquid State | 211 |
| 10.3 | An Introduction to Structures and Types of Solids | 212 |
| 10.4 | Structure and Bonding in Metals | 213 |
| 10.5 | Carbon and Silicon:  Network Atomic Solids | 215 |
| 10.6 | Molecular Solids | 215 |
| 10.7 | Ionic Solids | 216 |
| 10.8 | Vapor Pressure and Change of State | 217 |
| 10.9 | Phase Diagrams | 219 |

Your textbook justifies **grouping solids and liquids** together, separate from gases, by **examining common properties**. The **densities, compressibilities** and **heats of phase change** all indicate that the forces that hold solids and liquids together are similar.  Gases have no such forces.  In this chapter, your textbook discusses the **bonding models**, **structure** and **other properties** of liquids and solids.

## 10.1 - Intermolecular Forces

When you finish this section you will be able to list intermolecular forces and describe some of the effects on liquids and solids.

Your textbook differentiates between **intra**molecular and **inter**molecular forces.

- **Intramolecular forces** mean **forces within a molecule**.  We dealt with these in chapter 8.
- **Intermolecular forces** mean **forces between molecules**.  These are the forces that hold molecules together as liquids and solids.

There are three kinds of forces that are discussed in this section.

- **Dipole-dipole forces** result when the partial positive and negative charges of neighboring polar covalent molecules attract.  These forces are about 1% as strong as intramolecular covalent bonds.

- **Hydrogen bonds** are a special case where dipoles in **small, highly electronegative atoms** (such as fluorine) form a **surprisingly strong** interaction with the small hydrogen, which has a highly positive charge per unit size.  Hydrogen bonding leads to substances with unusually high intermolecular bond energies.  This is shown in Figure 10.3 of your textbook.

- **London dispersion forces** are caused by the instantaneous dipoles which arise in a molecule as a result of momentary imbalances in electron distribution.  These are very weak forces that become more important as the size of the atom of interest increases (see the discussion on **polarizability** in your textbook.)

## Example 10.1 A - London Dispersion Forces

The boiling point of Argon is -189.4°C.

    a.   Why is it so low?
    b.   How does this boiling point help prove that London dispersion forces exist?
    c.   The boiling point of Xenon is -111.9°C. Why is it higher than that of Argon?

### Solution

    a.   Argon does not interact with other substances because it is small and has a complete octet of valence electrons. Argon must be made quite cool to allow liquefication via London dispersion forces.

    b.   If these forces did not exist, Argon would never liquefy.

    c.   Xenon is bigger and has more electrons than Argon. The likelihood of momentary dipoles is thus greater. (It has a greater polarizability than Argon.)

## Example 10.1 B - The Effect of Intermolecular Forces

Put the following substances in order form lowest to highest boiling point.

$$C_2H_6, NH_3, F_2$$

### Solution

$$F_2, C_2H_6, NH_3$$

$F_2$ can only develop intermolecular London dispersion forces. $C_2H_6$ is not especially polar, but it does have a very slight electronegativity difference between the carbons and hydrogens. $NH_3$ exhibits hydrogen bonding, thus giving it a relatively high boiling point.

Problems 1-5 at the end of this study chapter cover material from this section.

## 10.2 - The Liquid State

The following review questions will serve to test your understanding of the material in this section.

1.   Why do liquids tend to bead up when on solid surfaces?
2.   What are **cohesive forces**? **Adhesive forces**? What causes these forces?
3.   What is surface tension? Why does it arise?
4.   Why does water form a concave meniscus when in a thin tube? Why does mercury form a convex m eniscus?
5.   What is **viscosity**? What is a requirement for a liquid to be viscous?
6.   Why do models of liquids tend to be more complex than those for either solids or gases?

## Example 10.2 - Properties of Liquids

Which would have a higher surface tension, $H_2O$ or $C_6H_{14}$?  Why?  Would the shape of the $H_2O$ meniscus in a glass tube be the same or different than $C_6H_{14}$?

## Solution

**Water**, having a dipole moment, has relatively **large cohesive forces**.  Hexane, $C_6H_{14}$, is essentially **nonpolar**.  It has no cohesive forces.  **Water** would therefore have the **higher surface tension**.

The water meniscus is concave because the adhesive forces of water to polar constituents on the surface of the glass are stronger than its cohesive forces.  **Hexane** would have a **convex meniscus**.  It has very small adhesive forces, and the slightly larger cohesive forces would dominate.

Problems 6-8 at the end of this study chapter cover material from this section.

## 10.3 - An Introduction to Structures and Types of Solids

When you finish this study section you will be able to:

- Define the basic terms relating to the structure of solids.
- Solve simple problems relating to x-ray diffraction by crystals.

This section in your textbook begins by contrasting **crystalline solids** (highly regular atomic arrangements) and **amorphous solids** (disordered atomic arrangements).  The focus of the section (and the bulk of the chapter) is on **crystalline solids**.

## Example 10.3 A - Basic Terms

Answer the following questions regarding the terms introduced in this section of your textbook.

    a.   Define **lattice**.
    b.   Define **unit cell**.
    c.   List three types of **cubic unit cells**.
    d.   What is diffraction?
    e.   Why are x-rays used in diffraction analyses of solids?
    f.   What is the difference between **ionic**, **molecular**, and **atomic solids**?

## Solution

    a.   **Lattice:  A three-dimensional system of points** designating the centers of the components in a crystal-line solid.
    b.   **Unit cell:  The smallest repeating unit** of the lattice.
    c.   As shown in **Figure 10.9 of your textbook**, the three types of cubic unit cells are **simple cubic, body-centered cubic** and **face-centered cubic**.
    d.   **Diffraction** is the **scattering of light** from a **regular array** of points or lines, where the spacings between the points or lines (in our case planes of atoms) are related to the wavelength of the light.

   e.   X-rays are used because their wavelengths are similar to the distances between atomic nuclei.

   f.   **Ionic solids** form electrolytes when dissolved in water. **Molecular solids** do not. An **atomic solid** contains atoms of only one kind of element. These atoms are covalently bonded to each other.

---

There are three types of atomic solids described in your textbook - metallic, network, and group VIII A solids. Note the description and examples of these (as given in Table 10.3) just before the start of Section 10.4. Review the meaning of each of the terms of the **Bragg equation** (Equation 10.3 in your textbook), which is used to determine the structures of crystalline solids. When you are comfortable with the terms, try the next example.

---

## Example 10.3 B - Bragg Equation

A topaz crystal has a lattice spacing (d) of 1.36 Å. Calculate the wavelength of x-ray that should be used if θ = 15.0° (assume n = 1).

## Solution

The Bragg equation is in the form

$$n\lambda = 2d \sin \theta$$

where **n** = the order of diffraction

     λ = the wavelength of the "incident energy" (the beam hitting the sample).

     **d** = the spacing between planes of atoms

     **sinθ** = the sine of the angle of reflection of the beam.

        sin(15°) = 0.26

$$\lambda = \frac{2d \sin \theta}{n} = \frac{2\,(1.36\text{Å})\,(0.26)}{1} = 0.71\text{Å}$$

---

Problems 9-12 at the end of this study chapter cover material from this section.

---

## 10.4 - Structure and Bonding in Metals

When you finish this section you will be able to:

- Discuss bonding in metals.
- Calculate the density of metals based on the structure of their unit cell.
- Define and give examples of substitutional and interstitial alloys.

This section of your textbook begins by listing **properties of metals**. These are

-   high thermal and electrical conductivity,
-   malleability (can form thin sheets), and
-   ductility (can be pulled into a wire).

A model that describes the structure of metals must be able to explain these properties.

The proposed model assumes that the atoms of metals are **uniform hard spheres** that are packed to best utilize available space (**closepacking**). Figures 10.13 - 10.15 in your textbook give examples of packing in layers.

- **hexagonal close packing** (hcp) has every **other** layer being spatially equivalent ("ababab..."). Examples include magnesium, zinc, cadmium, cobalt and lithium.

- **cubic close packing** (ccp) or **a face-centered cubic structure** has every **third** layer being spatially equivalent ("abcabcabc..."). Examples include silver, aluminum, nickel, lead and platinum.

Each sphere in each structure has **12 equivalent nearest neighbors (the coordination number is 12)** - 6 in its layer, 3 in the layer above and 3 in the layer below (see Figure 10.16 in your textbook). The exception to the above list is

- **body-centered cubic structure** (bcc). In this structure, spheres are **not** close packed. The **coordination number** for such a sphere is **8** (see Figure 10.9(b) in your textbook). Iron and alkali metals have bcc structures.

The calculation of properties such as **density** from crystal structures is possible if you know the type of unit cell that a metal (or ionic solid, as we shall see later) forms. Your textbook deals only with the **ccp** structure in this regard. The following example shows how density is calculated using the atomic radius and the ccp structure.

---

## Example 10.4 A - Crystal Structure and Density

The radius of a nickel atom is 1.24 Å (1 Å = $1 \times 10^{-8}$ cm). Nickel crystallizes with a cubic close packed structure (face-centered cubic). Calculate the **density** of solid nickel.

### Strategy

The goal is to calculate density, which is **mass/volume**. The **volume** in this case can be represented by the **unit cell**. As shown in your textbook, the ccp unit cell contains **4 atoms** of nickel.

Use your understanding of stoichiometry to calculate the **mass of nickel** in the unit cell.

The volume for the unit cell can be calculated using the Pythagorean Theorem, as illustrated in Sample Exercise 10.2 in your textbook. The volume of the unit cell, a cube, is

$$V = d^3$$

where d = the length on a side.

According to the theorem, the square of the hypotenuse is equal to the sum of the squares of the sides of a right triangle, or in our case,

$$h^2 = d^2 + d^2$$

h = 4r (r = the radius of nickel), as illustrated in the sample exercise.

$$(4r)^2 = 2d^2, \text{ or } 16r^2 = 2d^2$$

$$8r^2 = d^2, \text{ or } \mathbf{d = r\sqrt{8}}$$

Therefore,

$$V = d^3 = r^3 8^{1.5} = \mathbf{22.63r^3}$$

We can now solve for density.

## Solution

$$\text{mass} = 4 \text{ atoms} \times \frac{58.70 \text{ g Ni}}{6.022 \times 10^{23} \text{ atoms}} = 3.900 \times 10^{-22} \text{ g}$$

$$\text{volume} = 22.63 \times (1.24 \text{ Å} \times 1 \times 10^{-8} \text{ cm/Å})^3 = 4.315 \times 10^{-23} \text{ cm}^3$$

$$\text{density} = 3.900 \times 10^{-22} \text{ g} / 4.315 \times 10^{-23} \text{ cm}^3 = 9.04 \text{ g/cm}^3$$

This is in good agreement with the actual value of **8.90 g/cm³**.

---

Your textbook justifies **conduction** of **electricity** and **heat** by metals by using the **band theory**. The discussion says that if you have a **few** molecular orbitals (MOs) when a couple of atoms covalently bond, then a covalently bonded system with **many** atoms should lead to MOs with many energies (a "continuum"). Electrons in filled MOs can be excited into empty MOs and thus travel throughout the metal. Notice how your knowledge of MO theory (Chapter 9) helps you to understand bonding in metals.

The section ends with a discussion of **metal alloys**. When you think you are comfortable with the material, try the following example.

---

## Example 10.4 B - Metal Alloys

Why are high-carbon steels **interstitial alloys** while brass is a **substitutional alloy**?

## Solution

An **interstitial alloy** is formed when holes in the close packed metal structure are occupied by **small atoms** (in high-carbon steels the iron holes are occupied by carbon).

A **substitutional alloy** contains **similar sized** atoms of more than one element (the holes are not occupied). An example is the combination of copper and zinc to form the brass alloy.

Note that the makeup of the alloy greatly affects its properties.

---

Problems 13-17 at the end of this study chapter cover material from this section.

---

## 10.5 & 10.6 - Carbon and Silicon:  Network Atomic Solids and Molecular Solids

The questions listed below will help you to review the material presented in these sections.

### Section 10.5

1. Define **network solid**.
2. List the **properties** of network solids.
3. Why does diamond have carbons with sp3 hybridization while those in graphite are sp2 hybridized?
4. Use MO theory to explain why graphite conducts electricity while diamond does not.

5.  Why can carbon form  bonds while silicon can not?
6.  What is the difference between **silica** and **silicate**?
7.  What property does glass share with ceramics?  How are they different?
8.  What can be done to make ceramics somewhat flexible?
9.  What is the purpose of adding arsenic to a semiconductor?
10. What is a **p-type** semiconductor?  How does it work?
11. Describe how a p-n junction acts as a rectifier.
12. How do electrons interact in the gallium arsenide laser to produce color?

## Section 10.6

11. Define **molecular solid**.
12. Give some **examples** of molecular solids.
13. Describe the relative bonding strength and bond distances **within** and **between** molecules of a molecular solid.
14. Why are some larger nonpolar molecular solids solid at 25°C?

---

Problems 18-21 at the end of this study chapter cover material from this section.

---

# 10.7 - Ionic Solids

When you finish this section you will be able to:

- Discuss the rationale for the packing that is observed in closest packed structures.
- Classify substances according to the types and properties of the solids they form.

Your textbook says that the structure of most binary ionic solids can be explained by the **closest packing of spheres**.  Anions, which are usually larger than cations with which they combine (see section 8.4 in your textbook), are packed in either an hcp or ccp arrangement.  Cations fill the holes within the packed anions.

**KEY IDEA:  The packing arrangement is done in such a way as to minimize anion-anion and cation-cation repulsions.**

The nature of the holes depends on the ratio of the anion to cation size.  Trigonal holes are smallest, followed by tetrahedral with octahedral being the largest.

---

## Example 10.7 A - Packing in Ionic Structures

Would AlP have a closest packed structure which is more like NaCl or ZnS?

$$\text{Ionic Radii are:}\quad Al^{3+} = 0.50 \text{ Å}, P^{3-} = 2.12 \text{ Å}$$
$$Zn^{2+} = 0.74 \text{ Å}, S^{2-} = 1.84 \text{ Å}$$
$$Na^+ = 0.95 \text{ Å}, Cl^- = 1.81 \text{ Å}$$

## Solution

We need to take the anion to cation radius ratio in each of the three cases.

$S^{2-}/Zn^{2+} = 2.49$ (tetrahedral holes)
$Cl^-/Na^+ = 1.90$ (octahedral holes)
$P^{3-}/Al^{3+} = 4.24$ (????)

The aluminum ion is very small compared to the phosphorus ion. Therefore, not much room is needed by the $Al^{3+}$ cations. **Tetrahedral holes** are adequate. **AlP is more like ZnS than NaCl.**

---

Study Table 10.7 in your textbook. Based on the information given in the table, as well as your overall knowledge of chemistry, try the next example, which deals with classifying solids.

---

## Example 10.7 B - Types of Solids

Based on their properties, classify each of the following substances as to the **type of solid** it forms.

   a.   Fe       b.   $C_2H_6$       c.   $CaCl_2$       d.   graphite       e.   $F_2$

### Solution

   a.   Solid iron is an **atomic solid** with metallic properties.
   b.   Solid $C_2H_6$, ethane, contains nonpolar molecules and is a **molecular solid**.
   c.   Solid $CaCl_2$ contains $Ca^{2+}$ and $Cl^-$ ions and is an **ionic solid**.
   d.   Graphite is made up of nonpolar carbon atoms covalently bonded in directional planes. It is a **network solid**.
   e.   Solid fluorine is made up of nonpolar fluorine molecules. It is a **molecular solid**.

---

Problems 22-26 at the end of this study chapter cover material from this section.

---

## 10.8 - Vapor Pressure and Changes of State

When you finish this section you will be able to:

   • Interconvert between **vapor pressure, temperature,** and **enthalpy of vaporization** of a liquid.
   • Perform calculations regarding the energy of phase changes.

Your textbook introduces some very useful terms in this section. You need to be able to define **vaporization, enthalpy of vaporization, condensation, sublimation, enthalpy of fusion, melting point** and **boiling point.**

**Dynamic equilibrium** is a concept that you will be using a great deal in the latter half of your chemistry studies. It means that *two opposing processes are occurring at the same rate.* The net effect is *no observable change.* But the system is not static. In this section, **the equilibrium vapor pressure** means that evaporation and condensation by a liquid are occurring at the same rate. The net effect is to have a **constant** vapor pressure exerted by the liquid.

**The vapor pressure of a liquid depends on** the molecular weight of the liquid.

A heavier substance will have a lower vapor pressure than a lighter substance, all other things being equal, because the atoms are more polarizable, leading to larger intermolecular forces. A substance with hydrogen bonding interactions will have a lower vapor pressure (will be less volatile) than a nonpolar substance. Your textbook introduces the Clausius-Clapeyron equation (equation 10.5 in your textbook) which interrelates the **vapor pressure, temperature** and **enthalpy** of a liquid.

$$\ln \frac{P_{T_1},vap}{P_{T_2},vap} = \frac{\Delta H_{vap}}{R} \left[ \frac{1}{T_2} - \frac{1}{T_1} \right]$$

where $T_1$ and $T_2$ are **temperatures** in Kelvins

$\Delta H_{vap}$ is the **enthalpy of vaporization** of the liquid

$P_{T_1,vap}$ and $P_{T_2,vap}$ are the **vapor pressures** of the liquid at temperatures $T_1$ and $T_2$.

---

## Example 10.8 A - Clausius-Clapeyron Equation

The vapor pressure of 1-propanol at 14.7°C is 10.0 torr.  The heat of vaporization is 47.2 kJ/mol.  Calculate the vapor pressure of 1-propanol at 52.8°C.

## Solution

Let's list what we are given.

$\Delta H_{vap} = 47.2$ kJ/mol
R = 8.314 J/K mol = 0.008314 kJ/K mol
$T_1 = 14.7°C = 287.9$ K
$T_2 = 52.8°C = 326.0$ K
$P_{T_1,vap} = 10.0$ torr

$P_{T_2,vap} = X$

Substituting,

$$\ln \left[ \frac{10.0}{X} \right] = \frac{47.2 \text{ kJ/mol}}{0.008314 \text{ kJ/K mol}} \left[ \frac{1}{326.0 \text{ K}} - \frac{1}{287.9 \text{ K}} \right]$$

$$\ln \left[ \frac{10.0}{X} \right] = 5677 \, (-4.06 \times 10^{-4})$$

or

$$\ln \left[ \frac{10.0}{X} \right] = 2.305$$

Taking the antilog of both sides,

X/10 = 10.02

$X = P_{T_2,vap} = 100.2 = $ **100 torr**

---

Look at Figure 10.42 in your textbook.  This illustrates the **heating curve** for water.  Note two important observations:

- The **temperature** of a substance **remains constant** during a **phase change**.
- The **temperature rises** when heat is input while a substance is in **one phase**.

You can find the amount of **energy required** to convert water from ice at $T_1$ to steam at $T_2$ by using the following information:

- heat capacity of ice (**2.1 J/g °C**)
- $\Delta H_{fusion}$ of water (**6.0 kJ/mol**)

- heat capacity of liquid water (**4.2 J/g °C**)
- $\Delta H_{vap}$ of water (**43.9 kJ/mol**)
- heat capacity of steam (**1.8 J/g °C**)

How much of the information will be used will depend on the problem you have to solve.

---

## Example 10.8 B - Heating Curve

How much energy does it take to convert 130 g of ice at -40°C to steam at 160°C?

### Strategy

There are **5 steps** involved in this conversion from ice to steam.

1. Heating ice from -40°C to the melting point.
2. Melting ice to form liquid water.
3. Heating liquid water to its boiling point.
4. Boiling liquid water to form steam.
5. Heating to 160°C.

The **total energy** required is the **sum** of the energy required in each of the 5 steps. The appropriate constants for each step are given in the discussion preceding this problem. The units of heat capacity contain °C because the temperature is rising in each of these steps. The units "enthalpy of fusion and vaporization" do not because the temperature is constant at a phase change.

### Solution

Energy used = sum of energies from individual steps. There are 7.22 mol of water in 130 g.

Step 1 = 40°C × 130 g × 2.1 J/g °C   = 10.92 kJ
Step 2 = 7.22 mol × 6.0 kJ/mol        = 43.3 kJ
Step 3 = 100°C × 130 g × 4.2 J/g °C  = 54.6 kJ
Step 4 = 7.22 mol × 43.9 kJ/mol       = 317.0 kJ
Step 5 = 60°C × 130 g × 1.8 J/g °C    = 14.04 kJ

**Total Energy** = 10.92 + 43.3 + 54.6 + 317.0 + 14.04 = **440 kJ**

---

Problems 27-39 at the end of this study chapter cover material from this section.

---

## 10.9 - Phase Diagrams

When you finish this section you will be able to extract information from simple phase diagrams.

The beauty of this section is that it helps explain a large number of real-world phenomena. (See the Chemical Impact on diamonds near the end of the chapter in your textbook.)

You should be able to define the following terms: **phase diagram, critical temperature, critical pressure, critical point** and **triple point.** You should be able to answer the following general questions regarding material presented in this section.

1.  Why does the solid/liquid line in the phase diagram of water have a **negative** slope?  Why is it positive for carbon dioxide?
2.  Why does it take longer to cook an egg in the Rocky Mountains than at sea level?
3.  How does the phase diagram for water help explain why your blades glide on a liquid layer when you ice skate?
4.  How does the phase diagram for carbon dioxide help explain how a $CO_2$ fire extinguisher works?
5.  Snow sometimes sublimes.  How can this be so in spite of the phase diagram?

---

## Example 10.9 - Phase Diagrams

What phase changes does water undergo (see <u>Figure 10.47 in your textbook</u>) as the pressure changes while the temperature is held constant at -12°C?

### Solution

At very low pressures, water exists as a gas at -12°C.  As the pressure is increased, it turns into a solid.  At very high pressures water will liquefy.

---

Problem 40 at the end of this study chapter covers material from this section.

---

**END OF CHAPTER TEN**

# SELF-TEST QUESTIONS

1.  Define the following terms in your own words:

    a.  dipole-dipole forces          c.  London dispersion forces
    b.  hydrogen bonding

2.  Define the following terms in your own words:

    a.  crystaline solids             c.  lattice
    b.  amorphous solids              d.  unit cell

3.  Of HF, HCl and HBr, which has the highest boiling point?  Why?  Which has the lowest boiling point?

4.  Propane, $C_3H_8$, is a gas at room temperature, hexane, $C_6H_{14}$, is a liquid, and dodecane, $C_{12}H_{26}$ is a solid. Explain.

5.  Which would you expect to have a lower melting point $C_3H_8$ or $CH_3OH$?  Why?

6.  Why does water "bead up" more on a car that is waxed than on one that isn't waxed?

7.  Would mercury bead up more on a waxed or unwaxed car?

8.  Which would have greater surface tension $N_2(l)$ or $Br_2(l)$?

9.  X-rays of wavelength 0.1541 nm produced a reflection angle $\theta = 15.55°$.  What is the spacing between crystal planes (n = 1)?  What would be the angle of reflection for n = 2?

10. X-rays of wavelength 2.63 Å were used to analyze a crystal.  A reflection was produced at $\theta = 13.9°$.  Assuming n = 1, calculate the spacing between the planes of the atoms producing this reflection.

11. The second order reflection (n = 2) for a gold crystal is an angle of 22.20° for x-rays of 1.54 Å, what is the spacing between these crystal planes?

12. Classify the following as either ionic, molecular or atomic crystalline solids.

    a.  dry ice, $CO_2(s)$        c.  CaF(s)           e.  $C_{10}H_8(s)$ (naphthalene)
    b.  graphite                 d.  $MnO_2(s)$

13. Tungsten crystallizes in a body-centered cubic structure with a unit cell edge length of 3.1583 Å.  The density of tungsten metal is 19.3 $g/cm^3$ and its atomic weight is 183.85 g/mol.  Calculate the value of Avogadro's number by this method.

14. Silver crystallizes in a face-centered cubic structure with a unit cell edge length of 4.0776 Å.  The density is determined to be 10.5 $g/cm^3$.  Calculate the atomic weight of silver.

15. The density of gold is 19.3 $g/cm^3$.  Gold crystallizes in a face-centered cubic unit cell.  What is the radius of a gold atom?

16. Potassium crystallizes in a body-centered cubic structure.  Calculate the atomic radius of potassium if the unit cell edge length is 5.333 Å.

17. Why is iron relatively soft, ductile and malleable, while high carbon steels are much harder, stronger and less malleable?

18. What accounts for the lubricating ability of graphite?

19. Which electrons account for the conductivity of graphite?

20. Explain why graphite conducts electricity parallel to the layers much better than it conducts electricity perpendicular to the layers.

21. How many bonds would each sulfur atom in an $S_8$ molecule have to make to have the nearest noble gas electron configuration?

22. Cobalt fluoride crystallizes in a closest packed array of fluoride ions with cobalt ions filling one half of the octahedral holes. What is the formula?

23. Explain why anions usually have larger radii than cations.

24. What is the formula for the compound that crystallizes with a closest packed array of sulfur ions, that contains zinc ions in 1/8 of the tetrahedral holes and aluminum ions in 1/2 on the octahedral holes?

25. Manganese and fluoride ions crystallize such that each cubic unit cell has manganese ions at the corners and fluoride ions at the center of each edge. What is the formula of this compound?

26. What is the maximum radius for a cation that will fit in an octahedral hole in a face-centered cubic arrangement of chloride ions if the chloride ions are to remain in contact? (Radius of $Cl^-$ is 1.81 Å.)

27. Using the following vapor pressure data for $CCl_4$, make a graph and determine the normal boiling point of the liquid.

| Temp (°C) | 20.0 | 40.0 | 60.0 | 70.0 | 80.0 | 90.0 | 100.0 |
|-----------|------|------|------|------|------|------|-------|
| V.P. (torr) | 91.0 | 213.3 | 444.3 | 617.4 | 836.0 | 1110 | 1459 |

28. The vapor pressure of water at 25°C is 23.8 torr. Confirm the value of 43.9 kJ/mol for the heat of vaporization of water. (Use data for the normal boiling point as well as the vapor pressure given.)

29. On top of one of the peaks in Rocky Mountain National Park the pressure of the atmosphere is 550 torr. Determine the boiling point of water at this location.

30. In Denver, Colorado the pressure of the atmosphere is 697 torr. Determine the boiling point of water at this location.

31. Calculate the vapor pressure of water at 0°C (in kilopascals, kP).

32. Isopropanol, $C_3H_8O$, is also known as rubbing alcohol. The heat of vaporization is 42.1 kJ/mol. How much heat is needed to evaporate 25 g of isopropanol?

33. How much isopropanol must evaporate to cool 1 kg from 25°C to 20°C? (Specific heat of isopropanol is 2.59 J/g°C.)

34. What quantity of heat is required to melt 1.0 kg of ice at its melting point?

35. What quantity of heat is required to vaporize 1.0 kg of water at its boiling point?

36. What is the final temperature when 10 g of water at 0°C is added to 100 g of water at 75°C?

37. If 10 g of ice at 0°C comes in contact with 10 g of water at 50°C, calculate the final temperature reached by the system at equilibrium.

38. What is the final temperature when 10 g of ice at 0°C is added to 100 g of water at 75°C? ($\Delta H_f = 6.0$ kJ/mol, heat capacity of water is 4.2 J/g°C.)

39. If 10 g of ice at 0°C comes in contact with 50 g of water at 10°C, calculate the final temperature reached by the system at equilibrium.

40. How is the change in density for a solid-to-liquid phase change related to the slope of the liquid-solid line of a phase diagram?

# ANSWERS TO SELF-TEST QUESTIONS

1. Student's own answers.

2. Student's own answers.

3. HF has the highest boiling point because of hydrogen bonding.  HCl has the lowest boiling point.

4. All other things being equal, the higher the molecular weight of nonpolar compounds, the greater the intermolecular London forces.

5. $C_3H_8$ has a lower melting point because $CH_3OH$ has hydrogen bonding.

6. There are fewer adhesive forces to the nonpolar wax.

7. An unwaxed car.

8. $Br_2(l)$

9. 0.2874 nm, 0.5748 nm

10. 5.47 Å

11. 4.08 Å

12. a. molecular    c. ionic      e. molecular
    b. atomic      d. ionic

13. $6.08 \times 10^{23}$

14. 107 g/mol

15. 1.44 Å

16. 2.309 Å

17. Carbon forms strong directional bonds.

18. Weak attractive forces between layers.

19. The electrons in the $\pi$ bonds.

20. Parallel conduction is favored because $\pi$ electrons are localized above and below the layers.

21. 2 bonds

22. $CoF_2$

23. Extra electrons repel each other and have a lower $z_{eff}$.

24. $ZnAl_2S_4$

25. $MnF_3$

26. 0.489 Å, 0.265 Å, 0.183 Å

27. 77°C

28. 5.86 torr (if normal boiling point data is used), 5.19 torr (if 25°C data is used)

29. 91°C

30. 97°C

31. 0.61 kP

32. 17.5 kJ

33. 18.5 g

34. 333 kJ

35. 2259 kJ

36. 68°C

37. 0°C

38. 38°C

39. 0°C

40. If solid is more dense - positive slope;  if solid is less dense - negative slope.

# CHAPTER 11

# Properties of Solutions

| Section | Contents | Page in this Study Guide |
|---|---|---|
| 11.1 | Solution Composition | 226 |
| 11.2 | The Energies of Solution Formation | 228 |
| 11.3 | Factors Affecting Solubility | 229 |
| 11.4 | The Vapor Pressures of Solutions | 231 |
| 11.5 | Boiling-Point Elevation and Freezing-Point Depression | 234 |
| 11.6 | Osmotic Pressure | 236 |
| 11.7 | Colligative Properties of Electrolyte Solutions | 237 |
| 11.8 | Colloids | 238 |

This chapter deals with **solutions - homogeneous mixtures** of solids, liquids, or gases, as shown in Table 11.1 of your textbook. A new socially important application of salt solutions is the focus of the *chemical impact* at the beginning of the chapter.

## 11.1 - Solution Composition

When you finish this section you will be able to solve problems relating to the **mass percent, mole fraction, molality** and **normality** of a solution.

This section in the textbook introduces you to four new concentration terms. **Mass percent, mole fraction and molality** are useful because they are **temperature independent.** As the temperature of a solution changes, the **molarity changes slightly** because the **solution volume changes.** As your textbook points out, these new terms depend only on mass which is temperature independent.

The formulas for the new concentration units are:

$$\text{mass percent} = \frac{\text{g solute}}{\text{g solution}} \times 100\%$$

$$\text{mole fraction of A } (\chi_A) = \frac{n_A}{n_{total}} = \frac{\text{number of moles of A}}{\text{total number of moles}}$$

$$\text{molality (m)} = \frac{\text{n solute}}{\text{kg solvent}}$$

Also, recall the definition of **molarity** (moles of solute/liter of solution) that we have used throughout the course.

## Example 11.1 A - Concentration Units

A solution is prepared by adding 5.84 g of formaldehyde, $H_2CO$, to 100.0 g of water. The final volume of the solution was 104.0 mL. Calculate the molarity, molality, mass percent, and mole fraction of the formaldehyde in the solution.

## Solution

**Molarity:** You have 5.84 g $H_2CO \times \dfrac{1 \text{ mole } H_2CO}{30.03 \text{ g } H_2CO} = 0.194$ mol $H_2CO$

$$M = \frac{0.194 \text{ moles } H_2CO}{0.104 \text{ L solution}} = \textbf{1.87 M } H_2CO$$

**Molality:** $m = \dfrac{0.194 \text{ mol } H_2CO}{0.100 \text{ kg } H_2O \text{ (solvent)}} = \textbf{1.94 m } H_2CO$

**Mass percent** $= \dfrac{5.84 \text{ g } H_2CO}{105.84 \text{ g } (H_2CO + H_2O)} \times 100\% = \textbf{5.52\% } H_2CO$

(The mass percent of $H_2O$ must be 94.48%. Can you see why?)

**Mole fraction:** We know that there is 0.194 mole of $H_2CO$.

$$n_{H_2O} = 100.0 \text{ g } H_2O \times \frac{1 \text{ mol } H_2O}{18.02 \text{ g } H_2O} = 5.55 \text{ mol } H_2O$$

$$\chi_{H_2CO} = \frac{0.194 \text{ mol } H_2CO}{(0.194 \text{ mol } H_2CO + 5.55 \text{ mol } H_2O)} = \textbf{0.0338}$$

(The mole fraction of $H_2O$ must be 0.9662. Can you see why?)

---

The molarity will be equal to the molality only in very dilute aqueous solutions. The final concentration term in this section is **normality (N)**.

$$N = \frac{\text{equivalents of solute}}{\text{liter of solution}}$$

An **equivalent** is related to a mole. The number of equivalents of a given solute will always be **greater than or equal to** the number of moles of the solute. **An equivalent is the mass of the solute that can furnish or accept one mole of protons ($H^+$ in acid/base) or one mole of electrons (redox).**

The **equivalent mass** of a substance is the **mass of one equivalent** of the substance. The number of equivalents per mole of solute (hence the equivalent mass and the normality) depends upon the reaction. This is why it is such an awkward term. We can only hope the unit will eventually fall into disuse.

If we look at the reaction involving $H_3PO_4$,

$$H_3PO_4 + 3NaOH \longrightarrow PO_4^{3-} + 3H_2O + 3Na^+$$

Each mole of $H_3PO_4$ (97.99 g/mol) can supply 3 moles of protons (there are 3 equivalents per mole). The equivalent mass (the mass that can supply one mole of protons) is

$$\frac{97.99 \text{ g/mol}}{3 \text{ eq/mol}} = \textbf{32.66 g/eq}$$

## Example 11.1 B - Normality

Given the reaction listed above, if we have 28.42 g $H_3PO_4$ in 800 mL of water, what are the normality and molarity of the solution?

## Solution

We know from the discussion preceding this example that phosphoric acid has a mass of **32.66 g/eq**.

$$\text{equivalents of } H_3PO_4 = \frac{1 \text{ eq}}{32.66 \text{ g}} \times 28.42 \text{ g} = \mathbf{0.870 \text{ eq}}$$

$$N = \frac{0.870 \text{ eq } H_3PO_4}{0.800 \text{ L solution}} = \mathbf{1.09 \text{ N } H_3PO_4}$$

$$M = \frac{28.42 \text{ g}}{0.800 \text{ L}} \times \frac{1 \text{ mol}}{97.99 \text{ g}} = \mathbf{0.363 \text{ M } H_3PO_4}$$

Note the normality, as expected, is 3 times the molarity.

Problems 1-11 at the end of this study chapter cover material from this section.

## 11.2 - The Energies of Solution Formation

When you finish this section you will be able to determine the miscibility of a variety of substances.

This section concerns **how**, and **under what conditions**, a solution will be formed from the interaction of two substances. Your textbook lists **three steps** that must occur for solution formation to occur.

1. Expanding the solute (endothermic, $\Delta H = +$)
2. Expanding the solvent (endothermic, $\Delta H = +$)
3. Combining the solute and solvent (often exothermic, $\Delta H = -$)

In solution formation the enthalpy of solution (the sum of steps 1, 2, and 3) is often "-" (exothermic). However, this is not the determinant of whether a reaction will occur. It is just one outcome of solution formation. We will learn how to specifically determine if a reaction will occur in **Chapter 16**.

Although we can not yet determine explicitly if solution formation will occur, we can make the generalization that **like dissolves like**. This is illustrated in Table 11.3 in your textbook.

The key to predicting whether two substances will mix (are "miscible") is to establish the polarity of each. If they are similar, you can probably form a solution. If they are very different, a solution is not likely to form. A more extended explanation is given in the paragraph just preceding Table 11.3 in your textbook.

# Example 11.2 - Predicting Solubility

The following substances are slowly added to a 250 mL graduated cylinder: 50 mL of carbon tetrachloride ($CCl_4$, density = 1.4 g/mL), 50 mL of water (density = 1.0 g/mL) and 50 mL of cyclohexane ($C_6H_{12}$, density = 0.8 g/mL). After cyclohexane has been added, **how would the liquids appear in the cylinder** (ie; would there be solution formation)? If solid $I_2$ flakes are added to the system, **in which layers (if any) would they dissolve?**

## Solution

$CCl_4$ has no dipole moment (the molecule is **nonpolar**).
$H_2O$ is a **polar** molecule.
$C_6H_{12}$ is essentially **nonpolar**.

Because of the different densities of the liquids, $CCl_4$ would be the bottom layer. Water would sit on top of $CCl_4$. $C_6H_{12}$ would be the top layer.

Iodine is nonpolar. When solid $I_2$ flakes are added to the layers, it would dissolve in our nonpolar layers, $CCl_4$ and $C_6H_{12}$. There would ultimately be purple $CCl_4 + I_2$ and $C_6H_{12} + I_2$ layers surrounding a clear water layer.

---

Problems 12-14 at the end of this study chapter cover material from this section.

---

# 11.3 - Factors Affecting Solubility

When you finish this section you will be able to:

- Predict relative solubilities of simple molecules based on structure.
- Solve problems relating to Henry's Law.

In considering **solubility effects** it is important to **separate the behavior of solids and liquids from gases.** The solubility of gases is for the most part independent of structure. Solid and liquid solubility is highly structure dependent.

Reread the discussion in your textbook on "structure effects" as it relates to the solubilities of vitamins A and C, then work the next example.

---

# Example 11.3 A - Structure and Solubility

Determine whether or not each of the following compounds is likely to be **water soluble.**

a.  $CH_3CHNH_2$
     |
     OH

b.  $H_3C-N-C_6H_5$ with H above N and double bond below
     ‖
     $NC_6H_5$

c.  $C_4H_9CH=CH_2$

d.   $CH_3(CH_2)_4CH_2NH_2$

e.   $H_2C\text{-}CH_2$
     $\quad |\quad\ |$
     $\ \ HO\ OH$

## Strategy

An important factor in water solubility is the **ratio of polar to nonpolar groups** in the covalent molecule. The higher the ratio (with hydrogen bonding groups being especially good), the more likely the covalent substance is to be water soluble, unless there is no net dipole moment (such as in $CCl_4$).

However, there are no absolutes. There are other factors involved in solubility that you will learn more about when you are introduced to **organic chemistry** (Chapter 22 in your textbook).

## Solution

a.   **water soluble** (small molecule with 3 polar bonds)
b.   **insoluble** (large molecule with only 1 polar bond)
c.   **insoluble** (nonpolar molecule)
d.   **insoluble** (large molecule with 2 polar bonds)
e.   **water soluble** (small molecule with 2 polar bonds)

Note how the answers here are consistent with example 11.2 in this study guide.
Your textbook points out that **pressure has little effect on the solubilities of liquids and solids.** The solubility of gases is, for the most part, **independent of structure** and is given by **Henry's Law,**

$$P = kC$$

where  P = partial pressure of the gaseous solute above the solution (in atm)
    C = concentration of the dissolved gas (in mol/L)
    k = a constant for a particular solution (in L atm/mol)

Keep in mind that Henry's Law applies for **low gas concentrations** and gases that **do not react with the solution.**

---

# Example 11.3 B - Henry's Law

The solubility of $O_2$ is $2.2 \times 10^{-4}$ M at 0°C and 0.10 atm. Calculate the solubility of $O_2$ at 0°C and 0.35 atm.

## Solution

Because the temperature is the same in both solutions, the **Henry's Law constant is the same.** Therefore, if k = P/C then

$$\frac{P_1}{C_1} = k = \frac{P_2}{C_2}$$

$$P_1 = 0.10 \text{ atm} \qquad P_2 = 0.35 \text{ atm}$$
$$C_1 = 2.2 \times 10^{-4} \text{ M} \qquad C_2 = \text{???}$$

$$\frac{0.10 \text{ atm}}{2.2 \times 10^{-4} \text{ M}} = \frac{0.35 \text{ atm}}{C_2}$$

$$C_2 = 7.7 \times 10^{-4} \text{ M } O_2$$

With regard to temperature effects, the solubility of most, but not all, solids increases with temperature. Your textbook points out that the solid dissolves **more rapidly** at higher temperatures. Also, review the discussion on **thermal pollution** and **boiler scale**.

Problems 15-21 at the end of this study chapter cover material from this section.

## 11.4 - The Vapor Pressures of Solutions

When you finish this section you will be able to calculate

- the vapor pressure of a variety of solutions, and
- the molar mass from vapor pressure information.

This section deals with the vapor pressure relationships of solutions containing both volatile and nonvolatile solutes. Let's deal with each of these cases separately.

**Nonvolatile Solutes**

This behavior is described by **Raoult's Law**,

$$P_{soln} = \chi_{solvent}P^°_{solvent}$$

where $P_{soln}$ = vapor pressure of the solution
$\chi_{solvent}$ = mole fraction of the solvent (see <u>section 5.5 in your textbook</u> for a review of mole fractions)
$P^°_{solvent}$ = vapor pressure of the pure solvent

This equation says that the addition of a nonvolatile solute will cause the vapor pressure of the solution to fall in direct proportion to the mole fraction of the solute.

## Example 11.4 A - Raoult's Law

Glycerin, $C_3H_8O_3$, is a nonvolatile liquid. What is the vapor pressure of a solution made by adding 164 g of glycerin to 338 mL of $H_2O$ at 39.8°C? The vapor pressure of pure water at 39.8°C is 54.74 torr and its density is 0.992 g/mL.

### Strategy

We know that the vapor pressure of the solution will be lowered upon addition of (nonvolatile) glycerine. To determine the new vapor pressure, we must calculate the **mole fraction of water**, then multiply by the vapor pressure of pure water.

### Solution

$$\text{moles of glycerin} = 164 \text{ g} \times \frac{1 \text{ mol}}{92.1 \text{ g}} = 1.78 \text{ mol}$$

$$\text{moles of water} = 338 \text{ mL} \times \frac{0.992 \text{ g}}{\text{mL}} \times \frac{1 \text{ mol}}{18.0 \text{ g}} = 18.63 \text{ mol}$$

$$\chi_{water} = \frac{n_{water}}{n_{water} + n_{gly}} = \frac{18.63}{18.63 + 1.78} = 0.913$$

$$\mathbf{P_{water}} = \chi_{water}P^\circ_{water} = 0.913 \times 54.74 \text{ torr} = \mathbf{50.0 \text{ torr}}$$

Let's extend the concept further using a **nonvolatile electrolyte** that **dissociates completely** in water. When solving the problem, we are interested in **moles of solute** actually present **in the solution** (after dissociation), not moles added.

---

## Example 11.4 B - Raoult's Law with Electrolytes

What is the vapor pressure of a solution made by adding 52.9 g of $CuCl_2$ a strong electrolyte, to 800.0 mL of water at 52.0°C?  The vapor pressure of water at 52.0°C is 102.1 torr and its density is 0.987 g/mL.

### Solution

The strategy is the same as in our previous example.  However, $CuCl_2$ dissociates in aqueous solution:

$$CuCl_2(s) \text{----}> Cu^{2+}(aq) + 2Cl^-(aq)$$

As a result, we have **3 moles of solute formed for every mole we add.**

$$\text{moles of } CuCl_2 = 52.9 \text{ g} \times \frac{1 \text{ mol}}{134.5 \text{ g}} = 0.393 \text{ mol}$$

$$\text{total moles of solute} = 0.393 \text{ mol} \times 3 = \mathbf{1.18 \text{ mol solute}}$$

$$\text{moles of water} = 800.0 \text{ mL} \times \frac{0.987 \text{ g}}{mL} \times \frac{1 \text{ mol}}{18.0 \text{ g}} = \mathbf{43.8_6 \text{ mol } H_2O}$$

$$\chi_{water} = \frac{n_{water}}{n_{water} + n_{solute}} = \frac{43.86}{43.86 + 1.18} = \mathbf{0.974}$$

$$\mathbf{P_{water}} = \chi_{water} P^\circ_{water} = 0.974 \times 102.1 \text{ torr} = \mathbf{99.4 \text{ torr}}$$

---

<u>Sample Exercise 11.6 in your textbook</u> shows how Raoult's Law can be used to determine the molar mass of a nonvolatile solute.  Review that example, then try this one.

---

## Example 11.4 C - Molar Mass Via Raoult's Law

At 29.6°C pure water has a vapor pressure of 31.1 torr.  A solution is prepared by adding 86.7 g of "Y," a non-volatile nonelectrolyte to 350.0 g of water.  The vapor pressure of the resulting solution is 28.6 torr.  Calculate the molar mass of Y.

### Strategy

In this problem, we know the vapor pressure drops due to the addition of Y.

If $P_{water} = \chi_{water} P^\circ_{water}$, then $\chi_{water} = \frac{P_{water}}{P^\circ_{water}}$.  You can thus determine $\chi_{water}$.

$$\chi_{water} = \frac{n_{water}}{n_Y + n_{water}}$$

You know $n_{water}$ and $\chi_{water}$. You can therefore solve for $n_Y$, which leads to the molar mass of Y.

## Solution

$$\chi_{water} = \frac{28.6 \text{ torr}}{31.1 \text{ torr}} = \mathbf{0.920}$$

$$n_{water} = 350.0 \text{ g} \times \frac{1 \text{ mol}}{18.0 \text{ g}} = \mathbf{19.4 \text{ mol}}$$

$$\chi_{water} = \frac{n_{water}}{n_Y + n_{water}} \Rightarrow 0.920 = \frac{19.4}{n_Y + 19.4} \Rightarrow$$

$$0.920 n_Y + 17.8 = 19.4 \Rightarrow 0.920 n_Y = 1.60 \Rightarrow n_Y = 1.74 \text{ mol}$$

$$\text{molar mass of Y} = \frac{86.7 \text{ g}}{1.74 \text{ mol}} = \mathbf{49.9 \text{ g/mol}}$$

---

We have thus far worked with nonvolatile solutes. **Volatile solutes** contribute to the vapor pressure such that

$$\mathbf{P_{total} = P_{solute} + P_{solvent}}$$

The vapor pressure of each component can be expressed by Raoult's Law:

$$P_{solute} = \chi_{solute} P^{\circ}_{solute}$$

$$P_{solvent} = \chi_{solvent} P^{\circ}_{solvent}$$

therefore,

$$P_{total} = \chi_{solute} P^{\circ}_{solute} + \chi_{solvent} P^{\circ}_{solvent}$$

By extension, for a solution that contains n components,

$$P_{total} = \chi_1 P^{\circ}_1 + \chi_2 P^{\circ}_2 + \ldots + \chi_n P^{\circ}_n$$

Let's apply this to the following example.

---

## Example 11.4 D - Volatile Solutes

The vapor pressure of pure hexane ($C_6H_{14}$) at 60.0°C is 573 torr. That of pure benzene ($C_6H_6$) at 60.0°C is 391 torr. What is the expected vapor pressure of a solution prepared by mixing 58.9 g of hexane with 44.0 g of benzene (assuming an ideal solution)?

## Solution

$$n_{hexane} = 58.9 \text{ g} \times \frac{1 \text{ mole}}{86.0 \text{ g}} = 0.685 \text{ mol}$$

$$n_{benzene} = 44.0 \text{ g} \times \frac{1 \text{ mole}}{78.0 \text{ g}} = 0.564 \text{ mol}$$

$$\chi_{hexane} = \frac{n_{hexane}}{n_{hexane} + n_{benzene}} = \frac{0.685 \text{ mol}}{0.685 \text{ mol} + 0.564 \text{ mol}} = \mathbf{0.548}$$

$$\chi_{benzene} = 1 - 0.548 = \mathbf{0.452}$$

$$\mathbf{P}_{total} = \chi_{hexane} \, P^\circ_{hexane} + \chi_{benzene} \, P^\circ_{benzene} = 0.548 \, (573) + 0.452 \, (391) = \mathbf{491 \text{ torr}}$$

---

Problems 22-30 at the end of this study chapter cover material from this section.

---

## 11.5 - Boiling Point Elevation and Freezing Point Depression

When you finish this section you will be able to solve problems relating to the **colligative properties** of **boiling point elevation** and **freezing point depression**.

This section in your textbook introduces us to **colligative properties**. These properties depend **only on the concentration of solute particles**, not on the nature of the particles.

Your textbook deals first with **boiling point elevation**, that is, **the increase in boiling point of a liquid due to the addition of a nonvolatile solute**. The boiling point elevation, $\Delta T$, is given by

$$\Delta T = K_b m_{solute}$$

where $\Delta T$ is the boiling point elevation (in °C)
$K_b$ is the molal boiling point elevation constant (in °C kg/mol)
m is the molality of the solute (in mol solute/kg solvent)

Table 11.5 in your textbook gives values of $K_b$ for some common solvents. One thing to keep in mind is that much useful information can be obtained from colligative property experiments, including molar mass of the solute, molality and mass percent.

---

## Example 11.5 A - Boiling Point Elevation

A solution is prepared by adding 31.65 g of sodium chloride to 220.0 mL of water at 34.0°C (density = 0.994 g/mL, $K_b$ for water is 0.51°C kg/mol). Calculate the boiling point of the solution, assuming complete dissociation of NaCl in solution (which is not true, strictly speaking - see section 11.7).

## Solution

In order to calculate $\Delta T$, we need to calculate the **molality of NaCl**. In this case,

$$\text{molality} = \frac{\text{mol NaCl}}{\text{kg water}}$$

$$\text{mol NaCl} = 31.65 \text{ mL} \times \frac{1 \text{ mol}}{58.0 \text{ g}} = \mathbf{0.546 \text{ mol}}$$

$$\text{kg water} = 220.0 \text{ mL} \times \frac{0.994 \text{ g}}{\text{mL}} \times \frac{1 \text{ kg}}{1000 \text{ g}} = \mathbf{0.219 \text{ kg}}$$

$$m_{NaCl} = \frac{0.546 \text{ mol}}{0.219 \text{ kg}} = \mathbf{2.49 \text{ m}}$$

But, NaCl **dissociates** into $Na^+$ and $Cl^-$, giving 2 moles of solute per 1 mole dissolved.  Therefore,

$$\mathbf{m_{total} = 2.49 \times 2 = 4.98\ m}$$

We can now calculate $\Delta T$ directly.

$$\Delta T = K_b m = 0.51°C\ kg/mol \times 4.98\ m = 2.54°C$$

**The boiling point** = $100 + 2.54 = 102.54°C = \mathbf{102.5°C}$

As discussed in your textbook, the effect of adding a nonvolatile solute to a liquid is to lower the freezing point by an amount, $\Delta T$, given by

$$\mathbf{\Delta T = K_f\, m_{solute}}$$

where $\Delta T$ is the **freezing point depression** (in °C)
$\quad$ $K_f$ is the **molal freezing point depression** constant (in °C kg/mol)
$\quad$ **m** is the **molality of the solute** (in mol solute/kg solvent)

The overall effect of adding a nonvolatile solute to a liquid is to **extend the liquid range of the solvent**.  The same information can be found from both freezing point depression and boiling point elevation problems.

---

# Example 11.5 B - Freezing Point Depression

How many grams of glycerin ($C_3H_8O_3$) must be added to 350.0 g of water in order to lower the freezing point to -3.84° ($K_f$ for water = 1.86°C kg/mol)?

## Solution

$\Delta T = 3.84°C$
$K_f = 1.86°C\ kg/mol$

$$\text{molality} = \frac{\text{moles of glycerin}}{\text{kg of water}}$$

The bottom line then is to solve for the **molality of glycerin**, which leads to **moles** and finally **grams of glycerin**.

$$m_{glycerin} = \frac{\Delta T}{K_f} = \frac{3.84C°}{1.86°C\ kg/mol} = \mathbf{2.06_4\ m}$$

$$\textbf{moles of glycerin} = \frac{2.064\ \text{mol glycerin}}{\text{kg water}} \times 0.350\ \text{kg water} = \textbf{0.723 mol}$$

$$\textbf{g of glycerin} = 0.723\ \text{mol} \times 92.1\ \text{g/mol} = \textbf{66.6 g of glycerin}$$

---

Problems 31-36 at the end of this study chapter cover material from this section.

---

## 11.6 - Osmotic Pressure

When you finish this section you will be able to solve problems regarding osmosis and molar mass.

This section in your textbook begins with the definition of **osmosis** as **the flow of solvent into a solution through a semipermeable membrane.**

**Osmotic pressure** is **the pressure that just stops the osmosis.** Osmotic pressure is a **colligative property** because its value depends on the **concentration of the solute**, not its nature. Your textbook points out that osmotic pressure is especially useful for determining molar masses experimentally because **small solute concentrations produce large osmotic pressures.**

The relationship between osmotic pressure and solution concentration is given by

$$\pi = MRT$$

where $\pi$ is the **osmotic pressure** (in atm)
    $M$ is the **molarity of the solute** (in mol/L)
    $R$ is the **universal gas constant** (0.08206 L atm/K mol)
    $T$ is the **temperature** (in K)

Let's try an example that uses osmotic pressure to determine the molar mass of a substance.

## Example 11.6 A - Osmotic Pressure

The osmotic pressure of a solution containing 26.5 mg of aspartame per liter is 1.70 torr at 30°C. Calculate the molar mass of aspartame.

## Solution

$$\pi = MRT$$

We want to solve for molar mass. This can be done through **molarity**.

$$\pi = 1.70 \text{ torr} \times \frac{1 \text{ atm}}{760 \text{ torr}} = 2.24 \times 10^{-3} \text{ atm}$$

$$R = 0.08206 \text{ L atm/K mol}$$

$$T = 303 \text{ K}$$

$$M = \frac{\pi}{RT} = \frac{2.24 \times 10^{-3} \text{ atm}}{0.08206 \text{ L atm/K mol } (303 \text{ K})} = \mathbf{9.01 \times 10^{-5} \text{ mol/L}}$$

$$\textbf{molar mass} = \frac{\text{g aspartame}}{\text{mol}} \left|\left| \frac{1 \text{ L}}{9.01 \times 10^{-5} \text{ mol}} \times \frac{0.0265 \text{ g}}{\text{L}} = \textbf{294 g/mol} \right.\right.$$

Several new terms are introduced in the discussion following sample <u>Exercise 11.12 in your textbook</u>. You should be able to define **dialysis**, **hypertonic**, **hypotonic**, **isotonic solutions**, **crenation** and **lysis**.

You should also be able to describe the concept of **reverse osmosis** and how this can be used to **desalinate water**.

Problems 37-43 at the end of this study chapter cover material from this section.

## 11.7 - Colligative Properties of Electrolyte Solutions

At the end of this section you will be able to solve colligative property problems requiring the van't Hoff factor.

Recall our example 11.5 A, where we calculated the boiling point elevation of a solution to which NaCl had been added. We made an **assumption** that **NaCl completely dissociates**. Thus, our 2.49 m solution dissociated into 2.49 m $Na^+$ and 2.49 m $Cl^-$.

This section of your textbook points out that this assumption is **not valid, especially at high solute concentrations**. The reason seems to be **ion pairing**, where some sodium and chloride ions encounter one another, pair and are counted as a single particle.

The equation which takes experimentally observed dissociation into account is

$$\Delta T = imK$$

where **i** is the **van't Hoff factor** representing electrolyte dissociation in solution. Keep in mind that for a given electrolyte, **i is concentration dependent**. Also, **i = 1 for all nonelectrolytes**. The expected and observed values for several electrolytes is given in Table 11.6 of your textbook.

## Example 11.7 - Van't Hoff Factor

Use data from Table 11.6 in your textbook to calculate the **freezing point** and **expected osmotic pressure** of a 0.050 m $FeCl_3$ solution at 25°C. (Assume the density of the final solution equals 1.0 g/mL and that the liquid volume is unchanged by the addition of $FeCl_3$.)

## Solution

a.  *Freezing point*

From Table 11.6, $i_{observed}$ = 3.4 for this 0.05 m solution.

$\Delta T = imK_f = 3.4 \times 0.050 \text{ m} \times 1.86°C \text{ kg/mol} = 0.32°C$
$\qquad\qquad\quad \uparrow \qquad\quad \uparrow \qquad\qquad \uparrow$
$\qquad\qquad\quad i \qquad\quad m \qquad\qquad K_f$

The **freezing point** would be **-0.32°C**.

b.  *Osmotic pressure*

Analogous to the formula for freezing point depression, the formula for osmotic pressure (itself a colligative property) is

$$\pi = iMRT$$

We assumed that the density of the solution is 1.0 g/mL, and the volume of the liquid was unchanged by the addition of $FeCl_3$. In this **unusual** case, the molarity = 0.05 M.

$$\pi = 3.4 \times 0.050 \text{ M} \times 0.08206 \text{ L atm/K mol} \times 298.2 \text{ K}$$

$$\begin{array}{cccc} \uparrow & \uparrow & \uparrow & \uparrow \\ i & M & R & T \end{array}$$

**Osmotic pressure ($\pi$) = 4.16 atm**

---

Problems 44-46 at the end of this study chapter cover material from this section.

---

## 11.8 - Colloids

The following review questions will help to guide your study of this section.

1. What is the **Tyndall effect**?
2. How can the Tyndall effect be used to distinguish between a suspension and a true solution?
3. Define **colloid**.
4. Give some examples of **colloids**.
5. How can we prove that **electrostatic repulsion** helps stabilize a colloid?
6. Define **coagulation**.
7. How does heating destroy a colloid?
8. How does the addition of an electrolyte destroy a colloid?

---

Problem 47 at the end of this study chapter covers material from this section.

---

**END OF CHAPTER ELEVEN**

# SELF-TEST QUESTIONS

1.  Rubbing alcohol contains 585 g of isopropanol ($C_3H_7OH$) per liter (aqueous solution). Calculate the molarity.

2.  The density of a 10% (by mass) solution of NaOH is 1.109 $g/cm^3$. Calculate the concentration of this solution in molarity, molality and mole fraction.

3.  Toluene ($C_6H_5CH_3$), an organic compound often used as a solvent in paints, is mixed with a similar organic compound, benzene ($C_6H_6$). Calculate the molarity, molality, mass percent, and mole fraction of toluene in 600 mL of solution that contains 75.8 g of toluene and 95.6 g of benzene. The density of the solution is 0.286$g/cm^3$.

4.  A hydrochloric acid solution was made by adding 59.26 g HCl to 100 g of $H_2O$. The density of the solution was 1.19 $g/cm^3$. Calculate the concentration of HCl in molarity, molality, mass percent and mole fraction.

5.  What mass of $H_2SO_4$ is required to make 250 mL of a 6.00 M solution?

6.  For the reaction:

    $$H_2SO_4 + 2NaOH \longrightarrow Na_2SO_4 + 2H_2O$$

    what mass of $H_2SO_4$ is required to make 250 mL of a 1.00 N solution?

7.  How many grams of NaOH and $H_2O$ are required to prepare 100 g of a 28.7% NaOH solution?

8.  Find the concentration of NaOH in molarity, molality and mole fraction in 500 mL of the solution prepared in problem 7. Also, determine the mole fraction of $H_2O$.

9.  Seawater contains 1.94% chlorine (by mass). How many grams of chlorine are there in 400 mL of seawater if the density of seawater is 1.025 $g/cm^3$.

10. The molal concentration of a solution of salicylic acid ($C_7H_6O_3$), the primary organic molecule used in aspirin production, and ethanol ($C_2H_5O_3$) is 17.5 m. If 42 g of ethanol were used to make the solution, how many grams of salicylic acid were needed?

11. The concentration of glucose, $C_6H_{12}O_6$, in a biological fluid is 75 mg/100 g. What is the molal concentration?

12. Rationalize the water solubilities for the gases listed below (units are g/100 $cm^3$).

    | Gas | Solubility |
    |-----|------------|
    | $H_2$ | 0.0191 |
    | $CO_2$ | 0.141 |
    | HCl | 82.3 |
    | $NH_3$ | 89.9 |

    What would you expect for relative solubilities in hexane ($C_6H_{14}$)?

13. Would boric acid, $B(OH)_3$, be more soluble in ethanol, $C_2H_5OH$ or in benzene, $C_6H_6$?

14. Predict which substance would be more soluble in water.

    a.  $NH_2CH_2CH_2OCH_3$           $NH_2CH_2COOH$

    b.  $CH_3CH - CH_2$              $CH_3CH_2CHCH_2CH_3$
           |     |                              |
           OH   OH                        $CH_2OH$

    c.  $C_4H_9OH$                   $C_4H_9SH$

15. Ammonium salts can be used to make chemical cold packs. When these salts are dissolved in water the solution gets quite cold. What can we say about the heat of solution?

16. The solubility of oxygen in water at 0°C and 1 atm is 0.073 g per liter. What is k in Henry's law for this Temperature?

17. The solubility of nitrogen at 0°C in water is $8.21 \times 10^{-4}$ mole/L if the $N_2$ pressure above the water is 0.790 atm. What is solubility at 1 atm of $N_2$ and 0°C?

18. What is the Henry's law k for nitrogen in problem 17?

19. The solubility of nitrogen in blood at 37°C and 0.80 atm is $5.6 \times 10^{-4}$ M. If a deep sea diver breathes compressed air from a tank at a partial pressure of 3 atm, what would the solubility of nitrogen be in the diver's blood at 37°C?

20. Why doesn't Henry's law work for ammonia?

21. The solubility of $CO_2$ in water at 25°C and 1 atm is 0.034 M. What is Henry's Law Constant (k)? What would the solubility of $CO_2$ in water be at 0.038 atm and 25°C?

22. The vapor pressure of ethanol at 40°C is 135.3 torr. Calculate the vapor pressure of a solution of 53.6 g of glycerin, $C_3H_8O_3$, in 133.7 g of ethanol, $C_2H_5OH$, at 40°C.

23. Antifreeze solutions are mainly ethylene glycol, $C_2H_6O_2$, in water. Calculate the vapor pressure of a solution made by adding 101.6 g of ethylene glycol to 139.6 g of water at 50.0°C. At this temperature ethylene glycol is essentially nonvolatile and the vapor pressure of water is 92.51 torr.

24. Sucrose, $C_{12}H_{22}O_{11}$ (a nonvolatile substance), is a sweetener. A solution was made with 35.2 g of sucrose and 78.0 g of water at 30°C. Calculate the vapor pressure of the solution if the vapor pressure of water at 30°C is 31.824 torr.

25. A solution was prepared by dissolving 16.8 g of camphor ($C_{10}H_{16}O$) in 86.1 g of benzene ($C_6H_6$). At 23°C, the vapor pressure of benzene is 86.0 torr. If the vapor pressure of the resulting solution was 78.2 torr, calculate the vapor pressure of camphor at 23°C (note: solid camphor has a low volatility).

26. How many grams of a nonvolatile compound B (molar mass = 97.8 g/mol) would need to be added to 250 g of water to produce a solution with a vapor pressure of 23.756 torr? The vapor pressure of water at this temperature is 42.362 torr.

27. Calculate the vapor pressure of a solution of 40.27 g $MgCl_2$ in 500.3 mL of water at 25.0°C. The density of water is 0.9971 g/mL and the vapor pressure is 23.756 torr.

28. A 12.97 g sample of a metal chloride, general formula MCl, is added to 100.0 g of water at 30.0°C. The vapor pressure of the solution is 20.47 torr. The vapor pressure of pure water at 30.0°C is 31.824 torr. What is the formula of the metal chloride?

29. Calculate the vapor pressure of a solution made with 321 g of toluene ($C_7H_8$) and 398 g of benzene ($C_6H_6$). At 60°C, the vapor pressures of the two volatile components are $1.4 \times 10^{-4}$ torr and 396 torr respectively.

30. What is the vapor pressure of a solution made by adding 26.93 g potassium sodium tartrate, $KNaC_4H_4O_6$, to 676.5 g of water at 25°C? The vapor pressure of pure water is 23.756 torr at 25°C. Assume that potassium sodium tartrate is nonelectrolyte (a crummy assumption).

31. An aqueous solution of sucrose, $C_{12}H_{22}O_{11}$, boils at 112°C. What is the molality?

32. 1.51 g of white phosphorus ($P_4$) was dissolved in 18.0 g of carbon disulfide ($CS_2$). The boling point elevation of the carbon disulfide solution was found to be 1.36°C. Calculate the molality of the white phosphorous. What is $K_b$ for carbon disulfide?

33. A solution of carbon tetrachloride ($CCl_4$) has a molality of 1.35 m. The freezing point depression constant for $CCl_4$ is 30°C kg/mol and its freezing point is -22.99. At what temperature would the solution freeze?

34. What are the boiling and freezing points of a solution of 50.3 g of $I_2$ in 350.0 g of chloroform? (See Table 11.5 in your textbook for constants.)

35. A 4.367 g sample of an unknown hydrocarbon is dissolved in 21.35 g benzene. The freezing point of the solution is observed to be -0.51°C. Calculate the molar mass of the unknown.

36. During a Wisconsin winter, the temperature can reach -25°C (or colder!). How many grams of antifreeze [ethylene glycol ($C_2H_6$)] would you need to add to your radiator to keep 7.5 liters of water from freezing? Assume that the density is 1.0 g/mL ($K_f$ for water = 1.86°C kg/mol).

37. Calculate the osmotic pressure of a solution made by adding 13.65 g of sucrose, $C_{12}H_{22}O_{11}$, to enough water to make 250.0 mL of solution at 25°C.

38. What is the osmotic pressure of a $1.38 \times 10^{-2}$ M KBr?

39. The osmotic pressure of blood is 7.65 atm at 37°C. How much glucose, $C_6H_{12}O_6$, should be used per liter for an intravenous injection having the same osmotic pressure as blood (in 2/L)?

40. The osmotic pressure of an aqueous solution of a polypeptide solution was determined to be 4.80 torr. How many grams of polypeptide would be in 2 L of solution? (The molar mass of the polypeptide is 269 g/mol.)

41. The osmotic pressure of a tryptophan solution containing 1.136 g per liter is 103.5 torr at 25°C. What is the molar mass of tryptophan?

42. What is the minimum pressure required to desalinate 1.0 M salt (NaCl) solution at 25°C?

43. Calculate the osmotic pressure (in atm) of a 0.225 M aqueous solution of urea that is isotonic with sea water at 10°C.

44. Table 11.6 in your textbook lists values of observed van't Hoff factors for 0.05 m solutions. Would you expect the van't Hoff factors for 0.5 m solutions to be greater or less than the values for 0.05 m?

45. Using the van't Hoff factor from Table 11.6 in your textbook calculate the freezing point, boiling point and osmotic pressure (at 25°C) of a 0.050 m $MgCl_2$ solution. (Assume that 0.050 m = 0.050 M.)

46. Calculate the osmotic pressure of a 0.05 M solution of NaCl at 30°C. The van't Hoff Factor for NaCl is 2.0.

47. Indicate the type of the colloid that each of the following represents (i.e. aerosal, foam, emulsion, sol, or gel):

    a.  MOM (milk of magnesia)
    b.  salad dressing
    c.  meringue
    d.  rain cloud

# ANSWERS TO SELF-TEST QUESTIONS

1.  9.75 M

2.  molarity = 2.52 M;  molality = 2.78 m;  mole fraction = 0.05

3.  molality = 8.62 m $C_6H_5CH_3$; molarity = 1.37 M $C_6H_4CH_3$; mass percent = 44.2%; mole fraction = 0.401

4.  molarity = 12.1 M;  molality = 16.3 m;  mass % = 37.2%;  mole fraction = 0.226

5.  147 g

6.  73.5 g

7.  28.7 g NaOH, 71.3 g $H_2O$

8.  molarity = 12.50 M; molality = 17.5 m; mole fraction of NaOH = 0.240; mole fraction of $H_2O$ = 0.760

9.  7.95 g Cl

10. 101.43 g $C_7H_6O_3$

11. $4.2 \times 10^{-3}$ m

12. $H_2$ and $CO_2$ are non-polar, therefore not very soluble in water.  HCl and $NH_3$ are polar, therefore more soluble in water.  In hexane it is expected that $H_2$ and $CO_2$ will be more soluble than HCl and $NH_3$.

13. Ethanol.  Ethanol is more polar than benzene.  Because boric acid is also a polar substance, it would be more soluble in ethanol.

14. a.   $NH_2CH_2COOH$          c.   $C_4H_9OH$
    b.   $CH_3CH - CH_2$
            |        |
            OH   OH

15. It is an endothermic reaction therefore $\Delta H_{sol} > 0$.

16. 438.4 L atm/mol

17. $1.04 \times 10^{-3}$ M

18. $1.04 \times 10^{-3}$ L atm/mol

19. $2.1 \times 10^{-3}$ M

20. $NH_3 + H_2O \longrightarrow NH_4^+ + OH^-$, ammonia reacts with water

21. k = 29.41 L atm/mol; $1.30 \times 10^{-3}$ M

22. 112 torr

23. 76.39 torr

24. 31.09 torr

25. 0.285 torr

26. $1.063 \times 10^3$ g compound B

27. 22.72 torr

28. NaCl

29. 235 torr

30. 23.66 atm

31. 23.5 m

32. $6.78 \times 10^{-1}$ m:  2.01°C kg/mol

33. -63.49°C

34. 59.1°C, -66.2°C

35. 174 g

36. $6.3 \times 10^3$ g (or 6.3 kg)

37. 3.904 atm

38. 0.675 atm, 513 torr

39. 54.3 g/L

40. $1.06 \times 10^2$ g

41. 204 g/mol

42. 49 atm

43. 5.23 atm

44. 0.5 m factors would be less than 0.05 m factors.  The more concentrated the solute the more chance for interactions.

45. Freezing pt. = -0.25°C; Boiling pt. = 100°C; osmotic pressure 3.3 atm.

46. 2.49 atm

47. a.  sol
    b.  emulsion
    c.  foam
    d.  aerosol

# CHAPTER 12

# Chemical Kinetics

| Section | Contents | Page in this Study Guide |
|---|---|---|
| 12.1 | Reaction Rates | 244 |
| 12.2 | Rate Laws: An Introduction | 246 |
| 12.3 | Determining the Form of the Rate Law | 247 |
| 12.4* | The Integrated Rate Law | 251 |
| 12.6 | Reaction Mechanisms | 255 |
| 12.7 | A Model for Chemical Kinetics | 258 |
| 12.8 | Catalysis | 260 |

In the introductory remarks to this chapter, your textbook makes the *critical distinction between kinetics and thermodynamics*. *Thermodynamics* can be used to *predict if a reaction will occur*. It says nothing about how fast it occurs. *Kinetics* describes the *rate and mechanism* of a reaction *given that it occurs*. Thermodynamics says "if." Kinetics says "how" and "how fast."

## 12.1 - Reaction Rates

When you finish this section you will be able to use concentration vs. time data to determine relative reaction rates.

Your textbook defines the **rate of a reaction** as

$$\text{rate} = \frac{\text{change in concentration of a substance}}{\text{change in time}}$$

Let's look at data for the reaction of hydrogen and oxygen to give water,

$$2H_2(g) + O_2(g) \longrightarrow 2H_2O(l)$$

| Time(s) | $[H_2]$ | $[O_2]$ | $[H_2O]$ |
|---|---|---|---|
| 0.0020 | 0.050 | 0.080 | 0 |
| 0.0040 | 0.025 | 0.0675 | 0.0250 |
| 0.0060 | 0.018 | 0.064 | 0.032 |
| 0.0080 | 0.0125 | 0.0612 | 0.0375 |

*Section 12.5 in your textbook summarizes the previous sections on rate laws.*

The rate of disappearance of $H_2$ must be *twice* the rate for $O_2$, and equal to the rate of appearance of $H_2O$. This is because the *coefficients of the chemical equation tell you the relative rates*. That is,

$$\frac{-\Delta[H_2]}{\Delta t} = \frac{-2\Delta[O_2]}{\Delta t} = \frac{\Delta[H_2O]}{\Delta t}$$

where "[]" means **concentration in moles/liter**. This says that for **every mole of $O_2$ that reacts**, per unit time, 2 moles of $H_2$ react and 2 moles of $H_2O$ are produced. Note that the *rate of disappearance* of the reactants is given a negative ("-") sign. That implies that we are *losing reactants*, while gaining the product ("+" rate). Note also that your textbook uses the standard that all reaction rates should be *positive*.

Using the data from our reaction, the rate of reaction of *hydrogen* between 0.004 and 0.008 seconds is

$$\text{rate} = \frac{-\Delta[H_2]}{\Delta t} = \frac{-(0.0125 - 0.025)}{(0.008 - 0.004)} = \frac{-(-0.0125)}{0.004}$$

**rate = 3.1 mol/L·s**

We are *losing* hydrogen at a rate of 3.1 mol/L·s.

---

# Example 12.1 A - Reaction Rates

Answer the following questions using the hydrogen and oxygen rate data just presented.

a. Based on the *coefficients of the chemical equation alone*, what is the rate of oxygen reaction between 0.004 and 0.008 seconds?
b. What is the rate of water production during the same time period?
c. Use the concentration vs. time data to prove your answers.

## Solution

a. One mole of oxygen reacts for every two moles of hydrogen. That is, *oxygen reacts at 1/2 the rate of hydrogen*. The rate of reaction of hydrogen is **3.1 mol/L·s**. The **rate of reaction of oxygen** is therefore (3.1)/2 = **1.6 mol/L·s** (rounded off to two significant figures).

b. The balanced chemical equation says that one mole of hydrogen reacts to produce one mole of water. Therefore, the *rate of consumption of hydrogen equals the rate of production of water*.

**rate of water production = 3.1 mol/L·s**

c. For oxygen,

$$\frac{-\Delta[O_2]}{\Delta t} = \frac{-(0.0612 - 0.0675)}{(0.008 - 0.004)} = 1.6 \text{ mol/L·s}$$

For water,

$$\frac{\Delta[H_2O]}{\Delta t} = \frac{(0.0375 - 0.0250)}{(0.008 - 0.004)} = 3.1 \text{ mol/L·s}$$

Remember the order of operations of your calculator. After subtracting 0.0250 from 0.0375, press the "=" key before dividing by $\Delta t$, otherwise you will divide 0.0250 by $\Delta t$.

Look at Table 12.2 in your textbook. Notice that the *rate of $NO_2$ decomposition decreases with time* until equilibrium is reached. **The rate of a reaction is not constant**.

## Example 12.1 B - Practice with Reaction Rates

Given the hypothetical reaction

$$2A + nB ----> qC + rD$$

If     $\dfrac{-\Delta[A]}{\Delta t} = 0.050$ mol/L·s

$\dfrac{-\Delta[B]}{\Delta t} = 0.150$ mol/L·s

$\dfrac{\Delta[C]}{\Delta t} = 0.075$ mol/L·s

$\dfrac{\Delta[D]}{\Delta t} = 0.025$ mol/L·s

What are the coefficients n, q and r?

## Solution

The rate of disappearance of B is 3 times that of A. Therefore, **n** must be $3 \times 2$ or **6**. The rate of formation of C is 3/2 that of the decomposition of A. So **q** = 3/2 (2) = **3**. D appears at 1/2 the rate of disappearance of A. Therefore, **r** = 1/2 (2) = **1**. The equation is, therefore,

$$2A + 6B ----> 3C + D$$

Problems 1-2 at the end of this study chapter cover material from this section.

## 12.2 - Rate Laws:  An Introduction

Answer the following review questions after you have read the entire section.

1.  What **simplifying assumption** do we make regarding the rates of forward and reverse reactions in our study of rate laws?
2.  Define **rate law**.
3.  What are "**k**" and "**n?**"
4.  Why don't **products** appear in a rate law?
5.  Why is it important to **specify the component** whose rate we are describing with the rate law?
6.  What is a **differential rate law**?
7.  What is an **integrated rate law**?
8.  Give a couple of **practical examples** of the importance of determining rate laws.

## 12.3 - Determining the Form of the Rate Law

When you finish this study section you will be able to use the **method of initial rates** to determine **differential rate laws and rate constants**.

Recall from the last section that the **differential rate law** deals with the **dependence of rate on concentration**. The important question is, "how will the rate of reaction change if we change the concentration of our reactants?"

Let's take the hypothetical example of reactant "A" going to product "P."

$$A \longrightarrow P$$

A plot of the concentration of A "[A]" vs. time is similar in shape to the one shown in <u>Figure 12.2 in your textbook</u>. Examine the following data, where the **initial rate** = $\Delta[A]/\Delta t$, and is taken as close to t = 0 as possible.

| $[A]_o$ | Initial Rate of Reaction (mol/L·s) |
|---|---|
| 0.35 | $7.20 \times 10^{-4}$ |
| 0.70 | $2.90 \times 10^{-3}$ |
| 1.05 | $6.45 \times 10^{-3}$ |

("$[A]_o$" means the **initial concentration** of A. The subscript "o" means "initial.")

### How does the rate of decomposition of A change as we change [A]?

As [A] doubles from 0.35 to 0.70, the rate goes from $7.20 \times 10^{-4}$ mol/L·s to $29.0 \times 10^{-4}$ mol/L·s. The rate has increased by a factor of 4.0. As [A] triples, the rate increases by a factor of 9. This 2:4, 3:9 relationship suggests that the rate depends on the square of the concentration, or

$$\text{rate} = \frac{-\Delta[A]}{\Delta t} = k[A]^2$$

where k = the rate constant for the reaction.

Your textbook points out that such differential rate laws can only be found by experiments, not theoretically.

To solve for k, substitute in for several values of the rate with the corresponding values for [A], and average the resultant values for k. Using [A] = 0.70 M as an example,

$$\text{rate} = k[A]^2$$

$$2.90 \times 10^{-3} \text{ mol/L·s} = k (0.70 \text{ M})^2$$

$$k = \frac{2.90 \times 10^{-3} \text{ mol/L·s}}{(0.70 \text{ M})^2} = 5.92 \times 10^{-3} \text{ L/mol·s}$$

While the rate is not constant, *the rate constant is constant* for a particular reaction at a particular temperature.

## Example 12.3 A - Differential Rate Law

Using the data given above, what would the rate of reaction for A be (rate = $-\Delta[A]/\Delta t$) if [A] = 0.16 M?

## Strategy

You know the rate law and the value for k (which you would check by doing further experiments). You can therefore solve for the rate by substitution.

## Solution

$$\text{rate} = k[A]^2 = (5.92 \times 10^{-3} \text{ L/mol·s}) (0.16 \text{ M})^2 = 1.52 \times 10^{-4} \text{ L/mol·s}$$

Note that the rate of reaction is *considerably slower* with the *lower concentration* than with [A] = 0.70 M. That makes sense because the rate equation says that *the rate will increase as the square of the concentration increases.* If the concentration decreases, the rate does as well.

---

Let's look at the more complex example of

$$A + B \text{ ----> products.}$$

We want to determine a rate law for the decomposition of "A" that takes into account both [A] and [B]. We can hypothesize that the rate of reaction is proportional to the concentrations of A and B. The more of each, the faster the reaction.

$$\text{rate} = \frac{-\Delta[A]}{\Delta t} = k[A]^n[B]^p$$

Gathering experimental data will allow us to determine the rate law. Our goal is to determine **n, p,** and **k.** The sum of the exponents, **n + p,** is called the **reaction order.** Let's say the data are:

| Reaction # | $[A]_o$ | $[B]_o$ | Initial Rate of Reaction (L/mol·s) |
|------------|---------|---------|-------------------------------------|
| 1 | 0.100 | 0.100 | $1.53 \times 10^{-4}$ |
| 2 | 0.100 | 0.300 | $4.59 \times 10^{-4}$ |
| 3 | 0.200 | 0.100 | $6.12 \times 10^{-4}$ |
| 4 | 0.100 | 0.200 | $3.06 \times 10^{-4}$ |
| 5 | 0.300 | 0.600 | $8.26 \times 10^{-3}$ |

The key process here is to **DETERMINE HOW THE RATE VARIES WITH VARYING CONCENTRA- TION OF ONE COMPONENT WHILE THE CONCENTRATION OF THE OTHER IS HELD CON- STANT.**

**What happens to the rate when you vary [B] while holding $[A]_o$ constant?**

In reactions #1, #2, and #4, $[A]_o$ is constant. Therefore, any increase in rate must be related to the increase in $[B]_o$. Using data from reactions #1 and #2,

$$\text{increase in } [B]_o = \frac{0.300 \text{ M}}{0.100 \text{ M}} = \textbf{factor of 3}$$

$$\text{increase in rate} = \frac{4.59 \times 10^{-4} \text{ mol/L·s}}{1.53 \times 10^{-4} \text{ mol/L·s}} = \textbf{factor of 3}$$

**The rate increases *linearly* with $[B]_o$.**

$$\text{rate} = k[A]^n[B]^1$$

**What happens to the rate when you vary $[A]_o$ while holding $[B]_o$ constant?**

In reactions #1 and #3, $[B]_o$ is constant.  Therefore, any increase in rate must be related to the increase in $[A]_o$.

$$\text{increase in } [A]_o = \frac{0.200 \text{ M}}{0.100 \text{ M}} = \textbf{factor of 2}$$

$$\text{increase in rate} = \frac{6.12 \times 10^{-4} \text{ L/mol·s}}{1.53 \times 10^{-4} \text{ mol/L·s}} = \textbf{factor of 4}$$

**The rate increases with $[A]_o$** *squared*.

$$\textbf{rate} = \textbf{k[A]}^2\textbf{[B]} \quad \text{(if no exponent is given, ``1'' is assumed)}$$

We now have a complete rate law.  However, we need to **solve for k** and use other data to verify our solution.  Using data set #4 (we should really average the values obtained from all 5 reactions),

$$k = \frac{\text{rate}}{[A]^2[B]} = \frac{3.06 \times 10^{-4} \text{ L/mol·s}}{(0.100 \text{ M})^2(0.200 \text{ M})} = \textbf{0.153 L}^2\textbf{/mol}^2\textbf{·s}$$

(We will address the unusual units in the next section.)

As a **double check**, let's solve for the rate in reaction #5.

$$\textbf{rate} = (0.153 \text{ L}^2/\text{mol}^2\text{·s}) (0.300 \text{ M})^2(0.600 \text{ M}) = \textbf{8.26} \times \textbf{10}^{-3} \textbf{ L/mol·s}$$

Our rate law properly reflects the data.

---

# Example 12.3 B - Rate Law Determination

Determine the rate law and solve for the order and value of the rate constant for the reaction

$$C + D + E \longrightarrow \text{Products}$$

given the following data:

| Reaction # | $[C]_o$ | $[D]_o$ | $[E]_o$ | Initial rate (L/mol·s) |
|:---:|:---:|:---:|:---:|:---:|
| 1 | 0.400 | 0.300 | 0.560 | $7.14 \times 10^{-4}$ |
| 2 | 0.100 | 0.500 | 0.200 | $4.55 \times 10^{-5}$ |
| 3 | 0.100 | 0.200 | 0.200 | $4.55 \times 10^{-5}$ |
| 4 | 0.400 | 0.300 | 0.750 | $1.28 \times 10^{-3}$ |
| 5 | 0.100 | 0.300 | 0.560 | $3.57 \times 10^{-4}$ |

## Solution

$$\textbf{rate} = \textbf{k[C]}^n\textbf{[D]}^p\textbf{[E]}^q$$

We need to see how each component varies while the others remain constant.

1.  *Regarding dependence on [C]*:  $[D]_o$ and $[E]_o$ remain constant in reactions #1 and #5.  In going from #5 to #1, $[C]_o$ increases.

$$[C]_o = \frac{0.400 \text{ M}}{0.100 \text{ M}} = \textbf{factor of 4}$$

$$\textbf{rate} = \frac{7.14 \times 10^{-4} \text{ L/mol·s}}{3.57 \times 10^{-4} \text{ L/mol·s}} = \textbf{factor of 2}$$

Therefore, rate increases as $[C]_o^{1/2}$!

$$rate = k[C]_o^{1/2}[D]_o^P[E]_o^q$$

2. *Regarding dependence on [D]*:  $[C]_o$ and $[E]_o$ remain constant in reactions #2 and #3.  In going from #3 to #2. $[D]_o$ increases.

$$[D]_o = \frac{0.500 \text{ M}}{0.200 \text{ M}} = \text{factor of 2.5}$$

$$rate = \frac{4.55 \times 10^{-5} \text{ L/mol·s}}{4.55 \times 10^{-5} \text{ L/mol·s}} = \text{NO CHANGE!}$$

Therefore, the rate is **not dependent** on $[D]_o$.

$$\textbf{rate} = \textbf{k}[C]_o^{1/2}[D]_o^{0}[E]_o^q$$

3. *Regarding dependence on [E]$_o$*:  $[C]_o$ and $[D]_o$ remain constant in reactions #1 and #4.  In going from #1 to #4, $[E]_o$ increases.

$$[E]_o = \frac{0.750 \text{ M}}{0.560 \text{ M}} = \text{factor of 1.34}$$

$$rate = \frac{1.28 \times 10^{-3} \text{ L/mol·s}}{7.14 \times 10^{-4} \text{ L/mol·s}} = \text{factor of 1.79}$$

The square of 1.34 is 1.79, or $(1.34)^2 = 1.79$.  Therefore, rate increases as $[E]_o^2$.

$$rate = k[C]_o^{1/2}[D]_o^{0}[E]_o^2$$

or    $$\textbf{rate} = \textbf{k}[C]_o^{1/2}[E]_o^2, \text{ because } [D]_o^{0} = 1$$

4. *Solving for the rate constant, k*:  Pick any of the equations (though it is best to average all 5).  Using equation #3,

$$k = \frac{rate}{[C]_o^{1/2}[E]_o^2} = \frac{4.55 \times 10^{-5} \text{ L/mol·s}}{(0.100 \text{ M})^{1/2}(0.200 \text{ M})^2} = 3.6 \times 10^{-3} \text{ L}^{3/2}/\text{mol}^{3/2}\text{·s}$$

The order of k equals the sum of the exponents $n + p + q = 1/2 + 0 + 2 = $ **2 1/2 order**.

## Checking Your Work

Use **data set #5** and solve for the rate.

$$\textbf{rate} = 3.6 \times 10^{-3} (0.100)^{1/2}(0.560)^2 = \textbf{3.57} \times \textbf{10}^{-4} \textbf{ L/mol·s}$$

This agrees with the observed rate.

---

Problems 3-6 at the end of this study chapter cover material from this section.

---

## 12.4 - The Integrated Rate Law

When you finish this section you will be able to:

- Determine the proper units of a rate constant for any order rate law.
- Determine the reaction order from concentration v. time data.
- Calculate the half-life of a first- or second-order reaction.

### *Units of k*

In the last section, we reviewed the development of a differential rate law from initial rate data. The general form of the rate equation was

$$rate = \frac{-\Delta[A]}{\Delta t} = k[A]^n$$

where A is a reactant.

If n = 1 (a first order reaction) then the units of k are

$$k = \frac{rate}{[A]} = \frac{mol/L \cdot s}{mol/L} = 1/s$$

If n = 2 (a second-order reaction) then the units of k are

$$k = \frac{rate}{[A]^2} = \frac{mol/L \cdot s}{mol^2/L^2} = \mathbf{L/mol \cdot s}$$

If n = 3 (a third-order reaction) then

$$k = \frac{rate}{[A]^3} = \frac{mol/L \cdot s}{mol^3/L^3} = \mathbf{L^2/mol^2 \cdot s}$$

Do you see a pattern emerging? For an **nth order reaction** the units of k will be $\mathbf{L^{3/2}/mol^{3/2} \cdot s}$. Thus, for our 2 1/2-order reaction in Example 12.3 B, **k** is in units of $M^{1-(5/2)}s^{-1} = \mathbf{M^{-3/2}\ s^{-1}} \equiv \mathbf{L^{3/2}/mol^{3/2} \cdot s}$

---

## Example 12.4 A - The Units of the Rate Constant

Determine the units of k for the following differential rate law.

$$rate = k[A]^4[C][D]^2$$

### Solution

The order of this reaction is 4 + 1 + 2 = 7. The units of k are $mol^{(1-7)}/L^{(1-7)} \cdot s = \mathbf{L^6/mol^6 \cdot s}$.

---

### *First-Order Rate Law*

Recall that the difference between a **differential rate law** and an **integrated rate law** is that the latter relates **concentration of a substance to reaction time**.

Your textbook presents two *mathematically identical* forms of the first-order integrated rate law (rate = k[A]).

- $\ln[A] = -kt + \ln[A]_0$

- $\ln \dfrac{[A]_0}{[A]} = kt$

where $[A]_0$ = initial concentration of A
$[A]$ = concentration of A at time t
k = rate constant.

I have added a third equation (by exponentiating both sides of the previous equations), which you may find useful.

- $[A] = [A]_0 e^{-kt}$

Your textbook makes several points regarding this concentration vs. time relationship. To reemphasize some of them,

- The equation can be expressed in the form $y = mx + b$, where $y = \ln[A]$, $m = -k$, $x = t$ and $b = \ln[A]_0$.

- As a result, if you plot $\ln[A]$ vs. time and get a straight line, your data is likely to be first-order. If you do not get a straight line, some other order may predominate.

- The form of the equation that you should use will depend upon what you are given. You generally will avoid equations where the unknown is part of a log term.

---

## Example 12.4 B - Practice with First-Order Kinetics

All radioactive elements have nuclei which follow a first-order rate law when decaying (we will learn more about this in Chapter 21). Radon decays to polonium according to the following equation,

$$^{222}\text{Rn} \longrightarrow {}^{218}\text{Po} + {}^{4}\text{He}$$

The first-order rate constant for the decay is **0.181 days$^{-1}$**. If you begin with a 5.28 g sample of pure $^{222}$Rn, how much will be left after **1.96 days**? **3.82 days**?

### Strategy

*Your goal is to solve for [A].* You are given $[A]_0$, k and t. Although any of the three forms of the integrated rate law is usable, the most useful one solves for [A] directly,

$$[A] = [A]_0 e^{-kt}$$

### Solution

Given: $[A]_0 = 5.28$ g,  k = 0.181 d$^{-1}$,  t = 1.96 d

$$[A] = 5.28 e^{-(0.181 \times 1.96)} = 5.28 \times 0.701 = \textbf{3.70 g } ^{222}\textbf{Rn}$$

When we **increase** t to 3.82 days,

$$[A] = 5.28 e^{-(0.181 \times 3.82)} = 5.28 \times 0.501 = \textbf{2.64 g } ^{222}\textbf{Rn}$$

## Does Your Answer Make Sense?

In each case, the amount we ended with was *less* than the amount we started with. That makes sense, as we are *decomposing* $^{222}$Rn. The longer time period in the second part of the problem allowed even more to decompose. This answer should be less than in the first part. That too makes sense.

---

Note that in the problem you just finished **3.82 days** were required to **decompose 1/2 of the reactant** (5.28 g ----> 2.64 g). The time it takes to lose half of your reactant is called the **half-life, or $t_{1/2}$**. For $^{222}$Rn, $t_{1/2}$ = 3.82 days. For a first-order reaction (see your textbook for the proof),

$$t_{1/2} = \frac{0.693}{k}$$

**IN A FIRST-ORDER REACTION, THE HALF-LIFE IS *INDEPENDENT* OF THE INITIAL CONCENTRATION OF THE REACTANT.**

Look at Figure 12.4 in your textbook. This says that you *lose 1/2 of the remaining amount of substance for each half-life* that occurs. In effect,

$$[A] = \frac{[A]_o}{2^n}$$

where n = the number of half-lives that have occurred.

---

## Example 12.4 C - First-Order Half-Life

Using the same $^{222}$Rn nuclear decay data as in the previous example, how much $^{222}$Rn would remain from a 5.82 g sample after **11.46 days? 21.01 days?**

(Solve for 21.01 days using the half-life method, then, for practice, by using the integrated rate law.)

## Solution

The key is to calculate the number of half-lives ("n") in the given time period. For t = 11.46 days,

$$n_{(11.46)} = \frac{1 \text{ half-life}}{3.82 \text{ days}} \times 11.46 \text{ days} = 3.0 \text{ half-lives}$$

$$[A] = \frac{[A]_o}{2^n} = \frac{5.82 \text{ g}}{2^3} = \frac{5.82}{8} = \textbf{0.728 g } ^{222}\textbf{Rn}$$

For t = 21.01 days (half-life method),

$$n_{(21.01)} = \frac{1 \text{ half-life}}{3.82 \text{ days}} \times 21.01 \text{ days} = 5.50 \text{ half-lives}$$

$$[A] = \frac{5.82 \text{ g}}{2^{5.5}} = \frac{5.82}{45.3} = \textbf{0.129 g } ^{222}\textbf{Rn}$$

For t = 21.01 days (integrated rate law, k = 0.181 d$^{-1}$)

$$[A] = [A]_o e^{-kt} = 5.82e^{-(0.181 \times 21.01)} = \textbf{0.130 g } ^{222}\textbf{Rn}^*$$

*The *slight difference* in the last two answers came as a result of rounding off when *converting $t_{1/2}$ to k*. Without round off, k = 0.1814, and [A] = 0.129 g.

*Second-Order Rate Law*

Your textbook gives the following equation relating to the **integrated rate law for second-order data.**

$$bB \longrightarrow \text{Products}$$

$$\frac{1}{[B]} = kt + \frac{1}{[B]_0}$$

The equation is in the form $y = mx + b$ where $y = 1/[B]$, $m = k$, $x = t$ and $b = 1/[B]_0$. Thus, if your data fit a second-order model, a plot of $1/[B]$ vs. t should give a **straight line** with a **slope = k** and **intercept = $1/[B]_0$**. (See Figure 12.5 in your textbook.)

The equation relating half-life to k is

$$t_{1/2} = \frac{1}{k[B]_0}$$

**NOTICE THAT FOR A SECOND-ORDER SYSTEM, THE HALF-LIFE DEPENDS ON THE INITIAL CONCENTRATION!**

---

# Example 12.4 D - Second-Order Rate Law

The differential rate law for the association of iodine atoms to molecular iodine,

$$2I(g) \longrightarrow I_2(g)$$

is given by

$$\textbf{rate} = \textbf{k[I]}^2, \quad \textbf{where k} = \textbf{7.0} \times \textbf{10}^9$$

  a.  What are the proper units for k?
  b.  If $[I]_0 = 0.40$ M, calculate $[I]$ at $t = 2.5 \times 10^{-7}$ seconds.
  c.  If $[I]_0 = 0.40$ M, calculate $t_{1/2}$.
  d.  If $[I]_0 = 0.80$ M, how much time would it take for 75% of the iodine atoms to react?

## Solution

  a.  The units for a second-order rate constant are L/mol·s.

  b.  $\dfrac{1}{[I]} = kt + \dfrac{1}{[I]_0} \quad \Rightarrow \quad \dfrac{1}{[I]} = 7.0 \times 10^9 \text{ L/mol·s } (2.5 \times 10^{-7}) + \dfrac{1}{0.40 \text{ mol/L}}$

  $\dfrac{1}{[I]} = 1.75 \times 10^3 \quad \Rightarrow \quad \textbf{[I]} = \textbf{5.7} \times \textbf{10}^{-4} \textbf{ M}$

  c.  $\textbf{t}_{1/2} = \dfrac{1}{k[I]_0} = \dfrac{1}{7.0 \times 10^9 (0.40)} = \textbf{3.6} \times \textbf{10}^{-10} \textbf{ s}$

  Note that $t_{1/2}$ is so short that if we let the reaction proceed for as short as $2.5 \times 10^{-7}$ seconds, most of the iodine atoms have already reacted.

  d.  $[I]_0 = 0.80$ M. After 75% has reacted, 25% is left. Therefore, $[I] = [I]_0/4 = 0.20$ M.

  $\dfrac{1}{0.20 \text{ mol/L}} = 7.0 \times 10^9 \text{ L/mol·s (t)} + \dfrac{1}{0.80 \text{ mol/L}}$

$$5 = 7.0 \times 10^9(t) + 1.25 \quad \Rightarrow \quad 3.75 = 7.0 \times 10^9(t)$$

$$t = 5.4 \times 10^{-10} \text{ s}$$

You might think that an alternative method is to recognize that 75% reacted is 2 half-lives. If $[I]_o = 0.80$ M

$$t_{1/2} = \frac{1}{7.0 \times 10^9 \, (0.80)} = 1.8 \times 10^{-10} \text{s}$$

BUT THIS METHOD IS NOT VALID HERE! The first half-life would occur in $1.8 \times 10^{-10}$ s. However, the **initial concentration** for the second half-life **is now 0.40 M** (1/2 of 0.80 M!). As a result, $t_{1/2}$ for the second half-life is **twice as long** ($3.6 \times 10^{-10}$ s). The sum of $t_{1/2}$ for both half-lives equals $1.8 \times 10^{-10}$ s + $3.6 \times 10^{-10}$ s = **$5.4 \times 10^{-10}$ s**. This agrees with our results, but is the difficult way of arriving at the answer.

---

The remainder of this section in your textbook discusses zero, multicomponent, and pseudo-first-order rate laws. The last one is especially important for you to read about because so many multicomponent kinetic analyses are done using the strategy where one component is in huge excess. Also, for review, look over Table 12.6 in your textbook. It is a summary of the rate laws that you should be familiar with.

---

# Problems 7-16 at the end of this study chapter cover material from this section.

---

## 12.6 - Reaction Mechanisms

When you finish this section you will be able to evaluate the validity of simple reaction mechanisms.

Your textbook begins this chapter by defining several new terms. You should know the definitions of:

- **reaction mechanism** - the series of steps by which a chemical reaction occurs.

- **intermediate** - a species that is neither a reactant nor product. It is formed and consumed in the reaction sequence.

- **molecularity** - the number of species that collide to produce the reaction indicated by an elementary step. Uni and bimolecular are the two important molecularities. (Why isn't a termolecular molecularity likely?)

- **rate-determining step** - the slowest step. The reaction can only go as fast as the rate-determining step.

One key idea in this section is that a **chemical equation** only tells us **what** reactants become what products. It does not tell us **how** these changes occur. This is the goal of reaction mechanism studies.

Another key idea is that the differential rate law is determined by the **rate-determining step**. This step may be in one direction or reversible (both directions). Your goal is to be able to determine whether mechanisms are consistent with experimentally determined rate laws.

Review Sample Exercise 12.6 in your textbook. Then try the following examples.

## Example 12.6 A - Reaction Mechanisms

The balanced equation for the reaction of nitric oxide with hydrogen is

$$2NO + 2H_2 \longrightarrow 2H_2O + N_2$$

The experimentally determined rate law is

$$rate = k[NO]^2[H_2]$$

The following mechanism has been proposed,

$$NO + H_2 \xrightarrow{k_1} N + H_2O \qquad \text{(slow)}$$

$$N + NO \xrightarrow{k_2} N_2O \qquad \text{(fast)}$$

$$N_2O + H_2 \xrightarrow{k_3} N_2 + H_2O \qquad \text{(fast)}$$

Is this mechanism consistent with the observed rate law?

## Strategy and Solution

Your textbook says that there are two criteria that must be met if a mechanism is to be considered acceptable.

1. **The sum of the steps must give the balanced equation.** If you add up the three steps, you will find that this is in fact true here.

2. **The mechanism must agree with the observed rate law.** According to the proposed mechanism, the first step is rate-determining. This means that the overall mechanism must be given by the first step.

$$rate = k_1[NO][H_2]$$

This does not agree with the experimentally determined rate law, therefore **our observed mechanism is incorrect**.

## Example 12.6 B - More Reaction Mechanisms

Consider the same balanced equation as in the previous example. Is the following mechanism consistent with the experimentally determined rate law?

$$NO + H_2 \underset{k_{-1}}{\overset{k_1}{\rightleftharpoons}} N + H_2O \quad \text{(fast, with equal rates)}$$

$$N + NO \xrightarrow{k_2} N_2O \qquad \text{(slow)}$$

$$N_2O + H_2 \xrightarrow{k_3} N_2 + H_2O \quad \text{(fast)}$$

## Solution

If you add up the individual steps, you will find that they equal the balanced chemical equation. So far, so good. We must now **evaluate the mechanism**. The rate-determining step is the second one,

$$N + NO \xrightarrow{k_2} N_2O$$

Based on this,

$$rate = k_2[N][NO]$$

However, **N is an intermediate**. You can't control the initial amount of nitrogen you add, because there is none present at the beginning. We always want to express rate laws in terms of reactants and/or products.

The **first step** allows us to solve for N in terms of reactants and products.

$$forward\ rate = k_1[NO][H_2]$$

$$reverse\ rate = k_{-1}[N][H_2O]$$

The rates are equal. Therefore,

$$k_1[NO][H_2] = k_{-1}[N][H_2O]$$

Solving for [N],

$$[N] = \frac{k_1[NO][H_2]}{k_{-1}[H_2O]}$$

Substituting for [N] in our rate-determining equation,

$$rate = \frac{k_2 k_1[NO][H_2][NO]}{[H_2O]} = \frac{k'[NO]^2[H_2]}{[H_2O]}$$

This is close, but again, **no banana**. If you got this far, you have done excellent work. For an extra pat on the back, try the next challenging problem.

---

## Example 12.6 C - "Son of" Reaction Mechanisms

Same problem as before, same data. Test this mechanism:

$$NO + H_2 \xrightarrow{k_1} N + H_2O \qquad \text{(fast)}$$

$$N + NO \xrightarrow{k_2} N_2O \qquad \text{(fast)}$$

$$N_2O + H_2 \xrightarrow{k_3} N_2 + H_2O \qquad \text{(slow)}$$

### Solution

Again, our mechanisms add up to give the balanced chemical equation. The rate determining step is the third.

$$rate = k_3[N_2O][H_2]$$

However, **$N_2O$ is an intermediate**. There are no reverse reactions here. We can't explicitly substitute for [$N_2O$] with rate equations. **However**, $N_2O$ is formed **very quickly**, and irreversibly, from the first two reactions. Using **mole-ratios**,

- **1 mole of $N_2O$ comes from the combination of 2 moles of NO (2 nitrogens from 2 nitrogens).**

Therefore, we can say

$$[N_2O] = 2[NO] \quad \text{(from stoichiometry)}$$

In the rate-determining step, we can then substitute

$$2NO + H_2 \longrightarrow \text{products}$$

because $[N_2O]$ is directly related to $[NO]$, and **the formation of $N_2O$ is fast**.  Our rate equation becomes

$$\text{rate} = k_3[NO]^2[H_2]$$

And *this agrees with the observed rate law*.

(Reference for 12.6 A, B, and C is W. T. Lippencott, A. B. Garret and F. H. Verhock, Chemistry, J. W. Wiley, New York, (1977))

---

Problems 17-20 at the end of this study chapter cover material from this section.

---

## 12.7 - A Model for Chemical Kinetics

When you finish this section you will be able to:

- State the criteria for reaction according to the **collision model**.
- Solve problems relating to **activation energy**.

Your textbook points out two empirical facts about the rate of chemical reactions.   All other things being equal,

1.  The more **concentrated** your reactants, the **faster** the reaction.
2.  The **higher the temperature**, the **faster** the reaction.

These trends can be explained using **collision theory**.  The theory says that there are 4 steps involved in having a successful reaction.

1.  Molecules must move toward each other (through random motion).
2.  Molecules must hit with the proper orientation (see Figure 12.11 in your textbook).
3.  Molecules must hit with sufficiently high energy.
4.  Molecules will separate after reaction occurs.

The combination of steps 2 and 3 (orientation and energy) makes it unlikely that reaction will occur even in the best of circumstances.  However, the odds of collision are increased with higher concentrations.  The odds of collision with a sufficiently high energy (kinetic energy is proportional to temperature) are increased with higher temperature.

The rate constant, **k**, is a measure of the **fraction** of collisions with **sufficient energy** to produce a reaction,

$$k = Ae^{-E_a/RT}$$

where **k** is the **rate constant.**

**A** is the **frequency factor** (related to the collision frequency and orientation of collisions).

$E_a$ is the **activation energy** (in J/mol).

**R** is the **universal gas constant** (8.3145 J/K mol).

**T** is the **temperature** (in Kelvins).

Your textbook puts the equation in "y = mx + b" form by taking the ln of both sides,

$$\ln(k) = \underset{\uparrow \atop m}{\frac{-E_a}{R}} \underset{\uparrow \atop x}{\left[\frac{1}{T}\right]} + \underset{\uparrow \atop b}{\ln(A)}$$

$$\underset{\uparrow \atop y}{}$$

ln(k) can be plotted graphically vs. 1/T for a series of data to determine the slope and intercept, or for quick work, two sets of data can be used mathematically, yielding Equation 12.11 in your textbook,

$$\ln\left[\frac{k_2}{k_1}\right] = \frac{E_a}{R}\left[\frac{1}{T_1} - \frac{1}{T_2}\right]$$

# Example 12.7 A - Activation Energy

Your textbook gives the activation energy for the reaction

$$H_2(g) + I_2(g) \longrightarrow 2HI(g)$$

as being **167 kJ/mol.** It also gives the rate constant at 302°C as being **$2.45 \times 10^{-4}$ L/mol.** What is the rate constant for the reaction at 205°C?

## Solution

You are given data at two temperatures. Your goal is to solve for $k_2$, the rate constant at the second temperature.

$k_1 = 2.45 \times 10^{-4}$ L/mol·s          $T_1 = 302 + 273.2 = 575.2$ K

$k_2 = ???$                                    $T_2 = 205 + 273.2 = 478.2$ K

$E_a = 167$ kJ/mol = 167,000 J/mol (the units (J) must match that in R)

$$\ln\left[\frac{k_2}{k_1}\right] = \frac{E_a}{R}\left[\frac{1}{T_1} - \frac{1}{T_2}\right]$$

$$\ln\left[\frac{k_2}{2.45 \times 10^{-4} \text{ L/mol·s}}\right] = \frac{1.67 \times 10^5 \text{ J/mol}}{8.3145 \text{ J/K·mol}}\left[\frac{1}{575.2 \text{ K}} - \frac{1}{478.2 \text{ K}}\right]$$

$$\ln\left[\frac{k_2}{2.45 \times 10^{-4}}\right] = -7.083$$

Taking the **inverse ln** of both sides,

$$\frac{k_2}{2.45 \times 10^{-4}} = e^{-7.083} = 8.394 \times 10^{-4}$$

$$k_2 = 2.01 \times 10^{-7} \text{ L/mol·s}$$

### Does the Answer Make Sense?

$T_2$ was lower than $T_1$. We would therefore expect $k_2$ to be lower than $k_1$. Also, for every 10°C, the rate should be cut in half. The temperature was reduced here by 100°C. The rate constant should be reduced substantially, as it was.

---

## Example 12.7 B - Practice with Activation Energy

A second-order reaction has rate constants of **8.9 × 10⁻³ L/mol** and **7.1 × 10⁻² L/mol** at 3°C and 35°C respectively. Calculate the value for activation energy for the reaction.

### Solution

$k_1 = 8.9 \times 10^{-3}$ L/mol·s    $T_1 = 276.2$ K
$k_2 = 7.1 \times 10^{-2}$ L/mol·s    $T_2 = 308.2$ K

Using Equation 12.11 from your textbook,

$$\ln \left[ \frac{7.1 \times 10^{-2} \text{ L/mol·s}}{8.9 \times 10^{-3} \text{ L/mol·s}} \right] = \frac{E_a}{8.3145 \text{ J/K·mol}} \left[ \frac{1}{276.2 \text{ K}} - \frac{1}{308.2 \text{ K}} \right]$$

$$0.9019 = 4.52 \times 10^{-5} E_a$$

$$E_a = 2.0 \times 10^4 \text{ J/mol} = \textbf{20 kJ/mol}$$

---

Problems 21-25 at the end of this study chapter cover material from this section.

---

## 12.8 - Catalysis

The following questions will help you review the material in this section.

1. Define **catalysis**.
2. What are **enzymes**?
3. Why are catalysts more useful for increasing the reaction rate than raising the temperature?
4. How do catalysts work? How does this reflect the collision theory model?
5. What is a **heterogeneous catalyst**? Give some examples.
6. Differentiate between **absorption** and **adsorption**.
7. Describe the involvement of a heterogeneous catalyst in hydrogenation.
8. Discuss the role of catalysis in the conversion of sulfur dioxide to acid rain.
9. What is a **homogeneous catalyst**? Give an example.
10. How does nitric oxide act as a catalyst in the production of ozone?
11. How can chlorine catalyze the decomposition of ozone?

---

**END OF CHAPTER TWELVE**

# SELF-TEST QUESTIONS

1.  At 40°C $H_2O_2$(aq) will decompose according to the following reaction:

$$2H_2O_2(aq) \text{ ----> } 2H_2O(l) + O_2(g)$$

The following data were collected for concentration of $H_2O_2$ at various times.

| time(s) | $[H_2O_2]$(M) |
|---|---|
| 0 | 1.000 |
| $2.16 \times 10^4$ | 0.500 |
| $4.32 \times 10^4$ | 0.250 |

a.   Calculate the average rate of decomposition of $H_2O_2$ between 0 and $2.16 \times 10^4$ s. Use this rate to calculate the rate of production of $O_2$(g).

b.   What are these rates for the time period $2.16 \times 10^4$ s to $4.32 \times 10^4$ s?

2.  Given the following hypothetical equation and data:

$$fA \rightarrow gB + 3C$$

$$-\Delta[A]/t = 0.156 \text{ mol/L·s}$$
$$-\Delta[B]/t = 0.026 \text{ mol/L·s}$$
$$-\Delta[C]/t = 0.078 \text{ mol/L·s}$$

find the coefficients f and g.

3.  Use the given data for the hypothetical reaction:

$$2A + B \text{ ----> products,}$$

to determine the rate law and to evaluate the rate constant at 30°C.

| Reaction # | [A] | [B] | Initial rate (mol/L·s) |
|---|---|---|---|
| 1 | 0.1 | 0.1 | $3 \times 10^{-2}$ |
| 2 | 0.1 | 0.3 | $3 \times 10^{-2}$ |
| 3 | 0.2 | 0.3 | $6 \times 10^{-2}$ |

4.  Derive the rate law expression and calculate the rate constant for the reaction:

$$A + B + 3C \text{ ----> products,}$$

given the following data for 15°C.

| Reaction # | [A] | [B] | [C] | Initial rate (mol/L·s) |
|---|---|---|---|---|
| 1 | 0.4 | 0.1 | 0.1 | $6.0 \times 10^{-3}$ |
| 2 | 0.4 | 0.2 | 0.1 | $6.0 \times 10^{-3}$ |
| 3 | 0.4 | 0.3 | 0.2 | $1.2 \times 10^{-2}$ |
| 4 | 1.2 | 0.4 | 0.2 | 0.11 |

5.  Indicate the overall order of reaction for each of the following rate laws.

a.   $R = k[NO_2][F_2]$
b.   $R = k[I]^2[H_2]$
c.   $R = k[H_2][Cl_2]^{1/2}$

6. The following hypothetical reaction was performed:

$$A + 2B + C + 4D \rightarrow products$$

Determine the rate law and calculate the rate constant for the following data collected at 20°C.

| Reaction # | [A] | [B] | [C] | [D] | Initial rate (mol/L·s) |
|---|---|---|---|---|---|
| 1 | 0.25 | 0.30 | 0.60 | 0.15 | $7.20 \times 10^{-5}$ |
| 2 | 0.75 | 0.30 | 0.60 | 0.15 | $2.17 \times 10^{-4}$ |
| 3 | 0.25 | 0.30 | 0.20 | 0.15 | $7.20 \times 10^{-5}$ |
| 4 | 0.75 | 0.30 | 0.60 | 0.45 | $6.51 \times 10^{-4}$ |
| 5 | 0.75 | 0.44 | 0.60 | 0.15 | $4.67 \times 10^{-4}$ |

7. The decomposition of $H_2O_2(aq)$ into $H_2O(l)$ and $O_2(g)$ is first order. From the data in problem 1, determine the rate constant and the half life.

8. Using the answer from problem 7, calculate the concentration of $H_2O_2$ remaining after 8 hours if the initial concentration was 5.0 M.

9. Determine the rate constant (k) and $t_{1/2}$ for the data pertaining to the decomposition of phosphine ($PH_3$).

$$4PH_3(g) \rightarrow P_4(g) + 6H_2(g)$$

| Reaction # | [PH₃] | Rate (mol/L·s) |
|---|---|---|
| 1 | 0.18 | $2.4 \times 10^{-3}$ |
| 2 | 0.54 | $7.2 \times 10^{-3}$ |
| 3 | 1.08 | $1.4 \times 10^{-2}$ |

10. Using the data and answers from the previous problem, calculate the amount of $PH_3$ remaining after 2 minutes. If the initial concentration of phosphine is 1.00 M, how much time would be needed for 90% of the phosphine to decompose?

11. Data for the decomposition of compound AB to give A and B is given below. Determine the rate expression, the rate constant, and the $t_{1/2}$ for a 1 M solution.

| Reaction # | [AB] | Rate (mol/L·s) |
|---|---|---|
| 1 | 0.2 | $3.2 \times 10^{-3}$ |
| 2 | 0.4 | $12.8 \times 10^{-3}$ |
| 3 | 0.6 | $28.8 \times 10^{-3}$ |

12. The rate law for the following second order reaction at 10°C can be written as:

$$2NOBr(g) \rightarrow 2NO(g) + Br_2(g)$$

$$rate = k[NOBr]^2, \text{ where } k = 0.80 \text{ L/mol·s}$$

a. Determine $t_{1/2}$ when $[NOBr]_o = 0.65M$.
b. Calculate [NOBr] at $t = 5.8 \times 10^{-3}s$ if $[NOBr]_o = 0.65M$.
c. If $[NOBr]_o = 1M$, how long would it take for 50% NOBr to react?

13. Using the information given in problem 12, calculate $[NOBr]_o$ if it took 4.31s for 75% to react.

14. The decomposition of NOCl is a second order reaction with $k = 4.0 \times 10^{-8}$ L/mol·s. Given an initial concentration 0.50 M, what is the half life? How much is left after $1 \times 10^8$ s? What is the half life for an initial concentration of 0.25 M?

15. IBr(g) decomposes to form $I_2$(g) and $Br_2$(g). A plot of 1/[IBr] v. time gave a straight line. Write the general rate law for the reaction.

16. The rate law for the reaction:

$$2NO(g) + Cl_2(g) ----> 2NOCl(g)$$

is $R = k[NO]^2[Cl]$. If an experiment was performed in which the partial pressure of NO(g) initially was 0.1 atm and the initial partial pressure of $Cl_2$(g) was 10 atm, what experimental data would give a straight line plot?

17. Write the rate laws for the following proposed mechanisms for the decomposition of IBr to $I_2$ and $Br_2$.

    a.  IBr(g) ----> I(g) + Br(g)                    (fast)
        IBr(g) + Br(g) ----> I(g) + $Br_2$(g)        (slow)
        I(g) + I(g) ----> $I_2$(g)                    (fast)

    b.  IBr(g) ----> I(g) + Br(g)                    (slow)
        I(g) + IBr(g) ----> $I_2$(g) + Br(g)          (fast)
        Br(g) + Br(g) ----> $Br_2$(g)                 (fast)

    c.  IBr(g) + IBr(g) ----> $I_2Br^+$(g) + $Br^-$(g)   (fast)
        $I_2Br^+$(g) ----> $Br^+$(g) + $I_2$(g)           (slow)
        $Br^-$(g) + $Br^+$(g) ----> $Br_2$(g)             (fast)

    d.  IBr(g) + IBr(g) ----> $I_2$(g) + $Br_2$(g)    (one step)

18. Which of the following mechanisms are consistent with the observed rate law,

$$R = [H_2O_2][I^-][H_3O^+],$$

for the reaction:

$$H_2O_2 + 3I^- + 2H^+ ----> 2H_2O + I_3^-.$$

If none are, write a rate equation that is consistent with the mechanism.

    a.  $H_3O^+$ + $I^-$ ----> HI + $H_2O$              (fast)
        $H_2O_2$ + HI ----> $H_2O$ + HOI                 (slow)
        HOI + $H_3O^+$ + $I^-$ ----> $2H_2O$ + $I_2$      (fast)
        $I^-$ + $I_2$ ----> $I_3^-$                        (fast)

    b.  $H_3O^+$ + $H_2O_2$ ----> $H_3O_2^+$ + $H_2O$     (fast)
        $H_3O_2^+$ + $I^-$ ----> $H_2O$ + HOI            (slow)
        HOI + $H_3O^+$ + $I^-$ ----> $2H_2O$ + $I_2$      (fast)
        $I_2$ + $I^-$ ----> $I_3^-$                        (fast)

    c.  $H_2O_2$ + $I^-$ ----> $H_2O$ + $OH^-$           (slow)
        $H_3O^+$ + $I^-$ ----> $H_2O$ + HI              (fast)
        $H_3O^+$ + $OI^-$ + HI ----> $2H_2O$ + $I_2$     (fast)
        $I_2$ + $I^-$ ----> $I_3^-$                        (fast)

19. Write the rate law for the following predicted mechanism for the production of nitrogen dioxide ($NO_2$):

$$NO + O_2 \rightleftharpoons NO_3 \quad \text{(fast)}$$
$$NO_3 + NO \rightarrow 2NO_2 \quad \text{(slow)}$$

20. Which of the following rate laws is consistent with the proposed mechanism for the reaction:

$$3ClO^- \rightarrow ClO_3^- + 2Cl^-$$

$$ClO^- + ClO^- \rightarrow ClO_3^- + 2Cl^- \text{ (slow)}$$
$$ClO^- + ClO_2^- \rightarrow ClO_3^- + 2Cl^- \text{ (fast)}$$

a.   $R = k[ClO^-][ClO_2^-]$
b.   $R = k[ClO^-]^3$
c.   $R = k[ClO^-]^2[ClO_2^-]$
d.   $R = k[ClO^-]^2$

21. The activation energy for the decomposition of $HI(g)$ to $H_2(g)$ and $I_2(g)$ is 186 kJ/mol. The rate constant at 555 K is $3.52 \times 10^{-7}$ L/mol·s. What is the rate constant at 645 K?

22. The rate constant for the decomposition of acetonecarbonic acid was determined to be $6.42 \times 10^{-5}$ L/mol·s at 10°C and $2.03 \times 10^{-1}$ at 78°C. Calculate the activation energy.

23. The rate constant for the reaction:

$$C_4H_8 \text{----}> 2C_2H_4$$

at 325°C is $6.1 \times 10^{-8}$ s$^{-1}$. At 525°C the rate constant is $3.16 \times 10^{-2}$ s$^{-1}$. Calculate the activation energy.

24. The rate of a reaction increases 2.21 times as the temperature changes from 70°C to 80°C. Calculate the activation energy.

25. The activation energy for the hypothetical reaction $2A + B \rightarrow C$ is $8.9 \times 10^4$ J/mol. If the original rate of the reaction is tripled at a temperature of 325K, at what temperature did the reaction begin?

# ANSWERS TO SELF-TEST QUESTIONS

1. a. Average rate of decomposition of $H_2O_2$ = $2.31 \times 10^{-5}$ mol/L·s
      Rate of production of $O_2$ = $1.16 \times 10^{-5}$ mol/L·s
   b. Average rate of decomposition of $H_2O_2$ = $1.16 \times 10^{-5}$ mol/L·s
      Rate of production of $O_2$ = $5.79 \times 10^{-6}$ mol/L·s

2. f = 6, g = 1

3. $R = k[A]$, $k = 0.3$ $s^{-1}$

4. $R = k[A]^2[C]$, $k = 0.38$ $L^2/mol^2 \cdot s$

5. a. 2
   b. 3
   c. 1 1/2

6. $R = k[A][B]^2[D]$, $k = 2.14 \times 10^{-2}$ $L^3/mol^3 \cdot s$

7. $k = 3.21 \times 10^{-5}$ $s^{-1}$, $t_{1/2} = 2.16 \times 10^4$ s

8. 2.0 M

9. $k = 1.3 \times 10^{-2} s^{-1}$, $t_{1/2}$ or 53.5s

10. $[PH_3] = 0.21M$, 177s = 2 min. 57 sec.

11. $R = k[AB]^2$, $k = 0.080$ L/mol·s, $t_{1/2} = 12.5$ s

12. a. $t_{1/2} = 1.92s$
    b. $[NOBr] = 0.648M$
    c. $t = 1.25s$

13. $[NOBr]_o = 0.87M$

14. $t_{1/2}(0.50$ M$) = 5 \times 10^7$ s, $[NOCl]_{10^8 \text{ s}} = 0.17$ M, $t_{1/2}(0.25M) = 1 \times 10^8$ s

15. $R = k[IBr]^2$

16. 1/[NO] vs. t (pseudo-second order)

17. a. $R = k[IBr]^2$     c. $R = k[IBr]^2$
    b. $R = k[IBr]$      d. $R = k[IBr]^2$

18. a and b; for c the rate law would be $R = k[H_2O_2][I^-]$

19. $R = k[NO]^2[O_2]$

20. d

21. $9.76 \times 10^{-5}$ L/mol·s

22. $9.79 \times 10^{-4}$ J/mol

23. $2.61 \times 10^5$ J/mol = 261 kJ/mol

24. 80 kJ/mol

25. 315K

# CHAPTER 13

# Chemical Equilibrium

| Section | Contents | Page in this Study Guide |
|---|---|---|
| 13.1 | The Equilibrium Condition | 267 |
| 13.2 | The Equilibrium Constant | 267 |
| 13.3 | Equilibrium Expressions Involving Pressures | 270 |
| 13.4 | Heterogeneous Equilibria | 272 |
| 13.5 | Applications of the Equilibrium Constant | 273 |
| 13.6 | Solving Equilibrium Problems | 276 |
| 13.7 | Le Chatelier's Principle | 278 |

Your textbook introduces the concept of **EQUILIBRIUM** by noting that **no reaction goes fully to completion**. Some reverse reaction, however small, always exists. The direction of equilibrium may lie "**far to the right**" (reaction almost complete), "**far to the left**" (negligible reaction), or somewhere in between. Whatever the extent of reaction, the key point is that equilibrium is reached when **the rates of the forward and reverse reactions are equal**.

## 13.1 and 13.2 - The Equilibirum Condition and The Equilibrium Constant

When you finish this section you will be able to:

- list some characteristics of reactions at equilibrium, and
- write a mass action expression for a given balanced chemical equation.

## Example 13.1 - Background Questions

1. What is **dynamic equilibrium**?
2. What is true about the **initial** rate of forward and reverse reactions in a system where only reactants are present?
3. What is true about the rates of forward and reverse reactions **at equilibrium**?
4. Why does equilibrium occur?
5. What are some of the factors that determine the equilibrium position of a reaction?

### Solution

1. Dynamic equilibrium is the process in which the rates of the forward and reverse reaction are equal. As a result, the overall reaction appears static, but is not.

2. Initially, there is only forward reaction. After some product builds up, the reverse reaction begins.
3. The rates will be **equal** at equilibrium.
4. Equilibrium occurs because products, once they are produced, can combine to form reactants at a rate that will eventually **equal** that of product formation by reactants.
5. Among the factors listed in your textbook are initial concentrations, relative energies of the reactants and products, and the energy and organization of the products.

---

Section 13.2 opens with the statement of the **Law of Mass Action**. For a reaction

$$nA + mB \rightleftharpoons pC + qD$$

the equilibrium constant K is given by

$$K = \frac{[C]^p [D]^q}{[A]^n [B]^m}$$

One way to understand the significance of K is to assume that the forward reaction involves nA and mB as the rate-determining step, and the reverse reaction involves as the rate-determining step pC and qD.

$$\text{rate}_{\text{forward}} = k_f [A]^n [B]^m$$

$$\text{rate}_{\text{reversee}} = k_r [C]^p [D]^q$$

At equilibrium, $\text{rate}_{\text{foward}} = \text{rate}_{\text{reverse}}$, or $k_f [A]^n [B]^m = k_r [C]^p [D]^q$. Rearranging,

$$\frac{k_f}{k_r} = K = \frac{[C]^p [D]^q}{[A]^n [B]^m}$$

K is thus the **ratio** of the forward to reverse **rate constants** (not the rate)! As with the kinetic rate-determining step, reactant or product **coefficients become exponents** when put in the mass action expression (or "equilibrium expression").

---

# Example 13.2 A - Equilibrium Expressions

Write the equilibrium expression for each of the following reactions:

a. $PCl_5(g) \rightleftharpoons PCl_3(g) + Cl_2(g)$
b. $S_8(g) \rightleftharpoons 8S(g)$
c. $Cl_2O_7(g) + 8H_2(g) \rightleftharpoons 2HCl(g) + 7H_2O(g)$

## Solution

Remember that **coefficients become exponents** and that K = products/reactants.

products
↓

a. $K = \dfrac{[PCl_3][Cl_2]}{[PCl_5]}$

↑
reactant

coefficient of S
↓

b.   $K = \dfrac{[S]^8}{[S_8]}$

coefficient   coefficient
of $H_2O$    of HCl
↓            ↓

c.   $K = \dfrac{[H_2O]^7 \, [HCl]^2}{[Cl_2O_7][H_2]^8}$
coefficient of $H_2$

---

There are several types of equilibrium problems that you will need to solve in this chapter:

- Solving for K (knowing equilibrium concentrations).
- Solving for equilibrium concentrations (knowing K).
- Solving for equilibrium concentrations (knowing K and initial concentrations).

We will work with the first two types of problems now.

---

## Example 13.2 B - Calculating an Equilibrium Constant

Calculate the equilibrium constant, K, for the following reaction at 25°C,

$$H_2(g) + I_2(g) \;\rightleftharpoons\; 2HI(g)$$

if the equilibrium concentrations are $[H_2] = 0.106$ M, $[I_2] = 0.022$ M, and $[HI] = 1.29$ M.

### Solution

The mass action (or equilibrium) expression is

$$K = \dfrac{[HI]^2}{[H_2][I_2]}$$

We are given the **equilibrium concentrations**. We can substitute to get

$$K = \dfrac{(1.29)^2}{(0.106)(0.022)} = 7.1 \times 10^2 \qquad \text{(dimensionless - the units cancel)}$$

---

## Example 13.2 C - Practice with Equilibrium Expressions

Using the same reaction as in our previous problem (with $K = 7.1 \times 10^2$ at 25°C), if the **equilibrium concentrations** of $H_2$ and $I_2$ are 0.81 M and 0.035 M respectively, calculate the equilibrium concentration of HI.

### Solution

As before, the equilibrium expression is valid.

$$K = \dfrac{[HI]^2}{[H_2][I_2]}$$

Substituting,

$$7.1 \times 10^2 = \frac{[HI]^2}{(0.81)(0.035)}$$

$$[HI]^2 = 20.13, \quad \textbf{[HI] = 4.49 M = 4.5 M (2 sig figs)}$$

## Checking Your Answer

In the majority of equilibrium problems, you will be *given K* and asked to find the concentration of a substance. After finding the solution, the best way of checking your answer is to *substitute concentration values into the equilibrium expression to make sure you get the known value of K*. If you do (with a reasonable round off error) your solution is correct. If you don't, there is probably an error. Checking by solving for K,

$$K = \frac{(4.49)^2}{(0.81)(0.035)} = \textbf{7.1} \times \textbf{10}^2$$

Our [HI] is therefore correct.

---

# Example 13.2 D - Changing K When Changing the Reaction Coefficients or Direction

Your textbook discusses certain conclusions regarding the value of K when you modify a balanced chemical equation. Reread that discussion. Given our balanced chemical equation and value for K from the past two problems, calculate the value of K for the following reactions.

## Solution

    a.    The equation is the **reverse** of the original.

$$K' = 1/K = 1/710 = \textbf{1.4} \times \textbf{10}^{-3}$$

    b.    The equation is **1/3** of the original.

$$K' = K^{1/3} = (710)^{1/3} = \textbf{8.9}$$

    c.    The equation is **reversed** and is **four times** the original.

$$K' = 1/K^4 = K^{-4} = (710)^{-4} = \textbf{3.9} \times \textbf{10}^{-12}$$

---

Problems 1-13 at the end of this study chapter cover material from this section.

---

## 13.3 - Equilibrium Expressions Involving Pressures

When you finish this section you will be able to solve for K or $K_p$ given appropriate data.

Your textbook points out that equilibria for gases can be expressed either in **concentrations** or in **pressure terms**. We use **K** for the equilibrium constant gotten by using **concentrations** and $K_p$ for that using **pressures**.

K and $K_p$ can be related. Before we explicitly show this relationship, let's solve a problem which uses pressures of gases rather than concentrations.

---

# Example 13.3 A - Calculating $K_p$

Calculate $K_p$ for the following reaction

$$CH_3OH(g) \rightleftharpoons CO(g) + 2H_2(g)$$

given the equilibrium pressures as follows:

$$P_{CH_3OH} = 6.10 \times 10^{-4} \text{ atm} \quad P_{CO} = 0.387 \text{ atm} \quad P_{H_2} = 1.34 \text{ atm}$$

## Solution

The **form** of the equilibrium expression is the **same** whether it involves concentration **or** pressure data. That is,

$$K = \frac{products}{reactants}$$

For this example,

$$K_p = \frac{P_{H_2}^2 \, P_{CO}}{P_{CH_3OH}} = \frac{(1.34 \text{ atm})^2 (0.387 \text{atm})}{(6.1 \times 10^{-4} \text{ atm})} = 1.14 \times 10^3 \text{ atm}^2$$

---

# Example 13.3 B - Relating K to $K_p$

**Derive an expression** that relates K to $K_p$ for the reaction given in the previous problem. Then **calculate K** for the reaction at 25°C using the value for $K_p$.

## Deriving the Expression

According to the ideal gas equation, PV = nRT. Concentration units can be expressed as moles/liter, or **n/V**. Relating this to the ideal gas equation,

$$\frac{n}{V} = \frac{P}{RT}$$

$$K = \frac{[H_2]^2[CO]}{[CH_3OH]} = \frac{\left[\frac{n}{V}H_2\right]^2 \left[\frac{n}{V}CO\right]}{\left[\frac{n}{V}CH_3OH\right]}$$

Substituting P/RT for n/V,

$$K = \frac{(P_{H_2}/RT)^2(P_{CO}/RT)}{(P_{CH_3OH}/RT)}$$

Factoring out P's,

$$K = \frac{P_{H_2}^2\, P_{CO}}{P_{CH_3OH}} \left[ \frac{(RT)^{-2}\,(RT)^{-1}}{(RT)^{-1}} \right]$$

$$K_p = \frac{P_{H_2}^2\, P_{CO}}{P_{CH_3OH}} \quad \text{Therefore, } \mathbf{K = K_p(RT)^{-2}}$$

Note how this agrees with the formula given in your textbook.

$$\mathbf{K = K_p(RT)^{-\Delta n}}$$

Using R = 0.08206 L atm K$^{-1}$ mol$^{-1}$
    T = 298.2 K,

$$K = 1.14 \times 10^3 \text{ atm}^2\, [(0.08206 \text{ L atm K}^{-1} \text{ mol}^{-1})\,(298.2 \text{ K})]^{-2}$$
$$= 1.14 \times 10^3 \text{ atm}^2\, (1.67 \times 10^{-3} \text{ L}^{-2} \text{ mol}^2 \text{ atm}^{-2})$$
$$\mathbf{K = 1.90 \text{ mol}^2 \text{ L}^{-2}}$$

---

Problems 14-19 at the end of this study chapter cover material from this section.

---

## 13.4 - Heterogeneous Equilibria

When you finish this section you will be able to write equilibrium expressions for reactions involving pure solids and liquids.

Equilibrium expressions involve concentrations (or pressures) of substances that **change** from initial to the equilibrium conditions. A pure substance, such as water, changes **amount**, but not **concentration**.

THE CONCENTRATIONS OF PURE SOLIDS AND LIQUIDS REMAIN CONSTANT. That means that pure solids and liquids **can be incorporated into the equilibrium constant.** For example, the equilibrium expression for

$$CaCO_3(s) \;\rightleftharpoons\; CaO(s) + CO_2(g)$$

IS NOT $K = \dfrac{[CaO][CO_2]}{[CaCO_3]}$

Rather, $[CaCO_3]$ and $[CaO]$ are constant. The **amounts** of each will change, but their **concentrations** (a <u>derivative</u> term) are constant. Thus, we can incorporate them into our equilibrium expression.

$$\mathbf{K = [CO_2]}$$

---

## Example 13.4 - Equilibrium Expressions Involving Pure Solids and Liquids

Write equilibrium expressions for each of the following reactions:

a.   $Ba(OH)_2(s) \rightleftharpoons Ba^{+2}(aq) + 2OH^-(aq)$

b.   $NH_4NO_2(s) \rightleftharpoons N_2(g) + 2H_2O(g)$

c.   $HCl(g) + NH_3(g) \rightleftharpoons NH_4Cl(s)$

d.   $Zn(OH)_2(s) + 2OH^-(aq) \rightleftharpoons Zn(OH)_4^{2-}(aq)$

e.   $CH_3CO_2H(aq) \rightleftharpoons H^+(aq) + CH_3CO_2^-(aq)$

f.   $Al(NO_3)_3(s) + 6H_2O(l) \rightleftharpoons [Al(H_2O)_6]^{3+}(aq) + 3NO_3^-(aq)$

## Solution

Remember that concentration of pure solids and liquids are not put into the equilibrium expression.  These are incorporated into the equilibrium constant.

a.   $K = [Ba^{2+}][OH^-]^2$

b.   $K = [N_2][H_2O]^2$,   or   $K_p = P_{N_2} P^2_{H_2O}$

c.   $K = \dfrac{1}{[HCl][NH_3]}$ or   $K_p = \dfrac{1}{P_{HCl} P_{NH_3}}$

d.   $K = \dfrac{[Zn(OH)_4^{2-}]}{[OH^-]^2}$

e.   $K = \dfrac{[H^+][CH_3CO_2^-]}{[CH_3CO_2H]}$

f.   $K = [NO_3^-]^3[[Al(H_2O)_6]^{3+}]$

---

Problems 2, 3, 7, and 10 at the end of this study chapter covers material from this section.

---

## 13.5 - Applications of the Equilibrium Constant

When you finish this section you will be able to:

- Use the **reaction quotient (Q)** to predict the direction of chemical reactions toward equilibrium.
- Use the reaction quotient to aid in solving simple equilibrium problems.

Your textbook begins this section by reviewing those things that K can and cannot indicate:

- K **does not** reflect **how fast** a reaction goes.
- A **large** value of K means that mostly **products** will be present at equilibrium.
- A **small** value of K means that mostly **reactants** will be present at equilibrium.

The **theme** of this section is the use of the **reaction quotient (Q)**.  Q predicts the direction in which the reaction will go to reach equilibrium.  You calculate Q by using the law of mass action on the **initial, not equilibrium**, concentrations (or pressures) of the reaction substances.

- If **Q is equal to K**, the system is **at equilibrium**.
- If **Q is greater than K**, mathematically, there is too much product present.  The system will **shift to the left** to reach equilibrium.
- If **Q is less than K**, there is too much reactant present.  The system will **shift to the right** to reach equilibrium.

## Example 13.5 A - Predicting the Direction of Equilibrium

Let's reexamine the reaction between hydrogen gas and iodine gas that we used in examples 13.2 B and C.

$$H_2(g) + I_2(g) \rightleftharpoons 2HI(g)$$

$$K = 7.1 \times 10^2 \text{ at } 25°C.$$

Predict the direction that the system will shift in order to reach equilibrium given each of the following initial conditions:

  a. $Q = 427$
  b. $Q = 1522$
  c. $[H_2]_o = 0.81$ M $\quad [I_2]_o = 0.44$ M $\quad [HI]_o = 0.58$ M
  d. $[H_2]_o = 0.078$ M $\quad [I_2]_o = 0.033$ M $\quad [HI]_o = 1.35$ M
  e. $[H_2]_o = 0.034$ M $\quad [I_2]_o = 0.035$ M $\quad [HI]_o = 1.50$ M

## Solution

  a. **$Q < K$**, so the reaction will form more products and **shift to the right**.

  b. **$Q > K$**, so the reaction will form more reactants and **shift to the left**.

  c. $Q = \dfrac{[HI]_o^2}{[H_2]_o[I_2]_o} = \dfrac{(0.58)^2}{(0.81)(0.44)} = 0.94$

  **$Q < K$**, so the reaction will **shift to the right**.

  d. $Q = \dfrac{(1.50)^2}{(0.078)(0.033)} = 7.1 \times 10^2$

  **$Q = K$**, so the system is **at equilibrium**.

  e. $Q = \dfrac{(1.50)^2}{(0.034)(0.035)} = 1.8 \times 10^3$

  **$Q > K$**, so the reaction will **shift to the left**.

---

Earlier in our discussion (see Section 13.2) we mentioned three types of equilibrium problems. The third, and by far most interesting, is the problem where we are *given initial concentrations and K* and are asked to *find equilibrium concentrations*.

In the remainder of this section, we will introduce this type of problem. We will examine it in some detail in the next section.

---

## Example 13.5 B - Equilibrium Concentrations - Large Value for K

Let's continue to use our data from our hydrogen and iodine reaction ($K = 7.1 \times 10^2$ at 25°C). Calculate the **equilibrium concentrations** if a 5.00 L vessel initially contains 15.7 g of $H_2$ and 294 g of $I_2$.

## Strategy and Solution

The key is to determine **the direction of the reaction**. Thus, we must **compare Q to K**. (Note that our data must be converted from g/L to mol/L.)

$$[H_2]_o = \frac{15.7\ g}{5.00\ L} \times \frac{1\ mol}{2.016\ g} = 1.56\ M$$

$$[I_2]_o = \frac{294\ g}{5.00\ L} \times \frac{1\ mol}{253.8\ g} = 0.232\ M$$

$$[HI]_o = 0.00\ M$$

$$Q = \frac{[HI]_o^2}{[H_2]_o[I_2]_o} = \frac{0^2}{1.56\,(0.232)} = 0$$

**Q < K**, so the reaction will **go to the right.** (Which makes sense, as there is no product!)

Our next step is critical - setting up the **table** of **initial and final conditions**. In order to do this well, we must *take K into account*. You cannot memorize the equilibrium table set-up. You must <u>evaluate</u> each problem separately.

The value of K is 710. We can *assume* that the reaction will go *essentially to completion*, although there will be some small amount that will be unreacted. If this is true, the *stoichiometry* of the reaction dictates that $I_2$ *is the limiting reactant* and that $H_2$ will be in excess.

|  | $H_2$ | + | $I_2$ | ⇌ | $2HI$ |
|---|---|---|---|---|---|
| initial | 1.56 M | | 0.23 M | | 0 M |
| final | 1.328 M | | 0 M | | 0.464 M |

However, the reaction does not go all the way. There will be some $I_2$ left over. Call the amount "+ X." An identical amount of $H_2$ will be left over (+ X). Twice the amount of HI will not be formed (2 moles of HI to 1 mole of $I_2$ or $H_2$), so the amount that will not form is "-2X." The "-" indicates that some small amount will remain on the reactant side. Summarizing in table form,

|  | initial | change | final |
|---|---|---|---|
| $H_2$ | 1.56 M | -0.232 + X | 1.328 + X |
| $I_2$ | 0.232 M | -0.232 + X | + X |
| HI | 0 M | +0.464 - 2X | 0.464 - 2X |

## Assumption

Since the reaction is fairly complete, neglect X **relative to** 1.328 and 0.464. If the assumption is valid, X will be < 5% of 1.328 and 0.464. Therefore, $[H_2] \approx 1.328\ M$, $[I_2] = X$, and $[HI] \approx 0.464\ M$. Solving,

$$K = \frac{[HI]^2}{[H_2][I_2]} \quad \Rightarrow \quad 710 = \frac{(0.464)^2}{(1.328)(X)}$$

$$X = [I_2] = 2.28 \times 10^{-4}\ M$$

$$[H_2] = 1.328\ M, \quad [I_2] = 2.28 \times 10^{-4}\ M, \quad [HI] = 0.464\ M.$$

## Testing our Assumption

Is $2.28 \times 10^{-4}$ less than 5% of 1.328? Yes. Is it less than 5% of 0.464? Yes. If it were not, we would solve the equation by the quadratic formula (see the next section).

## Double Check

Solve for K.

$$K = \frac{(0.464)^2}{1.328 \, (2.28 \times 10^{-4})} = 7.1 \times 10^2$$

Remember that the key to solving equilibrium problems lies with **making and testing valid assumptions.**

---

**Problems 20-27 at the end of this study chapter cover material from this section.**

---

## 13.6 - Solving Equilibrium Problems

When you finish this section you will be able to solve many equilibrium problems, and test any assumptions you make.

At the beginning of this section, your textbook presents a "**Procedure for Solving Equilibrium Problems.**" You may follow that procedure, however, you must always think about what assumptions can make your problem solving easier and then **test the validity** of your assumptions.

We have already dealt with **large values of K.** Now let's do an example which involves a **small value of K.** (What assumptions might you make?)

---

## Example 13.6 A - Equilibrium Calculations - Small Value for K

The reaction between nitrogen and oxygen to form nitric oxide has a value for the equilibrium constant at 2000 K of **K = 4.1 × 10⁻⁴.** If 0.50 moles of $N_2$ and 0.86 moles of $O_2$ are put in a 2.0 L container at 2000 K, what would the equilibrium concentrations of all species be?

$$N_2(g) + O_2(g) \rightleftharpoons 2NO(g)$$

## Strategy and Solution

The **MOST IMPORTANT** question you must raise in doing equilibrium problems is "*What does the value of K tell me about the extent of reaction?*"

With **very small values of K,** such as this one, the reaction will **stay far to the left.** Only a small amount of $N_2$ and $O_2$ (call the amount "X") will react. The small amount of NO that will be formed will be "**2X**," because of the 2:1 stoichiometry between NO and $N_2$ or $O_2$. We may now set up our equilibrium table.

$$[N_2]_o = \frac{0.50 \text{ moles}}{2.0 \text{ L}} = 0.25 \text{ M} \qquad [O_2]_o = \frac{0.86 \text{ moles}}{2.0 \text{ L}} = 0.43 \text{ M}$$

|        | initial(M) | change | final(M)  |
|--------|-----------|--------|-----------|
| $N_2$  | 0.25      | -X     | 0.25 - X  |
| $O_2$  | 0.43      | -X     | 0.43 - X  |
| NO     | 0         | +2X    | 2X        |

Because K is small, **let us assume that X is negligible** compared to 0.25 and 0.43. That is, 0.25 - X ≈ 0.25 and 0.43 - X ≈ 0.43. We can now solve for X (and test the validity of our assumption).

$$K = \frac{[NO]^2}{[N_2][O_2]} \quad \Rightarrow \quad 4.1 \times 10^{-4} = \frac{(2X)^2}{(0.25)(0.43)}$$

$$4X^2 = 4.41 \times 10^{-5} \quad \Rightarrow \quad X = 3.3 \times 10^{-3} \, M$$

$$[NO] = 2X = 6.6 \times 10^{-3} \, M, \quad [N_2] = 0.25 \, M, \quad [O_2] = 0.43 \, M$$

We are not finished yet. We must **test our assumption** and **check the answer**. Is X less than 5% of 0.25 M? Yes (it is about 1.3%). Is X less than 5% of 0.43 M? Yes (it is about 0.8%). Our assumption was O.K.

We can check our math by **solving for K** using our equilibrium concentrations,

$$K = \frac{(6.6 \times 10^{-3})^2}{(0.25)(0.43)} = 4.0_5 \times 10^{-4}$$

Acceptably close, considering round-off.

---

We have solved problems with very large K's and very small K's. The majority of our problems will have these kinds of equilibrium constants. Occasionally you will have to solve a problem where the *value for K is intermediate* (roughly between 0.01 and 100). But remember, there are *no absolutes* in solving these kinds of problems. Assumptions that you make must *always be tested*.

In problems with intermediate K's, you *cannot (in general) make assumptions* based on the extent of reaction. It is too uncertain. The trade off is that you must explicitly solve for X, often using the *quadratic formula*.

---

## Example 13.6 B - Equilibrium Calculations - Intermediate Value for K

Sulfurous acid dissociates in water as follows:

$$H_2SO_3(aq) \rightleftharpoons H^+(aq) + HSO_3^-(aq)$$

If $[H_2SO_3]_0 = 1.50 \, M$ and $[H^+] = [HSO_3^-] = 0 \, M$, calculate the equilibrium concentrations of all species at 25°C if $K = 1.20 \times 10^{-2}$ for this reaction.

### Solution with Assumptions

The value of K is small. Let us **assume** that we can **neglect dissociation of $H_2SO_3$** (ie; X is negligible compared to 1.5 M).

|  | initial(M) | change | final(M) |  |
|---|---|---|---|---|
| $H_2SO_3$ | 1.50 | -X | 1.50 - X | ($\approx$ 1.50??) |
| $H^+$ | 0 | +X | X |  |
| $HSO_3^-$ | 0 | +X | X |  |

$$K = \frac{[H^+][HSO_3^-]}{[H_2SO_3]} \quad \Rightarrow \quad 1.20 \times 10^{-2} = \frac{X^2}{1.50}$$

$$X = 0.130 \, M = [H^+] = [HSO_3^-]$$

Let's test our assumption that X is negligible compared to 1.5 M. In fact, **0.130 is 9% of 1.5! Our assumption is not valid!** We must solve the problem explicitly.

## Explicit Solution

$$1.20 \times 10^{-2} = \frac{X^2}{1.50 - X}$$

You may use the quadratic equation (see section 13.6 in your textbook). Multiplying out,

$$0.0180 - 0.0120X = X^2$$

Setting our equation equal to zero,

$$X^2 + 0.0120X - 0.0180 = 0$$

The quadratic formula is

$$X = \frac{-b \pm (b^2 - 4ac)^{1/2}}{2a}$$

Because "-b" is negative (-0.012), the "$\pm$" in the quadratic must be "+," or one of our solutions will be a negative concentration, a physical impossibility. That is,

$$X = \frac{-b + (b^2 - 4ac)^{1/2}}{2a}$$

$$a = 1 \quad b = 0.0120 \quad c = -0.0180$$

$$X = \frac{-0.0120 + [(0.0120)^2 - 4(1)(-0.0180)]^{1/2}}{2(1)}$$

$$X = \frac{-0.0120 + 0.2686}{2} = 0.128 \text{ M}$$

$$[H_2SO_3] = 1.50 - 0.13 = 1.37 \text{ M}, \quad [H^+] = [HSO_3^-] = 0.128 \text{ M}$$

$$\text{Checking, } K = \frac{(0.128)^2}{1.37} = 1.20 \times 10^{-2} \cdot$$

---

The key to doing equilibrium problems is to **make assumptions when you can** to simplify the math. But you must **test all assumptions that you make.**

---

Problems 28-35 at the end of this study chapter cover material from this section.

---

## 13.7 - Le Chatelier's Principle

When you finish this section you will be able to predict the response of a system to stresses placed on it.

Your textbook defines **Le Chatelier's Principle**: "If a change is imposed on a system **at equilibrium** the position of the equilibrium will shift in a direction that tends to reduce that change."

Your textbook uses the **Haber process,**

$$N_2 + 3H_2 \rightleftharpoons 2NH_3$$

to conceptualize the following effects on the position of equilibrium:
- concentration
- pressure
- temperature

Look over the effects of each on the direction of equilibrium, then try the following problem.

---

## Example 13.7 A - Le Chatelier's Principle

Nitrogen gas and oxygen gas combine at 25°C in a closed container to form nitric oxide as follows:

$$N_2(g) + O_2(g) \rightleftharpoons 2NO(g) \quad \Delta H = +1.81 \text{ kJ}$$
$$K_p = 3.3 \times 10^{30}$$

What would be the effect on the direction of equilibrium (ie; would it shift to the **left**, **right** or **not at all**) if the following changes are made to the system?

a. $N_2$ is added
b. He is added
c. the container is made larger
d. the system is cooled

### Solution

a. From a mathematical point of view, if $P_{N_2}$ is larger, $Q_p$ will be smaller than $K_p$. To compensate more product must be formed. The reaction will shift **to the right**.

b. The addition of an **inert gas** such as He does not alter the partial pressure of any of the gases. Therefore, $Q_p$ is still equal to $K_p$, and the position of equilibrium **remains the same**.

c. **Enlarging** the container will favor the side with **more gas molecules**, because they can exist with fewer collisions per unit time. In this case, both sides have the same number of molecules. Therefore, there will be **no change** in the position of equilibrium.

d. This is an **endothermic** reaction ($\Delta H$ = "+"). It **requires heat** to go. Cooling the system will thus shift the equilibrium **to the left**.

---

## Example 13.7 B - Practice with Le Chatelier's Principle

The combination of hydrogen gas and oxygen gas to give water vapor can be expressed by

$$2H_2(g) + O_2(g) \rightleftharpoons 2H_2O(g) \quad \Delta H = -484 \text{ kJ}$$

Predict the effect of each of the following changes to the system on the direction of equilibrium.

a. $H_2O$ is removed as it is being generated
b. $H_2$ is added
c. the system is cooled

### Solution

a. moves **to the right** (removal of product forces Q to be less than K, and system compensates by making more product).

b.  moves **to the right** (same as above, except Q is less than K because $[H_2]_o$ is larger than at previous equilibrium).

c.  moves **to the right** because an **exothermic** reaction gives off heat. Cooling the system increases the temperature gradient between the system and surroundings thus allowing a continued flow of heat to the surroundings.

---

Problems 36-40 at the end of this study chapter cover material from this section.

---

**END OF CHAPTER THIRTEEN**

# SELF-TEST QUESTIONS

1. Write the equilibrium expression for each of the following reactions:

   a. $2H_2(g) + O_2(g) \rightleftharpoons 2H_2O(g)$
   b. $Cl_2(g) + 2Fe^{2+}(aq) \rightleftharpoons 2Fe^{3+}(aq) + 2Cl^-(aq)$
   c. $Cu^{2+}(aq) + 4NH_3(aq) \rightleftharpoons Cu(NH_3)_4^{2+}(aq)$

2. Write the equilibrium expression for each of the following reactions:

   a. $2NO(g) + O_2(g) \rightleftharpoons 2NO_2(g)$
   b. $Ag^+(aq) + I^-(aq) \rightleftharpoons AgI(s)$
   c. $Fe^{3+}(aq) + 3OH^-(aq) \rightleftharpoons Fe(OH)_3(s)$

3. Write the equilibrium expression for each of the following reactions:

   a. $Zn_2Fe(CN)_6(s) \rightleftharpoons 2Zn^{2+}(aq) + Fe(CN)_6^{4-}(aq)$
   b. $H^+(aq) + OH^-(aq) \rightleftharpoons H_2O(l)$

4. Calculate the equilibrium constant, K, at 25°C for the reaction:

$$2NO(g) + O_2(g) \rightleftharpoons 2NO_2(g)$$

   If the equilibrium concentrations are $NO_2 = 0.55$ atm, $NO = 6.5 \times 10^{-5}$ atm, $O_2 = 4.5 \times 10^{-5}$ atm.

5. Calculate the equilibrium constant, K, at 25°C for the Haber process:

$$3H_2(g) + N_2(g) \rightleftharpoons 2NH_3(g)$$

   If the equilibrium concentrations are $[H_2] = 0.85$ M, $[N_2] = 1.33$ M, $[NH_3] = 0.22$ M.

6. Calculate the value of K for the reaction: $2NO(g) + Cl_2(g) \rightleftharpoons 2NOCl(g)$ if the equilibrium concentrations are:

$$[NOCl] = 4.2 \times 10^{-2} \text{ mol/L}, \quad [NO] = 6.7 \times 10^{-1} \text{ mol/L}, \quad [Cl_2] = 2.9 \times 10^{-3} \text{ mol/L}$$

7. Write the equilibrium expression for each of the following reactions:

   a. $N_2(g) + 3H_2(g) \rightleftharpoons 2NH_3(g)$
   b. $I_2(s) + Cl_2(g) \rightleftharpoons 2ICl(g)$
   c. $2B(s) + 3F_2(g) \rightleftharpoons 2BF_3(g)$

8. Given your answer from **problem #4**, calculate the value for K at 25°C for each of the following reactions:

   a. $1/2NO(g) + 1/4O_2(g) \rightleftharpoons 1/2NO_2(g)$
   b. $2NO_2(g) \rightleftharpoons 2NO(g) + O_2(g)$
   c. $NO_2(g) \rightleftharpoons NO(g) + 1/2O_2(g)$

9. Given your answer from **problem #5**, calculate the value for K at 25°C for each of the following reactions:

   a. $2NH_3(g) \rightleftharpoons 3H_2(g) + N_2(g)$
   b. $NH_3(g) \rightleftharpoons 3/2H_2(g) + 1/2N_2(g)$
   c. $2NH_3(g) \rightleftharpoons 3H_2(g) + N_2(g)$

10. Write equilibruim expressions for each of the following reactions:

    a.   $CaCO_3(s) \rightleftharpoons CaO(s) + CO_2(g)$

    b.   $2HCN(aq) + Zn(s) \rightleftharpoons H_2(g) + 2CN^-(aq) + Zn^{+2}(aq)$

    c.   $2NaHCO_3(s) + 2CaHPO_4(s) \rightleftharpoons 2H_2O(g) + 2CO_2(g) + 2CaNaPO_4(s)$

11. The dissociation of acetic acid, $CH_3COH$, has an equilibrium constant at 25°C of $1.8 \times 10^{-5}$. The reaction is

$$CH_3COOH(aq) \rightleftharpoons CH_3COO^-(aq) + H^+(aq)$$

If the equilibrium concentration of $CH_3COOH$ is 0.46 moles in 0.500 L of water and that of $CH_3COO^-$ is $8.1 \times 10^{-3}$ moles in that same 0.500 L, calculate $[H^+]$ for the reaction.

12. Calculate the equilibrium constant, K, for the following reaction at 25°C if the equilibruim concentrations are $[Cl_2] = 0.371M$, $[F_2] = 0.194M$, and $[ClF] = 1.02M$.

$$Cl_2(g) + F_2(g) \rightleftharpoons 2ClF(g)$$

13. Write the equation for the reaction in **problem #11** if the value of K is $5.6 \times 10^4$ ($1/K_{original}$).

14. Derive an expression that relates K to $K_p$, and calculate the value of K at 25°C, for the reaction given in **problem #4**.

15. At 700K, the measured values for the partial pressures of ammonia, hydrogen, and nitrogen are 0.400 atm, 7.20 atm and 2.40 atm, respectively. Calculate $K_p$ and $K_c$ at 700K for the ammonia synthesis: $N_2(g) + 3H_2(g) \rightleftharpoons 2NH_3(g)$.

16. For the following process at 700°C, what is the partial pressure of each gas at equilibrium if the total pressure is 0.750 atm?

$$C(s) + CO_2(g) \rightleftharpoons 2CO(g) \quad K_p = 1.50$$

17. Calculate the value for $K_p$ at 25°C if the value for K is $3.7 \times 10^9$ °C mol$^{-1}$ for the reaction

$$CO(g) + Cl_2(g) \rightleftharpoons COCl_2(g)$$

18. Given the initial partial pressures of $p(PCl_5) = 0.0500$ atm, $p(PCl_3) = 0.150$ atm, and $p(Cl_2) = 0.250$ atm at 250°C for the following reaction, what must the equilibrium partial pressures be?

$$PCl_5(g) \rightleftharpoons PCl_3(g) + Cl_2(g) \quad K_p = 2.15$$

19. Consider the dimerization of nitrogen dioxide:

$$2NO_2(g) \rightleftharpoons N_2O_4(g) \quad {}^!K_p = 8.8$$

If the temperature is 298K and the total pressure is 0.220 atm, what are the equilibrium partial pressures?

20. The reaction of methane with water is given by the following equation:

$$CH_4(g) + H_2O(g) \rightleftharpoons CO(l) + 3H_2(g) \qquad K = 5.67$$

Predict the direction that the system will shift in order to reach equilibrium given the following initial values of Q.

a. Q = 11.85
b. Q = $3.8 \times 10^{-4}$
c. Q = 5.67

21. Determine what the system will do to reach equilibrium given the following values:

a.   K = $2.9 \times 10^2$; Q = $3.1 \times 10^1$
b.   K = 0.621; Q = $6.21 \times 10^{-1}$
c.   K = $7.3 \times 10^2$;  Q = $8.2 \times 10^2$

22. $K_p$ = 0.133 atm at a particular temperature for the reaction:

$$N_2O_4(g) \;\rightleftharpoons\; 2NO_2(g)$$

Calculate the reaction quotient, Q, if $pN_2O_4$ = 0.048 atm and $pNO_2$ = 0.056 atm.

23. Using the same reaction as in **problem #20** determine the direction that the system will shift in order to reach equilibrium given the following **initial concentrations**.

|     | $[CH_4]$ | $[H_2O]$ | $[CO]$ | $[H_2]$ |
|-----|----------|----------|--------|---------|
| a.  | 0.500    | 0.300    | 0.620  | 0.100   |
| b.  | $4.6 \times 10^{-3}$ | 0.800 | 0.200 | 1.00 |
| c.  | 0.818    | 0.750    | 0.650  | 2.00    |

24. The reaction,

$$2NO(g) \;\rightleftharpoons\; N_2(g) + O_2(g)$$

has a value of K = $2.4 \times 10^3$ at 2000 K.  If 0.61 g of NO are put in a previously empty 3.00 L vessel, calculate the equilibrium concentrations of NO, $N_2$ and $O_2$.

25. At 250°C the equilibrium constant for the reaction

$$PCl_5(g) \;\rightleftharpoons\; PCl_3(g) + Cl_2(g)$$

is 2.15.  If $PCl_5$ was initially the only gas present in the reaction vessel at 0.012500 atm, calculate the partial pressures of all of the gases after equilibrium has been reached.

26. Using the same reaction at 2000 K as in **problem #24**, calculate the equilibrium concentrations of NO, $N_2$ and $O_2$ if the initial concentrations of each species are:  [NO] = 0 M, $[N_2]$ = 0.850 M, $[O_2]$ = 0.560 M.

27. Using the reaction in **problem #18**, calculate the equilibrium partial pressures if the initial partial pressures are $p(PCl_5)$ = 0.850 atm, $p(PCl_3)$ = 0.440 atm, and $p(Cl_2)$ = 0.935 atm.

28. Hypobromous acid, HOBr, dissociates in water according to the following reaction:

$$HOBr(aq) \;\rightleftharpoons\; OBr^-(aq) + H^+(aq) \qquad K = 2.06 \times 10^{-9} \text{ at } 25°C$$

Calculate $[H^+]$ of a solution originally 1.25 M in HOBr.

29. The following reaction has an equilibrium constant of $6.2 \times 10^2$ at a certain temperature. Calculate the equilibrium concentrations of all species if 4.5 mol of each component were added to a 3 L flask.

$$H_2(g) + F_2(g) \rightleftharpoons 2HF(g)$$

30. Ammonia undergoes hydrolysis according to the following reaction:

$$NH_3(aq) + H_2O(l) \rightleftharpoons NH_4^+(aq) + OH^-(aq) \quad K = 1.8 \times 10^{-5} \text{ at } 25°C$$

Calculate $[NH_3]$, $[NH_4^+]$ and $[OH^-]$ in a solution originally 0.200 M in $NH_3$.

31. Using the same reaction and value for K as in **problem #30**, determine $[OH^-]$ if $[NH_3]$ = 0.500 M and $[NH_4^+]$ = 0.750 M.

32. Given the following reaction at 25°C:

$$2SO_2Cl(g) \rightleftharpoons 2SO_2(g) + Cl_2(g)$$

Calculate the equilibrium constant if the equilibrium concentrations are $[SO_2Cl]$ = 0.037 M, $[SO_2]$ = 0.591 M, and $[Cl_2]$ = 1.24 M.

33. Calculate the equilibrium concentration of $Cl_2$ for the following reaction at 25°C. The equilibrium constant is $2.3 \times 10^2$ and the equilibrium concentrations of $PCl_3$ and $PCl_5$ are 0.193 M and 0.615 M, respectively.

$$PCl_5(g) \rightleftharpoons PCl_3(g) + Cl_2(g)$$

34. The equilibrium constant for the reaction:

$$SbCl_3(g) + Cl_2(g) \rightleftharpoons SbCl_5(g)$$

at 448°C is 40. What are the equilibrium concentrations of $SbCl_3$, $Cl_2$ and $SbCl_5$ if $[Cl_2]_o$ = 0.620 M and $[SbCl_5]_o$ = 0.180 M?

35. Using the same reaction and K as in **problem #34**, calculate the equilibrium concentrations of all species in $[SbCl_5]_o$ = 1.25 M and $[Cl_2]_o$ = $[SbCl_3]_o$ = 0.

36. The following chemical process is at equilibrium:

$$2H_2(g) + O_2(g) \rightleftharpoons 2H_2O(g)$$

How would the process respond if the pressure were increased at constant temperature?

37. The reaction of carbon disulfide with chlorine is as follows:

$$CS_2(g) + 3Cl_2(g) \rightleftharpoons CCl_4(g) + S_2Cl_2(g) \qquad \Delta H° = -238 \text{ kJ}$$

Predict the effect of each of the following changes to the system on the direction of equilibrium:

a. The pressure on the system is doubled by halving the volume.
b. $CCl_4$ is removed as it is generated.
c. Heat is added to the system.

38. Given the following reaction at equilibrium,

$$3F_2(g) + Cl_2(g) \rightleftharpoons 2ClF_3(g)$$

    a.  Predict the effect if:  the pressure were reduced at constant temperature?
    b.  Predict the effect if:  the volume were reduced by increasing the pressure at constant temperature?

39. The reaction of nitrogen gas with hydrogen chloride is as follows:

$$N_2(g) + 6HCl(g) \rightleftharpoons 2NH_3(g) + 3Cl_2(g) \qquad \Delta H = +461 \text{ kJ}$$

Predict the effect of each of the following changes to the system on the direction of equilibrium:

    a. Triple the volume of the system
    b. The amount of nitrogen is doubled.
    c. Heat is added to the system.

40. Using the following equation, what is the effect on the equilibrium when the partial pressure of ammonia is increased?

$$NH_4Cl(s) \rightleftharpoons NH_3(g) + HCl(g)$$

# ANSWERS TO SELF-TEST QUESTIONS

1. a. $K = \dfrac{[H_2O]^2}{[H_2]^2[O_2]}$

   b. $K = \dfrac{[Cl^-]^2[Fe^{3+}]^2}{[Cl_2][Fe^{2+}]^2}$

   c. $K = \dfrac{[Cu(NH_3)_4^{2+}]}{[Cu^{2+}][NH_3]^4}$

2. a. $K = \dfrac{[NO^2]^2}{[NO]^2[O_2]}$

   b. $K = \dfrac{1}{[Ag^+][I^-]}$

   c. $K = \dfrac{1}{[Fe^{3+}][OH^-]^3}$

3. a. $K = [Zn^{2+}]^2[Fe(CN)_6^{4-}]$

   b. $K = \dfrac{1}{[H^+][OH^-]}$

4. $K_p = 1.6 \times 10^{12}$ atm$^{-1}$

5. $K = 0.059$

6. 1.4

7. a. $K = \dfrac{[NH_3]^2}{[N_2][H_2]^3}$

   b. $K = \dfrac{[ICl]^2}{[Cl_2]}$

   c. $K = \dfrac{[BF_3]^2}{[F_2]^3}$

8. a. $K_p = 1.1 \times 10^3$ atm$^{1/4}$
   b. $K_p = 6.2 \times 10^{-13}$ atm
   c. $K_p = 7.9 \times 10^{-7}$ atm$^{1/2}$

9. a. $K = 17$
   b. $K = 4.1$
   c. $K = 16.9$

10. a. $K = [CO_2]$

    b. $K = \dfrac{[H_2][CN^-]^2[Zn^{+2}]}{[HCN]^2}$

    c. $K = [H_2O]^2[CO_2]^2$

11. $[H^+] = 1.0 \times 10^{-3}$ M

12. $K = 1.45 \times 10^1$

13. $H^+ + CH_3COO^- \rightleftharpoons CH_3COOH$

14. $K = K_p(RT)^{-\Delta n} = K_p(RT) \Rightarrow K = 3.9 \times 10^{13}$

15. $K_p = 1.79 \times 10^{-4}$; $K_c = 5.4 \times 10^{-8}$

16. $p(CO_2) = 0.201$ atm; $p(CO) = 0.549$ atm

17. $K_p = 1.5 \times 10^8$ atm$^{-1}$

18. $p(PCl_5) = 0.023$ atm, $p(PCl_3) = 0.177$ atm, $p(Cl_2) = 0.277$ atm

19. $p(N_2O_4) = 0.109$ atm, $p(NO_2) = 0.111$ atm

20. a.   to the left
    b.   to the right
    c.   no shift - the system is at equilibrium.

21. a.   Q < K, system will shift to the right
    b.   Q = K, system is at equilibirium
    c.   Q > K, system will shift to the left

22. Q = 0.065

23. a.   $Q = 4 \times 10^{-3}$, system shifts to the right.
    b.   Q = 54, system shifts to the left.
    c.   Q = 8.5, system shifts to the left.

24. $[NO] = 6.9 \times 10^{-5}$ M; $[N_2] = [O_2] = 3.39 \times 10^{-3}$ M

25. $p(PCl_3) = p(Cl_2) = 0.0125$ atm; $p(PCl_5) = 7.2 \times 10^{-5}$ atm

26. $[NO] = 1.4 \times 10^{-2}$ M; $[N_2] = 0.850$ M; $[O_2] = 0.560$ M

27. $p(PCl_5) = 0.485$ atm, $p(PCl_3) = 0.855$ atm, $p(Cl_2) = 1.30$ atm

28. $[H^+] = 5.1 \times 10^{-5}$ M

29. $[H_2] = [F_2] = 0.167$M; $[HF] = 4.167$ M

30. $[NH_3] = 0.198$ M; $[OH^-] = [NH_4^+] = 1.9 \times 10^{-3}$ M

31. $[OH^-] = 1.2 \times 10^{-5}$ M

32. $3.16 \times 10^2$

33. $7.76 \times 10^2$ M

34. $[SbCl_3] = 7.7 \times 10^{-3}$ M; $[Cl_2] = 0.612$ M; $[SbCl_5] = 0.188$ M

35. $[SbCl_3] = [Cl_2] = 0.164$ M; $[SbCl_5] = 1.09$ M

36. The reaction would move left to right (fewer moles on the right side).

37. a.   The reaction shifts to the right.
    b.   The reaction shifts to the right.
    c.   The reaction shifts to the left.

38. a.   the reaction moves to the left
    b.   the reaction moves to the right

39. a.   The reactin shifts to the left.
    b.   The reaction shifts to the right.
    c.   The reaction shifts to the right.

40. Adding $NH_3(g)$ drives the reaction to the left.

# CHAPTER 14

# Acids and Bases

|  |  | Page in this |
| --- | --- | --- |
| Section | Contents | Study Guide |
| 14.1 | The Nature of Acids and Bases | 289 |
| 14.2 | Acid Strength | 291 |
| 14.3 | The pH Scale | 292 |
| 14.4 | Calculating the pH of Strong Acid Solutions | 295 |
| 14.5 | Calculating the pH of Weak Acid Solutions | 296 |
| 14.6 | Bases | 301 |
| 14.7 | Polyprotic Acids | 303 |
| 14.8 | Acid-Base Properties of Salts | 306 |
| 14.9 | The Effect of Structure of Acid-Base Properties | 309 |
| 14.10 | Acid-Base Properties of Oxides | 309 |
| 14.11* | The Lewis Acid-Base Model | 310 |

## 14.1 - The Nature of Acids and Bases

When you finish this section you will be able to:

- define acids and bases using the Brønsted-Lowry model,
- identify conjugate acid-base pairs, and
- write equilibrium expressions for acid dissociations.

Your textbook introduces two concepts of acids and bases. The **Arrhenius concept** says that an **acid supplies $H^+$ to an aqueous solution. A base supplies $OH^-$ to an aqueous solution.** This concept is limiting because there are many bases that do not contain $OH^-$.

A more global description of acids and bases is the **Brønsted-Lowry concept.** An **acid is a proton ($H^+$) donor. A base is a proton acceptor.**

The remainder of this section and the next several sections are devoted to the reaction of acids in water. The general reaction for an acid, HA, in water is

$$HA \quad + \quad H_2O \quad \rightleftharpoons \quad A^- \quad + \quad H_3O^+$$

$$\underset{\text{acid 1}}{\uparrow} \qquad \underset{\text{base 2}}{\uparrow} \qquad \underset{\substack{\text{conjugate} \\ \text{base 1}}}{\uparrow} \qquad \underset{\substack{\text{conjugate} \\ \text{acid 2}}}{\uparrow}$$

*14.12 contains a review of problem solving strategy, and is covered in earlier sections.

HA donates a proton to $H_2O$ giving $A^-$ and $H_3O^+$. HA and $A^-$ are **conjugate pairs**. $H_2O$ and $H_3O^+$ are also conjugate pairs.

---

# Example 14.1 A - Conjugate Pairs

Write the dissociation reaction for each of the following acids in water and identify the conjugate acid-base pairs.

   a.   formic acid (HCOOH)
   b.   perchloric acid ($HClO_4$)
   c.   the hydrated iron (III) ion $[Fe(H_2O)_6]^{+3}$

## Solution

   a.   $HCOOH(aq) + H_2O(l) \rightleftharpoons COOH(aq) + H_3O^+(aq)$

         acid 1          base 2      conjugate         conjugate
                                      base 1             acid 2

   Your textbook points out that it is common practice to eliminate the solvent, water, from the equation (it has a constant concentration). This leads to the more common form,

   $HCOOH(aq) \rightleftharpoons COOH(aq) + H^+(aq)$

   b.   long form:  $HClO_4(aq) + H_2O(l) \rightleftharpoons ClO_4^-(aq) + H_3O^+(aq)$

                     acid 1          base 2      conjugate      conjugate
                                                  base 1         acid 2

   common form:  $HClO_4(aq) \rightleftharpoons ClO_4^-(aq) + H^+(aq)$

   c.   long form:  $[Fe(H_2O)_6]^{3+}(aq) + H_2O(l) \rightleftharpoons [Fe(H_2O)_5OH]^{2+}(aq) + H_3O^+(aq)$

                     acid 1                 base 2        conjugate              conjugate
                                                           base 1                 acid 2

   common form:  $[Fe(H_2O)_6]^{3+}(aq) \rightleftharpoons [Fe(H_2O)_5OH]^{2+}(aq) + H^+(aq)$

---

Equilibrium expression for acid dissociations are written using the same concepts as for any other chemical equations. However, we use the **short form** of the acid dissociation when writing these expressions. Recall from your textbook that the equilibrium constant, K, is known as $K_a$ for acid dissociations.

---

# Example 14.1 B - Equilibruim Expressions For Acid Dissociations

Write an equilibrium expression for each of the equations in example 14.1 A.

## Solution

As always, $K = \dfrac{[products]}{[reactants]}$

   a.   $K_a = \dfrac{[COOH^-][H^+]}{[HCOOH]}$

b.  $K_a = \dfrac{[ClO_4^-][H^+]}{[HClO_4]}$

c.  $K_a = \dfrac{[[Fe(H_2O)_5OH]^{2+}][H^+]}{[[Fe(H_2O)_6]^{3+}]}$

---

Problems 1-5 at the end of this study chapter cover material from this section.

---

## 14.2 - Acid Strength

When you finish this study section you will be able to:

- compare relative strengths of acids, and
- solve problems regarding the autoionization of water.

The key idea of this section is that **the strength of an acid is indicated by the equilibrium position of the dissociation reaction.** If the equilibrium lies **far to the left** (as indicated by the value of $K_a$), the acid **does not dissociate** very much and is called **weak.** If the equilibrium lies **far to the right**, the acid **strongly dissociates** and is called **strong.** This idea is summarized in Table 14.1 of your textbook. **You should memorize the strong acids given below Table 14.1.** Note that with the strongest acids, the position of equilibrium lies so far to the right that $K_a$ cannot be measured.

Sample Exercise 14.2 in your textbook illustrates that:

THE STRONGER THE ACID, THE WEAKER ITS CONJUGATE BASE.
THE STRONGER THE BASE, THE WEAKER ITS CONJUGATE ACID.

---

## Example 14.2 A - Relative Acid and Base Strengths

Using Table 14.2 in your textbook, and knowing that $K_a$ for $H_2O = 1 \times 10^{-14}$ at 25°C, arrange the following acids in order of their strength. Then arrange their conjugate bases in order.

$$HOC_6H_5, \quad H_2O, \quad HSO_4^-, \quad NH_4^+, \quad HNO_3$$

### Solution

Acid strength is reflected by $K_a$ (except for very strong acids, where $K_a$ is too large to measure accurately). Also, $K_a$ for all but very strong acids is <<1, indicating that the equilibrium lies far to the left. $HNO_3$ is a very strong acid. According to the values of $K_a$ from Table 14.2, the list of acid strengths should read

$$HNO_3, \quad HSO_4^-, \quad NH_4^+, \quad HOC_6H_5, \quad H_2O$$
strongest <------------------------------------------> weakest

The list of conjugate base strength must be just the opposite:

$$OH^-, \quad OC_6H_5^-, \quad NH_3, \quad SO_4^{2-}, \quad NO_3^-$$
strongest <------------------------------------>weakest

Your textbook points out that water is an **amphoteric** substance (it can act as an acid OR a base). For the autoionization reaction

$$2H_2O \rightleftharpoons H_3O^+ + OH^-$$

$$K_a = K_w = 1 \times 10^{-14} \text{ at } 25°C$$

**Critical point: IN AQUEOUS SOLUTION, THE ION PRODUCT [H⁺][OH⁻] IS ALWAYS EQUAL TO $1 \times 10^{-14}$ AT 25°C. ALWAYS.** Therefore, if you know [H⁺] in a solution, you always know [OH⁻]. The reverse must also hold true.

Review the relationship given in your textbook between [H⁺] and [OH⁻] for an acidic, basic and neutral solution, then try the next example.

## Example 14.2 B - Conversion Between [H⁺] and [OH⁻]

At **10°C**, $K_w$ for the autoionization of water equals $2.9 \times 10^{-15}$. Calculate either [H⁺] or [OH⁻] as necessary under each of the following conditions.

   a.   Calculate [H⁺] if [OH⁻] = $9.3 \times 10^{-4}$ M. Is the solution acidic or basic?

   b.   Calculate [H⁺] and [OH⁻] for a neutral solution.

   c.   Calculate [OH⁻] if [H⁺] = $6.7 \times 10^{-1}$ M. Is the solution acidic or basic?

## Solution

$$K_w = [H^+][OH^-]$$

(Remember, [H₂O] is essentially constant, so it is incorporated into the equilibrium expression.)

   a.   [H⁺] = $K_w$/[OH⁻] = $2.9 \times 10^{-15} / 9.3 \times 10^{-4}$ = **$3.1 \times 10^{-12}$ M**
        [OH⁻] > [H⁺], so this solution is **basic**.

   b.   For a neutral solution, [H⁺] = [OH⁻]. If we let "X" = [H⁺] = [OH⁻], then

$$K_w = (X)(X) = X^2 = 2.9 \times 10^{-15}$$

$$X = [H^+] = [OH^-] = \textbf{5.4} \times \textbf{10}^{-8} \textbf{ M}.$$

   c.   [OH⁻] = $K_w$/[H⁺] = $2.9 \times 10^{-15} / 6.7 \times 10^{-11}$ = **$4.3 \times 10^{-5}$ M**
        [OH⁻] > [H⁺], so this solution is **basic**.

Problems 6-8 at the end of this study chapter cover material from this section.

## 14.3 - The pH Scale

When you finish this section you will be able to convert between pH, pOH, (OH⁻), and (H⁺).

This is the first time in our chemistry course that we need to make use of **logarithms**.  (See <u>Appendix 1.2 in your textbook</u> if you need a review.)  As your textbook points out,

$$pH = -log[H^+]$$

$$pOH = -log[OH^-]$$

The operator "p" means "**-log of the concentration of.**"  This is true for **anything**.

---

## Example 14.3 A - Understanding the Meaning of "p"

Translate the meaning of each of the following uses of "p," or convert to p, as necessary.  (Forgive the puns to follow.)

   a.   What is "-log[soup]" in terms of "p?"
   b.   What is "pCabin" in terms of "-log?"
   c.   What is "-log[ter Pan]" in terms of "p?"
   d.   What is "pRolling" in terms of "-log?"

### Solution

   a.   pSoup  (Aarrgh!)
   b.   -log[cabin]  (Oy!)
   c.   pter Pan  (Enough!)
   d.   -log[rolling]  (Isn't chemistry wonderful?)

---

Note the discussion on the **number of significant figures** present at the beginning of the section in your textbook. When you are comfortable with that, and you understand

$$\cdot \quad pX = -log[X]$$

$$\cdot \quad [X] = 10^{-pX}$$

Look over the discussion in **Sample Exercise 14.6** in your textbook.  Then try the following problem.

---

## Example 14.3 B - Converting Between p and Concentration

Calculate "p" or "[ ]" as necessary for each of the following.  (Remember to use the proper number of significant figures!)

   a.   calculate $[Cl^-]$ if $pCl = 7.32$
   b.   calculate $pAg$ if $[Ag^+] = 0.034$ M
   c.   calculate $pNO_3$ if $[NO_3^-] = 15$ M
   d.   calculate $[NH_4^+]$ if $pNH_4 = 11.87$

### Solution

   a.   $[Cl^-] = 10^{-pCl} = 10^{-7.32}$  (Enter **7.32** into your calculator, press the +/- key and press either **$10^x$** or **inv log**.)

$$[Cl^-] = 4.8 \times 10^{-8} \, M$$

b.   $pAg = -\log(0.034)$  (Enter **0.034** into your calculator, press the **log** key and press the **+/-** key.)

$$pAg = 1.47$$

c.   $pNO_3 = -\log(15) = -1.18$

d.   $[NH_4^+] = 10^{-11.87} = 1.3 \times 10^{-12} \, M$

---

For the next problem, recall that $K_w = [H^+][OH^-]$, or $pK_w = pH + pOH$. At 25°C, $K_w = 1.0 \times 10^{-14}$ and $pK_w =$ **14.** Therefore, at 25°C, **the sum of pH and pOH must always equal 14.0.**

---

# Example 14.3 C - Converting Among pH, pOH, [H⁺] and [OH⁻]

Fill in the blanks in the following table.

|            | pH   | pOH  | $[H^+]$ | $[OH^-]$ | acid, base, or neutral? |
|------------|------|------|---------|----------|-------------------------|
| solution a | 6.88 |      |         |          |                         |
| solution b |      |      |         | $8.4 \times 10^{-14}$ |            |
| solution c |      | 3.11 |         |          |                         |
| solution d |      |      | $1.0 \times 10^{-7}$ |    |                         |

## Solution

There are several ways of converting. You can convert to concentration or to "p". I will work each part a **different** way to demonstrate the possibilities.

a.   $[H^+] = 10^{-pH} = 10^{-6.88} = 1.3 \times 10^{-7} \, M$
     $[OH^-] = K_w/H^+ = 1.0 \times 10^{-14} / 1.3 \times 10^{-7} = 7.7 \times 10^{-8} \, M$
     $pOH = -\log[OH^-] = -\log(7.6 \times 10^{-8}) = 7.12$

b.   $pOH = -\log(8.4 \times 10^{-14}) = 13.08$
     $pH = 14 - pOH = 14 - 13.08 = 0.92$
     $[H^+] = 10^{-pH} = 10^{-0.92} = 0.12 \, M$

c.   $pH + pOH = 14 \Rightarrow 14 - 3.11 = pH = 10.89$
     $[OH^-] = 10^{-3.11} = 7.8 \times 10^{-4} \, M$
     $[H^+] = K_w/7.8 \times 10^{-4} = 1.3 \times 10^{-11} \, M$

d.   $[OH^-] = 1.0 \times 10^{-14} / 1.0 \times 10^{-7} = 1.0 \times 10^{-7} \, M$
     $pOH = -\log(1.0 \times 10^{-7}) = 7.00$
     $pH = 14.00 - 7.00 = 7.00$

The completed table is

|  | pH | pOH | $[H^+]$ | $[OH^-]$ | acid, base, or neutral? |
|---|---|---|---|---|---|
| solution a | 6.88 | 7.12 | $1.3 \times 10^{-7}$ | $7.7 \times 10^{-8}$ | acid |
| solution b | 0.92 | 13.08 | 0.12 | $8.4 \times 10^{-14}$ | acid |
| solution c | 10.89 | 3.11 | $1.3 \times 10^{-11}$ | $7.8 \times 10^{-4}$ | base |
| solution d | 7.00 | 7.00 | $1.0 \times 10^{-7}$ | $1.0 \times 10^{-7}$ | neutral |

Problems 9-16 at the end of this study chapter cover material from this section.

## 14.4 - Calculating the pH of Strong Acid Solutions

When you finish this section you will be able to calculate the pH of strong acid solutions.

Your textbook introduces you to the idea of **multiple equilibria** in this section. In order to properly assess acid-base problems in aqueous solution, you must always

a.  **recognize that autoprotolysis of water is ALWAYS occurring in an aqueous solution, and**
b.  **be able to determine whether autoprotolysis will contribute significantly to the acid-base character of a solution.**

Calculating the pH of strong acid solutions is in general, fairly straight forward, because the dissociation equilibrium lies so far to the right - that is, **the acid completely dissociates.** The autoprotolysis is negligible as a contributor of $H^+$ to the solution. (See the discussion regarding Le Chatelier's principle in this section of your textbook.) The rare exception to this is when your concentrated acid is exceptionally dilute ($< 10^{-6}$ M). In that case water can contribute a **relatively** large proportion of $H^+$ to the solution.

The bottom line is that $[H^+]$ at equilibrium is $\approx$ [strong acid]$_o$, except in very dilute solutions.

## Example 14.4 A - The pH of a Strong Acid

Calculate the **pH** and $[OH^-]$ of a $5 \times 10^{-3}$ M $HClO_4$ solution.

## Solution

$HClO_4$ is a very strong acid. It will completely dissociate. The autoprotolysis of water will not contribute significantly to $[H^+]$.

$$pH = -\log[HClO_4]_o = -\log[H^+] = -\log(5 \times 10^{-3} \text{ M})$$
$$\textbf{pH = 2.30}$$

$$[OH^-] = K_w/[H^+] = 1 \times 10^{-14} / 5 \times 10^{-3} = \textbf{2} \times \textbf{10}^{-12} \textbf{ M}$$

## Example 14.4 B - Practice with Strong Acids

A solution is prepared by adding 15.8 g of HCl to enough water to make a total volume of 400 mL. What is the pH of the solution? How much hydrogen ion is contributed by the autoprotolysis of water?

### Solution

Let's first find $[HCl]_o$.

$$\frac{\text{mol HCl}}{\text{L}} \;\bigg\|\; \frac{15.8\ g}{0.400\ L} \times \frac{1\ \text{mol}}{36.5\ g} = \textbf{1.08 M}$$

HCl is a strong acid, so it completely dissociates.

$$pH = -\log[H^+] = -\log[HCl]_o = -\log(1.08) = \textbf{-0.03}$$

The next question had to do with $[H^+]$ due to the autoprotolysis of water. We know that **THE ONLY SOURCE OF OH⁻ IS FROM THE AUTOPROTOLYSIS OF WATER.** We can find $[OH^-]$ because

$$[OH^-] = K_w/[H^+] = 1.0 \times 10^{-14} / 1.08 = \textbf{9.3} \times \textbf{10}^{\textbf{-15}} \textbf{ M}$$

In addition, $[H^+]$ due to $H_2O$ autoprotolysis must equal $[OH^-]$ due to $H_2O$ autoprotolysis (1:1 stoichiometry). Therefore,

$$[H^+]_{H_2O} = \textbf{9.3} \times \textbf{10}^{\textbf{-15}} \textbf{ M}$$

We can see Le Chatelier's principle at work here. If the autoprotolysis of water was not suppressed, $[H^+]_{H_2O} = 1.0 \times 10^{-7}$.

---

## Problems 17-23 at the end of this study chapter cover material from this section.

---

## 14.5 - Calculating the pH of Weak Acid Solutions

When you finish this section you will be able to calculate the pH and percent dissociation of weak acid solutions.

A succinct strategy for solving weak acid problems is proposed just above Sample Exercise 14.8 in your textbook. The key points of the strategy are

- Although there are often several reactions that can produce $H^+$, usually **only one predominates**. You can make the proper judgment based on the values of the equilibrium constants for the reactions.
- You must test any assumptions that you make regarding the extent of dissociation of a weak acid (ie; $[HA] \approx [HA]_o$).

Let's work the following problem using the procedure given in your textbook.

# Example 14.5 A - pH of a Weak Acid

Calculate the pH of a 0.500 M aqueous solution of formic acid, HCOOH ($K_a = 1.77 \times 10^{-4}$).

## Solution

Step 1: We recognize that HCOOH is a weak acid. The dissociation equilibrium lies far to the left. The same is true for the autoprotolysis of $H_2O$ ($K_w = 1.0 \times 10^{-14}$). The major species in solution are therefore **HCOOH and $H_2O$.**

Step 2: Both HCOOH and $H_2O$ can produce $H^+$.

$$HCOOH(aq) \rightleftharpoons COOH^-(aq) + H^+(aq) \quad K_a = 1.77 \times 10^{-4}$$

$$H_2O(l) \rightleftharpoons H^+(aq) + OH^-(aq) \quad K_w = 1.0 \times 10^{-14}$$

Step 3: The value of $K_a$ for the dissociation of HCOOH is far greater than the value of $K_w$ for the autoprotolysis of water. Therefore, HCOOH dissociation will predominate as a source of $H^+$. (Technically, we have **assumed** that $H_2O$ contributes a negligible amount of $H^+$. We will not test this assumption.)

Step 4: $K_a = \dfrac{[COOH^-][H^+]}{[HCOOH]}$

Steps 5, 6, and 7:

|  | initial(M) | change | final(M) |
|---|---|---|---|
| HCOOH | 0.500 | -X | 0.500 - X |
| $H^+$ | 0 | +X | X |
| $COOH^-$ | 0 | +X | X |

Step 8: $K_a = 1.77 \times 10^{-4} = \dfrac{X^2}{0.500 - X}$

Step 9: The equilibrium lies far to the left. Assume that the extent of dissociation, "X," is **negligible relative to 0.500 M.**

$$(0.500 - X \approx 0.500)$$

$$1.77 \times 10^{-4} = \frac{X^2}{0.500}$$

$$X = [H^+] = [COOH^-] = 9.41 \times 10^{-3} \text{ M}$$

Step 10: Comparing X to $[HCOOH]_o$,

$$\frac{X}{[HCOOH]_o} \times 100\% = \frac{9.41 \times 10^{-3}}{0.500} \times 100\% = \textbf{1.9\%}$$

This value is less than 5%, so **our assumption of negligible dissociation is valid.**

Step 11: $[H^+] = 9.41 \times 10^{-3}$ M $\Rightarrow$ **pH = 2.03**

## Example 14.5 B - Practice with Weak Acids

The **value for $K_a = 7.45 \times 10^{-4}$** for citric acid ($C_6H_{10}O_8$) (we'll call it "HCA"). Calculate the pH of a 0.200 M HCA solution.

## Solution

Our reactions of interest are

$$HCA(aq) \rightleftharpoons CA^-(aq) + H^+(aq)$$

$$H_2O(l) \rightleftharpoons H^+(aq) + OH^-(aq)$$

$K_a$ for HCA is much larger than $K_w$ for water. Therefore, the dissociation of HCA is the significant equilibrium. This equilibrium lies far to the left.

$$K_a = \frac{[CA^-][H^+]}{[HCA]}$$

|     | initial(M) | change | final(M) |
|-----|-----------|--------|----------|
| HCA | 0.200 | -X | 0.200 - X ($\approx$0.200) |
| $H^+$ | 0 | +X | X |
| $CA^-$ | 0 | +X | X |

$$7.45 \times 10^{-4} = \frac{X^2}{0.200}$$

$$X = [H^+] = [CA^-] = 1.22 \times 10^{-2} \text{ M}$$

$$\frac{X}{HCA} \times 100\% = \frac{1.22 \times 10^{-2}}{0.200} \times 100\% = 6.1\%$$

The percent of dissociation is greater than 5%, therefore, we **CANNOT "neglect X in comparison to 0.200."** (ie; We are NOT making this assumption, therefore, we do not have to test it.) We must therefore solve the equation **using the quadratic equation.**

$$7.45 \times 10^{-4} = \frac{X^2}{0.200 - X}$$

Clearing the denominator,

$$1.49 \times 10^{-4} - 7.45 \times 10^{-4}X = X^2$$

Setting the equation equal to zero,

$$\underset{a}{1X^2} + \underset{b}{7.45 \times 10^{-4}X} - \underset{c}{1.49 \times 10^{-4}} = 0$$

$$X = \frac{-b + (b^2 - 4ac)^{1/2}}{2a}$$

$$X = \frac{-7.45 \times 10^{-4} + [(7.45 \times 10^{-4})^2 - 4(1)(-1.49 \times 10^{-4})]^{1/2}}{2(1)}$$

$$X = \frac{-7.45 \times 10^{-4} + 0.02442}{2} = 1.18 \times 10^{-2} \, M$$

$$[H^+] = 1.18 \times 10^{-2} \, M \Rightarrow pH = 1.93$$

We may wish to check our results:

$$K_a = \frac{(1.18 \times 10^{-2})^2}{(0.200 - 1.18 \times 10^{-2})} = \frac{1.392 \times 10^{-4}}{0.1882} = 7.4 \times 10^{-4}$$

This is o.k. within round-off error.

---

With regard to calculating the pH of a mixture of weak acids, the basic question remains: **Which is the dominant equilibrium among the several that are followed?**

If you can resolve that, then the problem reduces to the pH of what is effectively one species in solution.

---

## Example 14.5 C - The pH of a Mixture of Weak Acids

Calculate the pH of a mixture of 2.00 M formic acid (HCOOH, $K_a = 1.77 \times 10^{-4}$) and 1.50 M hypobromous acid (HOBr, $K_a = 2.06 \times 10^{-9}$).  What is the concentration of both the hypobromite ion ($OBr^-$) and hydroxide ($OH^-$) ion at equilibrium?

## Solution

Like most other problems in equilibrium chemistry, this one is solved by **making proper assumptions.**  Looking at the three sources of $H^+$,

$$HCOOH(aq) \rightleftharpoons H^+(aq) + COOH^-(aq) \quad K_a = 1.77 \times 10^{-4}$$

$$HOBr(aq) \rightleftharpoons H^+(aq) + OBr^-(aq) \quad K_a = 2.06 \times 10^{-9}$$

$$H_2O(l) \rightleftharpoons H^+(aq) + OH^-(aq) \quad K_w = 1.0 \times 10^{-14}$$

We see that **the dissociation of formic acid is by far the most important as a supplier of $H^+$.**  Le Chatelier's principle will dictate that the dissociation of hypobromous acid and the autoprotolysis of water will be suppressed.  We have therefore reduced the problem to finding the pH of 2.00 M formic acid.

$$K_a = 1.77 \times 10^{-4} = \frac{[H^+][COOH^-]}{[HCOOH]}$$

|         | initial(M) | change | final(M)           |
|---------|------------|--------|--------------------|
| HCOOH   | 2.00       | -X     | 2.00 - X ($\approx 2.00$) |
| $H^+$   | 0          | +X     | X                  |
| $COOH^-$| 0          | +X     | X                  |

$$1.77 \times 10^{-4} = \frac{X^2}{2.00}$$

$$X = 1.88 \times 10^{-2} M = [H^+] = [COOH^-]$$

To test the 5% rule,

$$\frac{0.0188}{2.00} \times 100\% = 0.94\%. \text{ Our assumption was O.K.}$$

$$[H^+] = 0.0188 \Rightarrow pH = 1.72$$

To find [OBr⁻], use its equilibrium expression:

$$K_a = \frac{[OBr^-][H^+]}{[HOBr]}$$

We know the $K_a$, [H⁺], and [HOBr] (which $\approx$ [HOBr]$_o$, because of negligible dissociation.)

$$2.06 \times 10^{-9} = \frac{X(0.0188)}{1.50}$$

$$X = [OBr^-] = 1.6 \times 10^{-7} M$$

To find [OH⁻], if pH = 1.72, pOH = 14 - 1.72 = 12.28

$$[OH^-] = 10^{-12.28} = 5.2 \times 10^{-13} M$$

---

## Example 14.5 D - Percent Dissociation

Determine the percent dissociation of the formic acid solution given in the previous problem.

### Solution

$$\text{Percent dissociated} = \frac{\text{amount dissociated (M)}}{\text{initial concentration (M)}} \times 100\%$$

The amount dissociated = "X" = [COOH⁻]

$$\% \text{ dissociated} = \frac{0.0188}{2.00} \times 100\% = 0.94\%$$

You will note that we determined this as part of the previous problem!

---

Finally, your textbook deals with determination of $K_a$ from the percent dissociated. This type of problem is straight forward if you recognize that **you can calculate equilibrium concentrations from the percent of dissociation**, as is shown in our next example.

---

## Example 14.5 E - $K_a$ From Percent Dissociation

In a 0.500 M solution, uric acid ($HC_5H_3N_4O_4$) is 1.6% dissociated. Calculate the value of $K_a$ for uric acid.

## Solution

$$\% \text{ dissociated} = \frac{\text{amount dissociated}}{\text{initial concentration}} \times 100\%$$

The amount dissociated = $[C_5H_3N_4O_4^-]$ = $[H^+]$

$$1.6\% = \frac{[C_5H_3N_4O_4^-]}{0.500} \times 100\%$$

$$[C_5H_3N_4O_4^-] = [H^+] = 8.0 \times 10^{-3} \text{ M}$$

We can now substitute into our equilibrium expression.

$$K_a = \frac{[C_5H_3N_4O_4^-][H^+]}{[HC_5H_3N_4O_4]} = \frac{(8.0 \times 10^{-3})^2}{0.500} = 1.3 \times 10^{-4}$$

---

## Problems 30-41 at the end of this study chapter cover material from this section.

---

## 14.6 - Bases

When you finish this section you will be able to calculate the pH of a variety of basic solutions.

The key to determining the pH of basic solutions is to recognize that, **in an equilibrium sense**, bases work in the same way that acids do. Just as there are both strong and weak acids, there are both strong and weak bases.

Strong bases completely dissociate. Using LiOH in water as an example,

$$LiOH(s) \text{----}> Li^+(aq) + OH^-(aq)$$

Therefore, one can consider that $[OH^-] \approx [LiOH]_o$. Once you know $[OH^-]$, you can use $K_w$ to calculate $[H^+]$ and pH.

Your textbook points out that all alkali hydroxides are strongly basic. Alkaline earth hydroxides are also strongly basic, but somewhat less soluble than alkali hydroxides.

---

## Example 14.6 A - pH of a Strong Base

Calculate the pH of a solution made by putting 4.63 g of LiOH into water and diluting to a total volume of 400 mL.

## Solution

As we discussed earlier, LiOH is a strong base. The equilibrium concentration of $OH^-$ will be equal to the initial concentration of LiOH.

$$[LiOH]_o = \frac{4.63 \text{ g LiOH}}{0.400 \text{ L}} \times \frac{1 \text{ mol LiOH}}{24.0 \text{ g}} = 0.482 \text{ M}$$

$$[LiOH]_o = [OH^-] = 0.482 \text{ M}$$

$$pOH = 0.317$$

Recall that $pK_w = 14.00 = pH + pOH$.  Therefore,

$$pH = 14 - pOH = 14.00 - 0.32 = 13.68$$

$[H^+] = 10^{-13.68} = 2.1 \times 10^{-14}$ M.  The only source of $H^+$ was the autoprotolysis of water.

---

Weak bases react with water ("undergo hydrolysis") as described by the following equation.

$$B(aq) + H_2O(l) \rightleftharpoons BH^+(aq) + OH^-(aq)$$

| base 1 | acid 2 | acid 1 | base 2 |

For **weak bases**, as with weak acids, **the position of equilibrium lies far to the left**.  The strategy for solving for the pH of weak bases (via pOH) is the same as for weak acids.  Same steps.  Same assumption.  Same "5% test", as is shown in the following example.

---

## Example14.6 B - pH of a Weak Base

Calculate the pH of a 0.350 M solution of methylamine, $CH_3NH_2$ ($K_b = 4.38 \times 10^{-4}$).

### Strategy and Solution

We must proceed with the same problem-solving strategy as with weak acids.

Given the value of $K_b$, the extent of equilibrium for the base hydrolysis will be far to the left.

$$CH_3NH_2(aq) + H_2O(l) \rightleftharpoons CH_3NH_3(aq) + OH^-(aq)$$

Although the autoprotolysis of water can supply $OH^-$ to the solution, the value of $K_w$ is small relative to the $K_b$ of methylamine, so $[OH^-]$ due to water can be neglected.

$$K_b = \frac{[CH_3NH_3^+][OH^-]}{[CH_3NH_2]}$$

|  | initial(M) | change | final(M) |
|---|---|---|---|
| $CH_3NH_2$ | 0.350 | -X | 0.350 - X  ($\approx$0.350) |
| $OH^-$ | 0 | +X | X |
| $CH_3NH_3^+$ | 0 | +X | X |

(The assumption, which must be tested, is that $[CH_3NH_2] \approx [CH_3NH_2]_o$.  We will use the "5% rule" to verify this later.)

$$K_b = 4.38 \times 10^{-4} = \frac{X^2}{0.350}$$

$$X = [OH^-] = [CH_3NH_3^+] = 0.0124 \text{ M}$$

$$\frac{0.0124}{0.350} \times 100\% = 3.5\% \text{ hydrolysis, which passes the 5\% test.}$$

$$\text{pOH} = -\log(0.0124) = \mathbf{1.91}$$

$$\text{pH} = 14.00 - 1.91 = \mathbf{12.09}$$

## Does Our Answer Make Sense

There are really two ways to answer the question. As a double check of our math,

$$K_b = \frac{(0.0124)^2}{0.350} = 4.39 \times 10^{-4}$$

So far so good. The more significant question relates to the value for pH. We needed to have a pH > 7 for this base. We have that in this case. Therefore, the answer makes sense.

---

Problems 24-29 and 45-50 at the end of this study chapter cover material from this section.

---

## 14.7 - Polyprotic Acids

When you finish this section you will be able to solve for the pH and concentrations of species of polyprotic acids in aqueous solution.

A **polyprotic acid** can furnish **more than one proton** to a solution. Table 14.4 in your textbook lists a number of examples along with $K_a$ values.

Please note that in every case, $\mathbf{K_{a_1} \gg K_{a_2}}$ for these acids. This means that for most of these polyprotic acids, the first proton comes off **relatively easily**. The second (and third, where applicable) does not. Another way of saying this is that the second and third dissociation reactions are generally so far "to the left" that we can neglect them. The beauty of this is that most polyprotic acid pH problems **reduce to finding the pH from a single, dominant equation**.

As you do the Sample Exercises in your text and the examples here, you will see that our problem-solving strategy is essentially the same as with our previous acid problems.

---

## Example 14.7 A - pH of Oxalic Acid

Using the information in Table 14.4 of your textbook, **calculate the pH** of a 1.40 M $H_2C_2O_4$ (oxalic acid) solution and the **equilibrium concentrations of** $\mathbf{H_2C_2O_4}$, $\mathbf{HC_2O_4^-}$, $\mathbf{C_2O_4^{2-}}$ and $\mathbf{OH^-}$.

## Strategy and Solution

   A.  pH, $[H_2C_2O_4]$ and $[HC_2O_4^-]$

   The major species in solution are $\mathbf{H_2C_2O_4}$ **and** $\mathbf{H_2O}$. There are a number of equilibria that will occur. Based on the values of $K_{a_1}$, $K_{a_2}$, and $K_w$, by far the **most significant equilibrium** will be the dissociation of $H_2C_2O_4$.

$$H_2C_2O_4(aq) \rightleftharpoons HC_2O_4^-(aq) + H^+(aq)$$

$$K_{a_1} = \frac{[H^+][HC_2O_4^-]}{[H_2C_2O_4]}$$

|          | initial(M) | change | final(M)           |
|----------|-----------|--------|--------------------|
| $H_2C_2O_4$ | 1.40      | -X     | 1.40 - X  ($\approx$1.40) |
| $H^+$    | 0         | +X     | X                  |
| $HC_2O_4^-$ | 0         | +X     | X                  |

As always, we will test our "negligible dissociation" assumption (1.40 - X $\approx$ 1.40) later on.

$$6.5 \times 10^{-2} = \frac{[H^+][HC_2O_4^-]}{[H_2C_2O_4]} = \frac{X^2}{1.40}$$

$$X = [H^+] = [HC_2O_4^-] = 0.302$$

Testing the 5% rule,

$$\frac{0.302}{1.40} \times 100\% = 21.5\%!$$

Therefore, $[H_2C_2O_4] \neq [H_2C_2O_4]_o$ but rather equals "1.40 - X."

$$6.5 \times 10^{-2} = \frac{X^2}{1.40 - X}$$

Clearing the fraction and setting equal to zero so that we can use the quadratic formula,

$$X^2 + 0.065X - 0.091 = 0, \quad a = 1, b = 0.065, c = -0.091$$

Solving,

$$X = \frac{-0.065 + [(0.065)^2 - 4(1)(-0.091)]^{1/2}}{2(1)}$$

$$X = \frac{-0.065 + 0.6068}{2} = 0.271 \text{ M}$$

$$[H^+] = [HC_2O_4^-] = 0.27 \text{ M}$$

$$[H_2C_2O_4] = 1.40 - 0.271 = 1.1 \text{ M}$$

$$pH = -\log(0.27) = 0.57$$

Checking our math,

$$K_{a_1} = \frac{(0.27)^2}{1.1} = 0.066, \text{ O.K. within round-off error.}$$

B.  $[C_2O_4^{2-}]$ and $[OH^-]$

We can find $[C_2O_4^{2-}]$ by using the second dissociation equilibrium of oxalic acid,

$$HC_2O_4^-(aq) \rightleftharpoons H^+(aq) + C_2O_4^{2-}(aq) \quad K_{a_2} = 6.1 \times 10^{-5}$$

We know $[H_2C_2O_4^-]$, $[H^+]$ and $K_{a_2}$. We can therefore substitute into our equilibrium expression,

$$K_{a_2} = \frac{[H^+][C_2O_4^{2-}]}{[HC_2O_4^-]} \quad \Rightarrow \quad 6.1 \times 10^{-5} = \frac{(0.27)[C_2O_4^{2-}]}{0.27}$$

$$[C_2O_4^{2-}] = K_{a_2} = 6.1 \times 10^{-5} \text{ M}$$

Note that $[C_2O_4^{2-}]$ is smaller than $[HC_2O_4^-]$ by a factor of $10^4$. We can therefore conclude that the dissociation of $HC_2O_4^-$ was not a significant contributor of $H^+$ to the reaction.

We can find $[OH^-]$ by using $pK_w = pH + pOH$.

$$pOH = 14 - pH \quad \Rightarrow \quad pOH = 14 - 0.57 = \textbf{13.43}$$

$$[OH^-] = 10^{-13.43} = \textbf{3.7} \times \textbf{10}^{-14} \text{ M}$$

The autoprotolysis of water was clearly not important here (as evidenced by the low $[OH^-]$).

---

Keep in mind that even though the problem was long, it was all *based on the assumption* that only the first acid dissociation equilibrium was important.

---

## Example 14.7 B - Phosphoric Acid

Using data from <u>Table 14.4 in your textbook</u>, calculate the pH, $[PO_4^{3-}]$ and $[OH^-]$ in a 6.0 M phosphoric acid ($H_3PO_4$) solution.

### <u>Strategy</u>

In order to calculate $[PO_4^{3-}]$, we must use $K_{a_3}$. We must therefore know $[HPO_4^{2-}]$ and $[H^+]$. To know $[HPO_4^{2-}]$, we must use $K_{a_2}$, which means knowing $[H_2PO_4^-]$. This can be determined using $K_{a_1}$. **The bottom line is that** we must proceed as in the previous problem, by determining pH and working our way down[*]. We can find $[OH^-]$ by determining pH and using $pK_w$, as always.

### <u>Solution</u>

We will make the usual simplifying assumption that leads to

$$K_{a_1} = \frac{[H^+][H_2PO_4^-]}{[H_3PO_4]} \quad \Rightarrow \quad 7.5 \times 10^{-3} = \frac{X^2}{6.0}$$

$$X = [H^+] = [H_2PO_4^-] = \textbf{0.212 M}$$

We must now run the 5% test,

$$\frac{0.212}{6.0} \times 100\% = 3.5\%,$$

which passes the test of "negligible dissociation."

---

[*]It can be shown that $(PO_4^{3-}) = \frac{K_{a_1}K_{a_2}K_{a_3}(H_3PO_4)_o}{(H^+)}$ but that is beyond the scope of this study guide.

To find $[HPO_4^{2-}]$,

$$K_{a_2} = \frac{[H^+][HPO_4^{2-}]}{[H_2PO_4^-]} \Rightarrow 6.2 \times 10^{-8} = \frac{(0.212)[HPO_4^{2-}]}{0.212}$$

$$[HPO_4^{2-}] = 6.2 \times 10^{-8} = K_{a_2}$$

To find $[PO_4^{3-}]$,

$$K_{a_3} = \frac{[H^+][PO_4^{3-}]}{[HPO_4^{2-}]} \Rightarrow 4.8 \times 10^{-13} = \frac{(0.212)[PO_4^{3-}]}{6.2 \times 10^{-8}}$$

$$[PO_4^{3-}] = 1.4 \times 10^{-19} \text{ M}$$

We can find $[OH^-]$ in the usual way,

$$[OH^-] = K_w/[H^+] = 1.0 \times 10^{-14}/0.212$$

$$[OH^-] = 4.7 \times 10^{-14} \text{ M}$$

---

Problems 51-55 at the end of this study chapter cover material from this section.

---

# 14.8 - Acid-Base Properties of Salts

When you finish this section you will be able to calculate the pH of a variety of salt solutions.

Salts are ionic compounds. They dissociate in water and may exhibit acid-base behavior. The **key question** in deciding whether a salt will act as an acidic, basic or neutral species in solution is **"What are the acid-base properties, and strengths, of each component of the salt?"**

For example, sodium nitrite, $NaNO_2$, completely dissociates to give $Na^+$ and $NO_2^-$ ions. These are the main species in solution in addition to $H_2O$. What are the acid-base properties of each of these species? Remember your conjugate acid-base relationships.

- Strong acids and bases have weak conjugates. **$Na^+$ and other alkali and alkaline earth metals exhibit no acid-base properties.**
- The nitrite ion, $NO_2^-$, is the conjugate base of the moderately weak acid $HNO_2$ ($K_a = 4.6 \times 10^{-4}$). Remember that "weak" is a relative term. $HNO_2$ is weak compared to $HNO_3$, which has a $K_a \gg 1$. But $HNO_2$ is far stronger than HCN ($K_a = 6.2 \times 10^{-10}$). **The $NO_2^-$ ion will act as a very weak base.** (Your textbook points out that it is competing for protons against $OH^-$!)
- Water will have relatively little acid-base effect. Therefore, **an aqueous solution of $NaNO_2$ should be slightly basic.**

Your textbook proves the relationship that for an aqueous solution

$$K_w = K_a \times K_b$$

You will generally find the $K_a$ for a weak acid or $K_b$ for a weak base in chemical data tables (such as <u>Tables 14.2, 3 and 4 in your textbook</u>). You will have to **calculate $K_a$ or $K_b$ for the ion of a salt** from given values for the conjugate neutral acid or base.

# Example 14.8 A - Calculating $K_a$ or $K_b$

Using data given below, calculate $K_a$ or $K_b$ (as required), and write the reaction with water for each of the following aqueous ions:

    a.   $NO_2^-$ ($K_a$ for $HNO_2 = 4.6 \times 10^{-4}$)

    b.   $I^-$ ($K_a$ for $HI = 0.169$)

    c.   $C_6H_5NH_3^+$ ($K_b$ for aniline, $C_6H_5NH_2 = 2.34 \times 10^{-5}$)

    d.   $HC_3H_4N_2^+$ ($K_b$ for imidazole, $C_3H_4N_2 = 1.11 \times 10^{-7}$)

## Solution

In each case, $K_w = K_a \times K_b$

    a.   $NO_2^-(aq) + H_2O(l) \rightleftharpoons HNO_2(aq) + OH^-(aq)$
          $K_b = K_w/K_a = 1.0 \times 10^{-14} / 4.6 \times 10^{-4} = \mathbf{2.2 \times 10^{-11}}$

    b.   $I^-(aq) + H_2O(l) \rightleftharpoons HI(aq) + OH^-(aq)$
          $K_b = K_w/K_a = 1.0 \times 10^{-14} / 0.169 = \mathbf{5.9 \times 10^{-14}}$

    c.   $C_6H_5NH_3^+(aq) + H_2O(l) \rightleftharpoons C_6H_5NH_2(aq) + H_3O^+(aq)$
                     OR
          $C_6H_5NH_3^+(aq) \rightleftharpoons C_6H_5NH_2(aq) + H^+(aq)$
          $K_a = K_w/K_b = 1.0 \times 10^{-14} / 2.34 \times 10^{-5} = \mathbf{4.3 \times 10^{-10}}$

    d.   $HC_3H_4N_2^+(aq) \rightleftharpoons C_3H_4N_2(aq) + H^+(aq)$ (short form)
          $K_a = K_w/K_b = 1.0 \times 10^{-14} / 1.11 \times 10^{-7} = \mathbf{9.0 \times 10^{-8}}$

---

**Recalling the bottom line**, you must decide how each part of the salt (cation and anion) will react with water. Your textbook points out a sticky case (such as $NH_4C_2H_3O_2$) where **both cation and anion exhibit acid-base behavior.** When this occurs, the overall pH of the solution is determined by the relative $K_a$ and $K_b$ values for the ions (see Table 14.5 in your textbook).

---

# Example 14.8 B - Predicting Acid-Base Behavior

Using data from Tables 14.2, 3 and 4 in your textbook, predict whether each of the following will create an acid, base, or neutral aqueous solution.

    a. $Na_3PO_4$   b. $KI$   c. $HC_5H_5NCl$ (pyridinium chloride)   d. $NH_4F$

## Solution

    a.   $Na^+$ exhibits no acid-base behavior. $PO_4^{3-}$ is the most basic form of $H_3PO_4$ ($K_{b_1} = K_w/K_{a_3}$). This solution will be a **fairly strong base.**

    b.   $K^+$ exhibits no acid-base behavior. $I^-$ also exhibits no acid-base behavior. This solution will be **neutral.**

c.  $Cl^-$ exhibits no acid-base behavior.  Pyridinium ion, $HC_5H_5N^+$, is the conjugate acid of pyridine, $C_5H_5N$ ($K_a = K_w/K_b$).  This solution will be **acidic**.

d.  $NH_4^+$ is the conjugate acid of $NH_3$ ($K_a = K_w/K_b = 5.6 \times 10^{-10}$).  $F^-$ is the conjugate base of HF ($K_b = K_w/K_a = 1.4 \times 10^{-11}1^1$).  $K_a > K_b$, therefore the solution will be a **weak acid**.

---

# Example 14.8 C - pH of a Salt

Calculate the pH of a 0.500 M $NaNO_2$ solution.  ($K_a$ for $HNO_2 = 4.0 \times 10^{-4}$)

## Solution

As discussed in the previous example, $NaNO_2$ will act as a base in water, due to the hydrolysis of $NO_2^-$.

$$NO_2^-(aq) + H_2O(l) \rightleftharpoons HNO_2(aq) + OH^-(aq)$$

$$K_b = \frac{[HNO_2][OH^-]}{[NO_2^-]}$$

**Testing Soil pH**

$$K_b = K_w/K_a = 1.0 \times 10^{-14} / 4.6 \times 10^{-4} = \mathbf{2.2 \times 10^{-11}}$$

We may now proceed as with any other weak base problem.

|         | initial(M) | change | final(M)             |
|---------|------------|--------|----------------------|
| $NO_2^-$| 0.500      | -X     | 0.500 - X  (≈0.500)  |
| $OH^-$  | 0·         | +X     | X                    |
| $HNO_2$ | 0          | +X     | X                    |

$$2.2 \times 10^{-11} = \frac{X^2}{0.500}$$

$$X = [OH^-] = [HNO_2] = 3.3 \times 10^{-6}$$

This clearly passes the 5% rule ($K_b$ is so small)!

$$pOH = -\log(3.3 \times 10^{-6}) = 5.48$$

$$pH = 14 - 5.48 = \mathbf{8.52}$$

---

Reproduced here are $K_a$ and $K_b$ tables for use in this chapter.

## Values of $K_b$ for Some Common Weak Bases

| Name          | Formula      | Conjugate acid  | $K_b$                |
|---------------|--------------|-----------------|----------------------|
| Ammonia       | $NH_3$       | $NH_4^+$        | $1.8 \times 10^{-5}$ |
| Methylamine   | $CH_3NH_2$   | $CH_3NH_3^+$    | $4.38 \times 10^{-4}$|
| Ethylamine    | $C_2H_5NH_2$ | $C_2H_5NH_3^+$  | $5.6 \times 10^{-4}$ |
| Hydroxylamine | $H_2ONH_2$   | $H_2ONH_3^+$    | $1.1 \times 10^{-8}$ |
| Aniline       | $C_6H_5NH_2$ | $C_6H_5NH_3^+$  | $3.8 \times 10^{-10}$|
| Pyridine      | $C_5H_5N$    | $C_5H_5NH^+$    | $1.7 \times 10^{-9}$ |

$$CN^-(aq) + H_2O(l) \rightleftharpoons HCN(aq) + OH^-(aq)$$

$$K_b = K_w/K_a = 1.0 \times 10^{-14} / 6.2 \times 10^{-10} = 1.6 \times 10^{-5}$$

**The $CN^-$ ion is a much stronger base than $NH_4^+$ is an acid.** Its equilibrium will dominate.

$$K_b = \frac{[HCN][OH^-]}{[CN^-]} \quad \Rightarrow \quad 1.6 \times 10^{-5} = \frac{X^2}{0.800}$$

$$X = [OH^-] = [HCN] = 3.6 \times 10^{-3}$$

The 5% rule holds here (test for yourself!), so that

$$pOH = 2.44 \text{ and } \mathbf{pH = 11.56}$$

---

# Problems 56-61 at the end of this study chapter cover material from this section.

---

## 14.9 and 14.10 - The Effect of Structure on Acid-Base Properties and Acid-BaseProperties of Oxides

After you have read these two sections, answer the questions below. They will help you review the material.

Section 14.9

1. What are the two structural properties that determine if a molecule will act as an acid?
2. Why is HF a weak acid compared to HCl?
3. Explain why increasing the number of oxygens in an H-O-X grouping increases acid strength. (See Figure 14.6 in your textbook.)
4. What is the relationship between the charge on a metal ion and the acidity of attached water molecules? Why?
5. Why is HOCl a stronger acid than HOI?

---

## Example 14.9 - Structure and Properties

Given the two acids HOI and $HIO_3$, and two values for $K_a$, 0.17 and $2 \times 10^{-11}$, which value goes with which acid?

## Solution

$HIO_3$, having more oxygens, will draw electrons away from the O-H (acidic) bond. Therefore, the O-H bond is weakened and acid strength is substantially increased.

$$\mathbf{K_a \text{ for HOI} = 2 \times 10^{-11}. \quad K_a \text{ for } HIO_3 = 0.17.}$$

Section 14.10

6. Why are some oxides covalent?
7. Under what circumstances will the O-X bond in H-O-X be ionic? What is the outcome in water of such an ionic bond?
8. Define **acidic oxide**.
9. Define **basic oxide**.
10. Which groups tend to produce acidic oxides? Basic oxides?

## Example 14.10 - Acidic and Basic Oxides

Determine whether each of the following oxides will give an acidic or basic solution when added to water:

a. $K_2O$   b. $SO_3$   c. $MgO$

### Solution

a.   The K-O bond is ionic.  This bond can break, so that an O-H bond can form.  The solution is **basic**.

b.   The S-O bond is covalent.  O-H bonds will break to produce an **acidic solution**.

c.   A **basic solution** is produced (see part "a").

---

Problems 62-63 at the end of this study chapter covers material from this section.

---

## 14.11 - The Lewis Acid-Base Model

When you finish this section, you will be able to identify the Lewis acids and bases in a reaction.

Your textbook points out that the **Lewis acid-base** concept is **the most wide-ranging** of the three concepts introduced in this chapter.

**Lewis acids are electron acceptors.**
**Lewis bases are electron donors.**

$Al^{3+}$, $H^+$, $BF_3$ are examples of Lewis acids.
$NO_2^-$, $NH_3$ and $H_2O$ are examples of Lewis bases.

Note that many species that do not contain $H^+$ are Lewis acids.  The all-encompassing nature of the Lewis concept is what makes it so useful.  For example, for the reaction to form $Ag(NH_3)_2^+$,

$$
\begin{array}{ccc}
H & & H \\
| & & | \\
H\text{-}N\text{:} \;+\; Ag^+ \;+\; \text{:}N\text{-}H & \longrightarrow & [H_3N\text{:}Ag\text{:}NH_3]^+ \\
| & & | \\
H & & H
\end{array}
$$

$Ag^+$ is the Lewis acid, because it **accepts electrons** from $NH_3$.
$NH_3$ is the Lewis base, because it **donates electrons** to $Ag^+$.

---

## Example 14.11 - Lewis Acids and Bases

Identify the Lewis acid and base in each of the following reactions:

a.   $Cu^{2+}(aq) + 4NH_3(aq) \rightleftharpoons Cu(NH_3)_4^{2+}(aq)$
b.   $I^-(aq) + I_2(aq) \rightleftharpoons I_3^-(aq)$
c.   $Fe^{3+}(aq) + 6H_2O(l) \rightleftharpoons Fe(H_2O)_6^{3+}(aq)$

## Solution

a. $Cu^{2+} + 4:NH_3 \rightleftharpoons Cu(NH_3)_4^{2+}$

   ↑         ↑

  Lewis   Lewis
  acid     base

b. $I^- + I_2 \rightleftharpoons I_3^-$

  ↑     ↑

 Lewis  Lewis
 acid   base

c. $Fe^{3+} + 6\left[:O\begin{smallmatrix}H\\H\end{smallmatrix}\right] \rightleftharpoons Fe(H_2O)_6^{3+}$

    ↑        ↑

  Lewis   Lewis
  acid    base

---

Problems 64-66 at the end of this study chapter covers material from this section.

---

**END OF CHAPTER FOUTREEN**

# SELF-TEST QUESTIONS

1. List four strong acids and four strong bases.

2. Write the dissociation reaction for each of the following acids in water. Identify the conjugate acid-base pairs in each case.

   a. $C_6H_5COOH$ (benzoic acid)
   b. $H_3BO_3$ (boric acid)
   c. $H_2PO_4^-$ (dihydrogen phosphate)
   d. $HNO_3$ (nitric acid)

3. Write the dissociation reaction for each of the following acids in water and identify the conjugate acid-base pairs:

   a. $HC_2H_2ClO_2$
   b. $HCN$
   c. $NH_4Cl$

4. Write the equilibrium expressed for each of the reactions in problem #2.

5. Write equilibrium expressions for each of the equations in problem #3.

6. The values for $K_a$ for the acids in problem #2 are:

   | substance | $K_a$ |
   |-----------|-------|
   | $C_6H_5COOH$ | $6.14 \times 10^{-5}$ |
   | $H_3BO_3$ | $5.83 \times 10^{-10}$ |
   | $H_2PO_4^-$ | $6.3 \times 10^{-8}$ |
   | $HNO_3$ | $>>1$ |

   Put the acids in order from strongest to weakest.

7. Predict which one of the bases in each pair is stronger:

   a. $HCO_3^-$ or $CO_3^{-2}$
   b. $NO_3^-$ or $NO_2^-$

8. Put the conjugated bases of the acids in problem #6 in order from strongest to weakest.

9. The pH of solution is 11.93. What is $[H^+]$?   $[OH^-]$?   pOH?

10. For each of the following solutions at 25°C, calculate $[H^+]$ given $[OH^-]$ or $[OH^-]$ given $[H^+]$. Is the solution an acid or a base?

    a. $[OH^-] = 1 \times 10^{-4}$
    b. $[H^+] = 1 \times 10^{-6}$
    c. $[H^+] = 1 \times 10^{-9}$

11. A solution has $[OH^-] = 3.6 \times 10^{-1}$ M. Is this solution strongly or weakly acidic or basic?

12. Calculate the pH for each of the following solutions at 25°C:

    a. $[H^+] = 3.8 \times 10^{-7}$ M
    b. $[H^+] = 7.2 \times 10^{-4}$ M
    c. $[H^+] = 4.1 \times 10^{-13}$ M

13. Calculate the pOH for each of the following solutions at 25°C:

    a.  $[OH^-] = 2.0 \times 10^{-12}\,M$
    b.  $[OH^-] = 3.4 \times 10^{-11}\,M$
    c.  $[OH^-] = 9.2 \times 10^{-3}\,M$

14. At 100°C, $K_w = 4.9 \times 10^{-13}$.

    a.  What is the pH of a neutral solution at 100°C?
    b.  Calculate the pH at 100°C if $[OH] = 6.3 \times 10^{-12}\,M$.

15. Calculate the pH and pOH for each of the following solutions at 25°C;

    a.  $[H^+] = 2.0 \times 10^{-12}$
    b.  $[OH^-] = 3.9 \times 10^{-3}$
    c.  $[H^+] = 7.7 \times 10^{-1}$
    d.  $[OH^-] = 5.3 \times 10^{-9}$

16. If [nut butter] = 89.8 M, what is pNut butter?

17. Calculate the pH of a $7.0 \times 10^{-2}\,M$ HCl solution.

18. Calculate the pH of a $2.8 \times 10^{-5}\,M$ $HNO_3$ solution.

19. Fill in the missing information in the following table:

| | pH | pOH | $[H^+]$ | $[OH^-]$ | acid, base, or neutral? |
|---|---|---|---|---|---|
| solution a | 5.64 | | | | |
| solution b | | | | $3.9 \times 10^{-6}\,M$ | |
| solution c | | | 0.027 M | | |
| solution d | | 1.7 | | | |

20. The pH of a solution of $HClO_4$ is 3.11. What is $[HClO_4]$?

21. The pOH of 400 mL of a solution of $HNO_3$ is 12.44. How many grams of $HNO_3$ are in the solution?

22. If 0.10 mol of HCl is added to enough water to produce 1.0L of solution, calculate the concentrations of $H^+$ and $OH^-$ and the pH of the solution.

23. One liter of solution was prepared from water and $3.5 \times 10^{-6}$ mol of HCl. Calculate the $[H^+]$ and $[OH^-]$, and the pH of the solution.

24. The pH of a 300 g solution of NaOH is 12.97. The density of the solution is 1.10 g/mL. How many grams of NaOH are in the solution?

25. Add 0.0150 mol of LiOH to sufficient water to make 1L of solution. Calculate the $[LiOH]_o$, $[H^+]$, and $[OH^-]$ and the pH of the solution.

26. How many grams of KOH are necessary to prepare 800 mL of a solution of pH = 11.56?

27. How many grams of $Ba(OH)_2$ are necessary to prepare 400 mL of a solution of pH = 12.46?

28. Calculate the pH of a $2.8 \times 10^{-4}\,M$ $Ba(OH)_2$ solution, assuming complete dissociation of the $Ba(OH)_2$.

29. How many grams of NaOH are needed to prepare a 546 mL solution with pH of 10.00?

30. A solution is made by dissolving 18.4 g of $HNO_3$ in enough water to make 662 mL of solution. Calculate the pH of the solution.

31. The $pK_a$ of chloroacetic acid ($ClCH_2COOH$) is $1.36 \times 10^{-3}$. Calculate the pH, pOH, $[H^+]$ and $[OH^-]$ of a 1.00 M solution of chloroacetic acid.

32. What is the amount of hydrogen ion due to water in problem #31? What is the percent dissociation of chloroacetic acid?

33. Calculate the pH of a 0.237 M solution of benzoic acid, $C_6H_5COOH$ ($K_a = 6.14 \times 10^{-5}$).

34. A total of 0.0560 g of acetic acid is added to enough water to make 50 mL of solution. Calculate $[H^+]$, $[CH_3COO^-]$, $[CH_3COOH]$, and pH at equilibrium. ($K_a$ for acetic acid is $1.8 \times 10^{-5}$.)

35. Calculate $K_a$ of a weak acid "HW" if a solution with an initial concentration of 0.200 M has a pH of 3.15.

36. Calculate the $K_a$ of a 0.060 M weak monoprotic acid with a pH of 3.44.

37. Calculate the pH of a solution that contains 0.250 M $H_2SO_4$ and 1.00 M $CH_3COOH$ ($K_a = 1.8 \times 10^{-5}$).

38. Calculate the original molarity of a solution of formic acid, HCOOH ($K_a = 1.9 \times 10^{-4}$) whose pH is 3.26 at equilibrium.

39. Calculate the pH of a solution that contains 1.23 M benzoic acid, $C_6H_5COOH$ ($K_a = 6.14 \times 10^{-5}$) and 0.713 M arsenious acid, $H_3AsO_3$ ($K_a = 6.0 \times 10^{-10}$). What is the concentration of the benzoate ion ($C_6H_5COO^-$), hydroxide ion ($OH^-$) and arsenite ion ($H_2AsO_3^-$)?

40. What is the percent dissociation of the benzoic acid in problem #39?

41. Calculate the percent dissociation of 0.20 M benzoic acid whose $K_a = 6.6 \times 10^{-5}$.

42. Calculate the pH of a 0.20 M $NH_4Cl$ solution. ($K_a = 5.6 \times 10^{-10}$)

43. The pH of a 0.30 M solution of a weak base is 10.66. Calculate the $K_b$ of the base.

44. A solution of ammonia, $K_b = 1.8 \times 10^{-5}$, has a pH of 11.22. Calculate the molarity of the solution.

45. Calculate the pH of a solution made by putting 11.17 g of KOH into water and diluting to a volume of 600 mL.

46. Calculate the pH of a 0.500 M solution of dimethylamine, $(CH_3)_2NH$ ($K_b = 5.9 \times 10^{-4}$).

47. What is the percent of hydrolysis of the base in problem #46? What is the concentration of hydrogen ion as a result of the autoprotolysis of water?

48. Calculate the pH of a 0.76 M KOH solution?

49. Calculate how many grams of $HONH_2$ are needed to dissolve in enough water to make 250 mL of solution with a pH of 10.00. ($K_b = 1.1 \times 10^{-8}$)

50. Calculate the pH of a 0.15 M solution of $CH_3COONa$. ($K_b = 5.6 \times 10^{-10}$)

51. What is the pH of a 0.100 M solution of arsenic acid?

$$H_3AsO_4 \rightleftharpoons H_2AsO_4^- + H^+ \qquad K_a = 6.0 \times 10^{-3}$$
$$H_2AsO_4^- \rightleftharpoons HAsO_4^{2-} + H^+ \qquad K_a = 1.05 \times 10^{-7}$$
$$HAsO_4^{2-} \rightleftharpoons AsO_4^{3-} + H^+ \qquad K_a = 3.0 \times 10^{-12}$$

What is the concenration of $AsO_4^{3-}$ ion? (Hint: use quadratic equation)

52. Calculte the concentrations of $CO_3^{-2}$, $HCO_3^-$, and $H^+$ in a 0.025 M $H_2CO_3$ solution (see <u>Table 14.4 in your text</u> for $K_a$ values of carbonic acid).

53. Calculate the equilibrium concentration of $PO_4^{3-}$ in a solution prepared by adding 716.2 g of $NH_3PO_4$ to water and diluting to total volume of 750 mL.  (See <u>Table 14.4 in your textbook</u> for $K_a$ values of phosphoric acid.)

54. Calculate the concentrations of $H_2CO_3$, $HCO_3^-$, $CO_3^{-2}$, $H^+$, and $OH^-$ and the pH at equilibrium in a solution that is initially 0.010 M in $H_2CO_3$.

55. Calculate the concentrations of $C_2H_2O_4$, $C_2HO_4^-$, $C_2O_4^{-2}$, and $H^+$ in a 0.10 M oxalic acid solution.

56. Using data from <u>Table 14.2 in your textbook</u>, determine the values for $K_b$ of

    a. $CN^-$
    b. $NH_3$
    c. $C_6H_5O^-$

57. Arrange the bases in problem #56 in order from strongest to weakest.

58. Calculate the pH of a 0.450 M solution of sodium propionate;  $NaOCH_2CH_3$ ($K_a$ for propionic acid, $HOCH_2CH_3 = 1.34 \times 10^{-5}$).

59. Calculate the pH of a 0.20 M $K_2S$ solution ($K_a = 1 \times 10^{-13}$).

60. Determine the equilibrium concentration of HCN in a 0.0500 M NaCN solution ($K_a$ for HCN = $6.2 \times 10^{-10}$).

61. Decide qualitatively whether each of the following will be acidic, basic or neutral in solution.

    a.  NaCN
    b.  KCl
    c.  $NH_4NO_3$

62. Determine whether each of the following oxides will give an acid or basic solution when added to water.

    a.  $SO_2$
    b.  CaO
    c.  $NO_2$
    d.  $Cl_2O_7$

63. Predict whether an aqueous solution of ammonium acetate ($CH_3COONH_4$) will be acidic, basic, or neutral.

64. Identify the Lewis acid in each of the following reactions:

    a.  $Co^{2+} + 6CN^-(aq) \rightleftharpoons Co(CN)_6^{4-}(aq)$
    b.  $HCOOH(aq) + OH^-(aq) \rightleftharpoons COOH^-(aq) + H_2O(l)$
    c.  $AgCl(s) + Cl^-(aq) \rightleftharpoons AgCl_2^-(aq)$

65. Identify the Lewis acid and Lewis base in each of the following reactions:

    a.  $SnCl_4(s) + 2Cl^-(aq) \rightleftharpoons SnCl_6^{-2}(aq)$
    b.  $Hg^{+2}(aq) + 4CN^-(aq) \rightleftharpoons Hg(CN)_4^{-2}(aq)$
    c.  $Co^{+3}(aq) + 6NH_3(aq) \rightleftharpoons Co(NH_3)_6^{+3}(aq)$

66. Identify the following as Lewis acids or Lewis bases:

    a.  $BF_3$
    b.  $NH_3$
    c.  $SO_3$

# ANSWERS TO SELF-TEST QUESTIONS

1. acids: $HCl$, $HNO_3$, $HClO_4$, $H_2SO_4$. bases: $NaOH$, $KOH$, $Ba(OH)_2$, $LiOH$.

2.

|   | acid 1 | base 2 |   | base 1 | acid 2 |
|---|--------|--------|---|--------|--------|
| a. | $C_6H_5COOH$ | $+ H_2O$ | $\rightleftharpoons$ | $C_6H_5COO^-$ | $+ H_3O^+$ |
| b. | $H_3BO_3$ | $+ H_2O$ | $\rightleftharpoons$ | $H_2BO_3^-$ | $+ H_3O^+$ |
| c. | $H_2PO_4^-$ | $+ H_2O$ | $\rightleftharpoons$ | $HPO_4^{2-}$ | $+ H_3O^+$ |
| d. | $HNO_3$ | $+ H_2O$ | $\rightleftharpoons$ | $NO_3^-$ | $+ H_3O^+$ |

3.  a.  $HC_2H_2ClO_2(aq) + H_2O(l) \rightleftharpoons C_2H_2ClO_2^-(aq) + H_3O^+(aq)$

          acid 1       base 2       base 1       acid 2

    b.  $HCN(aq) + H_2O(l) \rightleftharpoons CN^-(aq) + H_3O^+(aq)$

          acid 1     base 2   base 1   acid 2

    c.  $NH_4Cl(aq) + H_2O(l) \rightleftharpoons NH_3Cl^-(aq) + H_3O^+(aq)$

          acid 1     base 2   base 1     acid 2

4.  a.  $K = \dfrac{[C_6H_5COO^-][H^+]}{[C_6H_5COOH]}$    c.  $K = \dfrac{[HPO_4^{2-}][H^+]}{[H_2PO_4^-]}$

    b.  $K = \dfrac{[H_2BO_3^-][H^+]}{[H_3BO_3]}$    d.  $K = \dfrac{[NO_3^-][H^+]}{[HNO_3]}$   $(K \gg 1)$

5.  a.  $K_a = \dfrac{[C_2H_2ClO_2^-][H^+]}{[HC_2H_2ClO_2]}$

    b.  $K_a = \dfrac{[CN^-][H^+]}{[HCN]}$

    c.  $K_a = \dfrac{[NH_3Cl^-][H^+]}{[NH_4Cl]}$

6.  $HNO_3$, $C_6H_5COOH$, $H_2PO_4^-$, $H_3BO_3$
    strongest--------------------->weakest

7.  a.  $CO_3^{-2}$
    b.  $NO_2^-$

8.  $H_2BO_3^-$, $HPO_4^{2-}$, $C_6H_5COO^-$, $NO_3^-$
    strongest------------------->weakest

9.  $[H^+] = 1.2 \times 10^{-12}$ M, $[OH^-] = 8.5 \times 10^{-3}$ M, pOH = 2.07

10. a.  $[H^+] = 1 \times 10^{-10}$, the solution is basic.
    b.  $[OH^-] = 1 \times 10^{-8}$, the solution is acidic.
    c.  $[OH^-] = 1 \times 10^{-5}$, the solution is basic.

11. strongly basic

12. a.  pH = 6.42
    b.  pH = 3.14
    c.  pH = 12.4

13.  a.    pOH = 11.70
     b.    pOH = 10.47
     c.    pOH = 2.04

14.  a.    pH = 6.15
     b.    pH = 1.11

15.  a.    pH = 11.70,   pOH = 2.30
     b.    pH = 11.59,   pOH = 2.41
     c.    pH = 0.11,    pOH = 13.89
     d.    pH = 5.72,    pOH = 8.28

16.  pNut butter = -1.95

17.  pH = 1.15

18.  pH = 4.55

19.

|  | pH | pOH | [H+] | [OH-] | acid, base, or neutral? |
|---|---|---|---|---|---|
| solution a | 5.64 | 8.36 | $2.3 \times 10^{-6}$ M | $4.4 \times 10^{-9}$ M | acid |
| solution b | 8.59 | 5.41 | $2.6 \times 10^{-9}$ M | $3.9 \times 10^{-6}$ M | base |
| solution c | 1.57 | 12.43 | 0.027 M | $3.7 \times 10^{-13}$ M | acid |
| solution d | 12.3 | 1.70 | $5.0 \times 10^{-13}$ M | $2.0 \times 10^{-2}$ M | base |

20.  $[HClO_4] = 7.8 \times 10^{-4}$ M

21.  0.69 g $HNO_3$

22.  $[H^+] = 0.010$ M,   $[OH^-] = 1.0 \times 10^{-13}$ M,   pH = 1

23.  $[H^+] = 3.5 \times 10^{-6}$ M,   $[OH^-] = 2.8 \times 10^{-9}$ M,   pH = 5.45

24.  1.02 g NaOH

25.  $[LiOH]_o = [OH^-] = 0.015$, $[H^+] = 6.7 \times 10^{-13}$, pH = 12.2

26.  0.16 g KOH

27.  2.0 g $Ba(OH)_2$

28.  pH = 10.45

29.  $2.2 \times 10^{-3}$ g

30.  pH = 0.355

31.  pH = 1.43,   pOH = 12.57,   $[OH^-] = 2.7 \times 10^{-13}$ M,   $[H^+] = 3.7 \times 10^{-2}$ M

32.  $[H^+]_{H_2O} = [OH^-]_{total} = 2.7 \times 10^{-13}$ M, 3.7%

33.  pH = 2.42

34.  $[H^+] = [CH_3COO^-] = 5.8 \times 10^{-4}$ M,   $[CH_3COOH] = 0.018$ M,   pH = 3.24

35.  $K_a = 2.52 \times 10^{-6}$

36. $K_a = 2.2 \times 10^{-6}$

37. $pH = 0.30$

38. $1.6 \times 10^{-3}$ M

39. $pH = 2.05$,   $[C_6H_5COO^-] = [H^+] = 9.0 \times 10^{-3}$ M,   $[OH^-] = 1.1 \times 10^{-12}$ M,   $[H_2AsO_3^-] = 4.8 \times 10^{-8}$ M

40. percent dissociation = 0.68%

41. 1.8%

42. $pH = 4.98$

43. $7.1 \times 10^{-7}$

44. 0.15 M

45. $pH = 13.52$

46. $pH = 12.23$

47. percent hydrolysis = 3.4%,   $[H^+]_{H_2O} = [H^+]_{total} = 5.9 \times 10^{-13}$ M

48. $pH = 13.88$

49. 7.71 g

50. $pH = 8.96$

51. $pH = 1.66$,   $[AsO_4^{3-}] = 1.5 \times 10^{-17}$ M

52. $[CO_3^{-2}] = 5.6 \times 10^{-11}$ M,   $[HCO_3^-] = [H^+] = 1.0 \times 10^{-4}$ M

53. Calculate $[H^+]$ and work your way down;   $[H^+] = 0.262$,   $(pH = 0.580)$;   $[PO_4^{3-}] = 1.0 \times 10^{-19}$ M

54. $[H_2CO_3] = 0.010$ M,   $[HCO_3^-] = [H^+] = 6.6 \times 10^{-5}$ M,   $[CO_3^{-2}] = 5.6 \times 10^{-11}$,   $[OH^-] = 1.5 \times 10^{-10}$ M, $pH = 4.18$

55. $[C_2H_2O_4] = 0.046$ M,   $[C_2HO_4^-] = [H^+] = 0.054$ M,   $[C_2O_4^{-2}] = 6.1 \times 10^{-5}$ M

56. a. $K_b = 1.6 \times 10^{-5}$
    b. $K_b = 1.8 \times 10^{-5}$
    c. $K_b = 6.3 \times 10^{-5}$

57. $C_6H_5O^- > NH_3 > CN^-$

58. $pH = 9.26$

59. $pH = 13.0$

60. $[HCN] = 8.9 \times 10^{-4}$ M

61. a. basic
    b. neutral
    c. acidic

62. a. acidic
    b. basic
    c. acidic
    d. acidic

63. neutral (pH = 7.0)

64. a. $Co^{2+}$
    b. HCOOH
    c. AgCl

65. a.   Lewis acid = $SnCl_4$,   Lewis base = $Cl^-$
    b.   Lewis acid = $Hg^{+2}$,   Lewis base = $CN^-$
    c.   Lewis acid = $Co^{+3}$,   Lewis base = $NH_3$

66. a.   Lewis acid
    b.   Lewis base
    c.   Lewis acid

# CHAPTER 15

# Applications of Aqueous Equilibria

<table>
<tr><td>Section</td><td>Contents</td><td>Page in this<br>Study Guide</td></tr>
<tr><td>15.1</td><td>Acid-Base Equilibria</td><td>320</td></tr>
<tr><td>15.2</td><td>Buffered Solutions</td><td>322</td></tr>
<tr><td>15.3</td><td>Buffer Capacity</td><td>327</td></tr>
<tr><td>15.4</td><td>Titrations and pH Curves</td><td>329</td></tr>
<tr><td>15.5</td><td>Acid-Base Indicators</td><td>335</td></tr>
<tr><td>15.6</td><td>Solubility Equilibria</td><td>337</td></tr>
<tr><td>15.7</td><td>Precipitation and Qualitative Analysis</td><td>341</td></tr>
<tr><td>15.8</td><td>Complex Ion Equilibria</td><td>342</td></tr>
</table>

This chapter covers a great deal of material. The beauty is that **you already have the knowledge to handle the material.** Very little is "new." Most of the material represents **applications of equilibrium** theory. If you **think through** the various problems, you should do just fine.

## 15.1 - Acid-Base Equilibria

When you finish this section you will be able to perform calculations on acidic solutions that involve a common ion.

The important concept in this section is how the addition of a "common ion" to a solution can depress the dissociation of an acid. This is based on Le Chatelier's principle. To illustrate the point, acetic acid ($HC_2H_3O_2$) dissociates in water as follows:

$$HC_2H_3O_2(aq) \overset{K_a}{\rightleftharpoons} H^+(aq) + C_2H_3O_2^-(aq)$$

Let us add sodium acetate ($NaC_2H_3O_2$), which completely dissociates to form $Na^+(aq)$ and $C_2H_3O_2^-(aq)$. The acetate ion, $C_2H_3O_2^-$, is *common* to the solution (see the above equation). It is called the *common ion. The common ion imposes a stress on the system. The system responds by shifting **to the left**. Therefore, there is much less dissociation then before.* The general effect of a common ion is to depress the **amount of dissociation.**

*Because the dissociation is depressed, we can assume that the equilibrium concentration of acetic acid equals the initial concentration. The same is true of the acetate ion.*

Let's see how this applies to the pH of a weak acid.

## Example 15.1 - The Common Ion Effect

Calculate the pH, and the percent dissociation of the acid, in each of the following solutions.

- a.   0.200 M $HC_2H_3O_2$  ($K_a = 1.8 \times 10^{-5}$)
- b.   0.200 M $HC_2H_3O_2$ in the presence of 0.500 M $NaC_2H_3O_2$

## <u>Solution</u>

a.   The major species are acetic acid, $HC_2H_3O_2$, and $H_2O$.  As we saw in the previous chapter, $K_w$ for the autoprotolysis of water is insignificant compared to the $K_a$ for acetic acid dissociation.  We need only calculate based on the acid dissociation.

$$HC_2H_3O_2(aq) \rightleftharpoons H^+(aq) + C_2H_3O_2^-(aq)$$

$$K_a = \frac{[H^+][C_2H_3O_2^-]}{[HC_2H_3O_2]}$$

Using techniques from Chapter 14,

$$1.8 \times 10^{-5} = \frac{X^2}{0.200}$$

$$X = [H^+] = [C_2H_3O_2^-] = 1.9 \times 10^{-3} \text{ M} \quad \textbf{pH = 2.72}$$

$$\textbf{\% dissociation} = \frac{1.9 \times 10^{-3}}{0.200} \times 100\% = \textbf{0.95\%}$$

b.   The difference here is that the presence of the acetate ion, $C_2H_3O_2^-$, ($NaC_2H_3O_2$ is a strong electrolyte, as discussed previously) will *depress the dissociation of $HC_2H_3O_2$* as predicted by Le Chatelier's principle.

We can assume (and test) that because there is so little dissociation,

$$[HC_2H_3O_2]_o \approx [HC_2H_3O_2] = 0.200 \text{ M}$$

and

$$[C_2H_3O_2^-] \approx [C_2H_3O_2^-] = 0.500 \text{ M}$$

This makes our problem remarkably easy to solve.  Only $[H^+]$ is unknown,

$$K_a = \frac{[H^+][C_2H_3O_2^-]}{[HC_2H_3O_2]} \quad \Rightarrow \quad 1.8 \times 10^{-5} = \frac{[H^+](0.500)}{0.200}$$

$$[H^+] = 7.2 \times 10^{-6} \text{ M, pH = 5.14}$$

Note that the **pH increased markedly** due to the presence of the **basic salt**.  The extent of dissociation is now reflected by $[H^+]$, because the only significant source of $H^+$ ions is the dissociation of $HC_2H_3O_2$.

$$\textbf{\% dissociation} = \frac{7.2 \times 10^{-6}}{0.200} \times 100\% = \textbf{3.6} \times \textbf{10}^{-3}\textbf{\%}$$

Therefore, the dissociation was substantially depressed, and our assumption was very good.

Problems 1-5 at the end of this study chapter cover material from this section.

## 15.2 - Buffered Solutions

When you finish this section you will be able to:

- Calculate the pH of buffer solutions.
- Determine the effect of the addition of acids and bases on the pH of buffer solutions.

**A buffered solution resists change in its pH** upon the addition of a strong acid or strong base.

The last section set us up for doing buffer problems by introducing us to the common ion effect on equilibrium. We saw that the **extent of equilibrium is depressed** by the addition of a common ion.

A buffered solution generally contains a weak acid and its salt **or** a weak base and its salt.

Try the following buffer calculation and think about

1. the common ion effect, and
2. the dominant equilibrium.

## Example 15.2 A - pH of a Buffered Solution

Calculate the pH of a solution that contains 0.250 M formic acid, HCOOH ($K_a = 1.8 \times 10^{-4}$) and 0.100 M sodium formate NaCOOH.

## Strategy

The major species in solution are **HCOOH, Na$^+$, COOH$^-$ and H$_2$O.**

| | |
|---|---|
| **Na$^+$** | has no acid-base properties. |
| **H$_2$O** | has $K_w = 1.0 \times 10^{-14}$, and is thus weak relative to HCOOH. |
| **COOH$^-$** | has a $K_b = K_w/K_a = 5.6 \times 10^{-11}$. There will be very little hydrolysis of COOH$^-$. It will be depressed further by the common ion effect. |
| **HCOOH** | has $K_a = 1.8 \times 10^{-4}$. Although this *acid dissociation is depressed* by the presence of the common ion COOH$^-$, *it will be the dominant equilibrium.* |

$$HCOOH(aq) \rightleftharpoons H^+(aq) + COOH^-(aq)$$

| | initial(M) | change | final(M) |
|---|---|---|---|
| HCOOH | 0.250 | -X | 0.250 - X ($\approx 0.250$ M) |
| H$^+$ | 0 | +X | X |
| COOH$^-$ | 0.100 | +X | 0.100 + X ($\approx 0.100$ M) |

As shown in our previous chapter, it is a virtual certainty that we can ignore "X" relative to the initial concentrations of HCOOH and COOH$^-$. We will test this with the 5% rule.

Keep in mind that buffer problems are among the easiest to solve, because the **initial and final concentrations** of the acid and conjugate bases **do not change significantly**, and they are usually known.

## Solution

$$K_a = \frac{[H^+][COOH^-]}{[HCOOH]} \quad \Rightarrow \quad 1.8 \times 10^{-4} = \frac{[H^+]\,0.100}{0.250}$$

$$[H^+] = 4.5 \times 10^{-4}\ M, \quad pH = 3.35$$

The actual % dissociation of HCOOH, which is equal to $[H^+]$, is small enough to easily pass the 5% rule.

---

The real utility of buffers comes when we add a strong acid or base. A buffered solution should *resist pH change* when a substantial amount of $H^+$ or $OH^-$ ions are added. ("Substantial amount" will vary from solution to solution, depending upon the "buffer capacity." We will consider that in section 15.3.)

If we add a strong base to a solution that contains a weak acid, the reaction is complete.

$$HA(aq) + OH^-(aq) \longrightarrow A^-(aq) + H_2O$$

Similarly, adding a strong acid to a weak base gives a complete reaction,

$$A^-(aq) + H^+(aq) \longrightarrow HA(aq)$$

Notice that in these equations, unless **excess** strong acid or base is added, **NO STRONG ACID OR BASE BUILDS UP**. Only weak species remain. This is why a buffer works so well. Let's use this idea in our next problem.

---

# Example 15.2 B - Addition of a Strong Base to a Buffer

Calculate the pH of the 0.250 M HCOOH/0.100 M NaCOOH buffer used in the last problem after the addition of 10.0 mL of 6.00 M NaOH to the **original buffered solution volume of 500 mL**.

## Strategy

As your textbook points out there are really two parts to this problem. The first is to *determine the moles* of weak acid and base existing *after the addition of the strong base ("stoichiometry part")*. The second part is to *calculate the pH of the buffered solution* after you have your new equilibrium concentrations *("equilibrium part")*.

## Solution

a.  **Stoichiometry Part**

We need to work in moles (or mmoles).

$$\text{mmoles HCOOH}_{initial} = 0.250\ \text{mmol/mL} \times 500\ \text{mL} = \textbf{125 mmol}$$
$$\text{mmoles COOH}^-_{initial} = 0.100\ \text{mmol/mL} \times 500\ \text{mL} = \textbf{50.0 mmol}$$
$$\text{mmol OH}^-\ \textbf{added} = 6.00\ \text{mmol/mL} \times 10.0\ \text{mL} = \textbf{60.0 mmol}$$

|  | HCOOH(aq) + | OH⁻(aq) ----> | COOH⁻(aq) |
|---|---|---|---|
| initial (mmol) | 125 | 60 | 50 |
| final (mmol) | 65 | ≈0 | 110 |

After addition of NaOH,

$$[HCOOH] = 65.0 \text{ mmol}/510 \text{ mL} = 0.128 \text{ M}$$

$$\uparrow \qquad \qquad \nwarrow$$

(500 + 10)                    equilibrium concentrations

$$\swarrow$$

$$[COOH^-] = 110 \text{ mmol}/510 \text{ mL} = 0.216 \text{ M}$$

b.  **Equilibrium Part**

As in the previous problem, the acid dissociation of HCOOH is the significant factor here.  As before,

$$K_a = \frac{[H^+][COOH^-]}{[HCOOH]} \qquad \Rightarrow \qquad 1.8 \times 10^{-4} = \frac{[H^+] \, 0.216}{0.128}$$

$$[H^+] = 1.06 \times 10^{-4}, \quad pH = 3.97 \text{ (The 5\% rule easily holds.)}$$

Comparing this result to that in the previous problem, we note that the *pH has gone from 3.35 to 3.97* after addition of a substantial quantity of OH⁻.  The *pH change of 0.62 units* indicates that our buffer has neutralized the base well.  If we just added *60 mmoles of OH⁻ to 500 mL of water*, [OH⁻] would equal 0.120 M, pOH would equal 0.92, and *pH would equal 13.08*.  The presence of the buffered solution allowed us to maintain an acid pH in spite of the strong base that was added.

---

# Example 15.2 C - Practice with Buffers

A solution is prepared by adding 31.56 g NaCN and 22.30 g HCN to 600.0 mL of water ($K_a$ for HCN = $6.2 \times 10^{-10}$).

a.  What is the pH of this solution?
b.  What is the pH after the addition of 50.0 mL of 3.00 M HCl?
c.  What is the pH after a **further** addition of 80.0 mL of 4.00 M NaOH?

## Strategy

We have the following major species in solution:  **Na⁺, H₂O, HCN, and CN⁻.**

HCN    is a weak acid, $K_a = 6.2 \times 10^{-10}$.

CN⁻,   its conjugate base, has $K_b = K_w/K_a = 1.0 \times 10^{-14} / 6.2 \times 10^{-10} = 1.6 \times 10^{-5}$

H₂O,   with its low $K_w$, does not affect the acid-base characteristics of the system.

Because **CN⁻ is the strongest acid-base substance in our solution**, its behavior will dominate.  The base hydrolysis reaction is

$$CN^-(aq) + H_2O(l) \rightleftharpoons HCN(aq) + OH^-(aq)$$

and

$$K_b = \frac{[HCN][OH^-]}{[CN^-]}$$

As always in problems involving buffered solutions, we can make our usual assumption regarding the lack of reaction of CN⁻ due to the presence of the common species CN⁻ (in the form of HCN).

After calculating the pH of this solution (part a), *we need to assess the pH after addition of HCl*, a strong acid, for *part b*.  In general terms, *the pH should go down, but only a small amount*, because we have a buffer.  The exception to this is if we *exceed* the buffer capacity. We must use our *two part approach*, taking into account the initial, complete reaction of CN⁻ with H⁺ (from HCl).

$$CN^-(aq) + H^+(aq) \longrightarrow HCN(aq)$$

As long as we have some CN⁻ ion left, we will maintain our buffered solution. We may then use our *equilibrium expression* to assess pH (via pOH).

In *part c, we must consider the addition of a strong base, OH⁻ ion, to the system*. It will react completely with our weak acid, HCN, to form CN⁻.

$$HCN(aq) + OH^-(aq) \longrightarrow CN^-(aq) + H_2O(l)$$

As long as we have some HCN left, we will maintain a buffered solution. *The pH of the solution should rise, but not much*, if we maintain our buffer. We may use our *equilibrium expression* to determine the pOH, and then pH.

## Solution

a.  Let's calculate [HCN] and [CN⁻].

$$[HCN] = \frac{\text{mol HCN}}{L} = \frac{1 \text{ mol}}{27.0 \text{ g}} \times \frac{22.30 \text{ g}}{0.600 \text{ L}} = \mathbf{1.37 \text{ M}}$$

$$[CN^-] = \frac{\text{mol NaCN}}{L} = \frac{1 \text{ mol}}{49.0 \text{ g}} \times \frac{31.56 \text{ g}}{0.600 \text{ L}} = \mathbf{1.07 \text{ M}}$$

Using the equilibrium expression for the buffered solution, as discussed above,

$$K_b = \frac{[HCN][OH^-]}{[CN^-]} \quad \Rightarrow \quad 1.6 \times 10^{-5} = \frac{1.37 \, [OH^-]}{1.07}$$

$$\mathbf{[OH^-] = 1.25 \times 10^{-5}, \quad pOH = 4.90, \quad pH = 9.10}$$

b.  **Stoichiometry Part**

$$\textbf{mmoles HCN}_{\text{initial}} = 1.37 \text{ mmol/mL} \times 600 \text{ mL} = \mathbf{822 \text{ mmol}}$$
$$\textbf{mmoles CN}^-_{\text{initial}} = 1.07 \text{ mmol/mL} \times 600 \text{ mL} = \mathbf{642 \text{ mmol}}$$
$$\textbf{mmoles H}^+ \textbf{ added} = 3.00 \text{ mmol/mL} \times 50.0 \text{ mL} = \mathbf{150 \text{ mmol}}$$

|  | CN⁻(aq) + | H⁺(aq) ----> | HCN(aq) |
|---|---|---|---|
| initial (mmol) | 642 | 150 | 822 |
| final (mmol) | 492 | ≈0 | 972 |

$$\mathbf{[CN^-]} = 492 \text{ mmol}/(600 + 50) \text{ mL} = \mathbf{0.757 \text{ M}}$$
$$\mathbf{[HCN]} = 972 \text{ mmol}/650 \text{ mL} = \mathbf{1.50 \text{ M}}$$

**Equilibrium Part** (Use the **equilibrium** expression)

$$K_b = \frac{[HCN][OH^-]}{[CN^-]} \quad \Rightarrow \quad 1.6 \times 10^{-5} = \frac{1.50 \, [OH^-]}{0.757}$$

$$\mathbf{[OH^-] = 8.1 \times 10^{-6} \text{ M}, \quad pOH = 5.09, \quad pH = 8.91}$$

c.  **Stoichiometry Part** (from part b)

$$\text{mmol HCN}_{\text{initial}} = 972 \text{ mmol}$$
$$\text{mmol CN}^-_{\text{initial}} = 492 \text{ mmol}$$

**We are adding** $80.0$ mL $\times$ $4.00$ mmol $H^+$/mL = **320 mmol OH⁻**. This will react with HCN, as discussed in our strategy section.

$$HCN(aq) + OH^-(aq) \longrightarrow CN^- + H_2O(l)$$

|  | 972 | 320 | 492 |
|---|---|---|---|
| initial (mmol) | 972 | 320 | 492 |
| final (mmol) | 652 | $\approx 0$ | 812 |

$$[HCN] = 652 \text{ mmol}/(650 + 80) \text{ mL} = \mathbf{0.893 \text{ M}}$$
$$[CN^-] = 812 \text{ mmol}/730 \text{ mL} = \mathbf{1.11 \text{ M}}$$

**Equilibrium Part** (use the **equilibrium** expression)

$$K_b = \frac{[HCN][OH^-]}{[CN^-]} \quad \Rightarrow \quad 1.6 \times 10^{-5} = \frac{0.893 \,[OH^-]}{1.11}$$

$$[OH^-] = 2.0 \times 10^{-5} \text{ M}, \quad pOH = 4.70, \quad pH = 9.30$$

## Does The Answer Make Sense?

The initial pH was basic (9.10). This makes sense because $CN^-$ is a stronger base than HCN is an acid.

The pH dropped (8.91) after addition of a strong acid. This makes sense.

The pH rose (9.30) after the subsequent addition of a strong base. This makes sense. In neither part b nor part c did we exceed our buffer capacity, so the pH changes were not drastic.

---

## Example 15.2 D - pH and pK$_a$

The $K_a$ of propionic acid, $HC_3H_5O_2$, is $1.34 \times 10^{-5}$ ($pK_a = 4.87$). What is the pH when $[HC_3H_5O_2] = [C_3H_5O_2^-]$?

## Solution

Let's look at the problem from a strictly mathematical point of view. If we use the equation for the acid dissociation of propionic acid,

$$HC_3H_5O_2(aq) \rightleftharpoons C_3H_5O_2^-(aq) + H^+(aq)$$

$$K_a = \frac{[C_3H_5O_2^-][H^+]}{[HC_3H_5O_2]}$$

If $[HC_3H_5O_2] = [C_3H_5O_2^-]$, $K_a = [H^+] = 1.34 \times 10^{-5}$ M and $pK_a = pH$, therefore **pH = 4.87**.

If we use the equation for the base hydrolysis of the propionate ion, $C_3H_5O_2^-$,

$$C_3H_5O_2^-(aq) + H_2O(l) \rightleftharpoons HC_3H_5O_2(aq) + OH^-(aq)$$

$$K_b = \frac{[HC_3H_5O_2][OH^-]}{[C_3H_5O_2^-]}$$

If $[C_3H_5O_2^-] = [HC_3H_5O_2]$,

$$[OH^-] = K_b = K_w/K_a = 1.0 \times 10^{-14} / 1.34 \times 10^{-5} = \mathbf{7.46 \times 10^{-10}}$$

$$pOH = 9.13, \quad \mathbf{pH = 4.87}$$

Although we would normally use the $K_a$ expression to solve for pH in this situation, two important points need to be made from this problem:

1.  **If $[HA] = [A^-]$, then $pH = pK_a$ and $pOH = pK_b$.**
2.  In buffer problems, **uniquely, either the $K_a$ or $K_b$ expressions can be used**, and the same pH will result. This is because **BOTH $[HA]$ and $[A^-]$ are explicitly known**.

---

# Problems 6-19 at the end of this study chapter cover material from this section.

---

## 15.3 - Buffer Capacity

When you finish this section you will be able to:

- Calculate the pH of a solution where the buffer capacity has been exceeded.
- Choose among alternatives, the best buffer system for a given pH.

Your textbook emphasizes the two important measures involved in buffered solutions.

1.  The pH is determined by the **ratio** of $[HA]/[A^-]$.
2.  The **buffer capacity** is determined by the **magnitudes** of $[HA]$ and $[A^-]$. The more you have, the more strong acid or base that can be neutralized.

Sample Exercise 15.7 in your textbook compares the pH change brought about by adding a strong acid to solutions with different buffering capacities. Let's try a problem where we **exceed** the buffering capacity of a solution.

---

## Example 15.3 A - Buffering Capacity

Calculate the pH of a 0.500 L solution that contains 0.15 M HCOOH ($K_a = 1.8 \times 10^{-4}$) and 0.20 M NaCOOH. Then, calculate the pH of the solution after the addition of 10.0 mL of 12.0 M NaOH.

### Solution

We can solve the problems using the same strategy as in the previous sections. The major species are **HCOOH, $Na^+$, $COOH^-$ and $H_2O$**. The dominant equilibrium will involve the acid dissociation of HCOOH.

$$HCOOH(aq) \;\rightleftharpoons\; H^+(aq) + COOH^-(aq)$$

$$K_a = \frac{[H^+][COOH^-]}{[HCOOH]} \quad \Rightarrow \quad 1.8 \times 10^{-4} = \frac{[H^+]\,0.20}{0.15}$$

$$[H^+] = 1.4 \times 10^{-4}\,M, \quad pH = 3.87$$

When we add the strong base, we have the reaction between $OH^-$ ions and HCOOH,

$$HCOOH(aq) + OH^-(aq) \longrightarrow COOH^- + H_2O(l)$$

We must proceed with our two step approach.

### 1. Stoichiometry Part

$$\text{mmoles HCOOH}_{\text{initial}} = 0.15 \text{ mmol/mL} \times 500 \text{ mL} = \textbf{75 mmol}$$
$$\text{mmoles COOH}^-_{\text{initial}} = 0.20 \text{ mmol/mL} \times 500 \text{ mL} = \textbf{100 mmol}$$
$$\text{mmoles OH}^- \text{ added} = 12.0 \text{ mmol/mL} \times 10.0 \text{ mL} = \textbf{120 mmol}$$

We have more $OH^-$ than is needed to neutralize the HCOOH. HCOOH is the **limiting reactant** in this solution.

|  | $HCOOH(aq)$ + | $OH^-(aq)$ ----> | $COOH^-(aq)$ + $H_2O(l)$ |
|---|---|---|---|
| initial (mmol) | 75 | 120 | 100 |
| final (mmol) | ≈0 | 45 | 175 |

We have 45 mmoles of excess $OH^-$! There is no **"Equilibrium part"** in this problem! $OH^-$ ion is a far stronger base than $COOH^-$, and will thus determine the pH.

$$[OH^-] = 45 \text{ mmol}/(500 + 10) \text{ mL} = \textbf{0.088 M}$$
$$pOH = 1.05 \quad \textbf{pH 12.95}$$

When we *exceeded the buffer capacity* of the solution, *the pH changed drastically*. If we would have doubled the amounts of formic acid and sodium formate we would not have exceeded the buffer capacity of the solution.

---

All other things being equal, the closer the $[HA]/[A^-]$ ratio is to 1, the better will be the buffer capacity of the solution. This will happen when $[H^+] = K_a$, or $pH = pK_a$. This means that when selecting a buffer that will be used to maintain the pH of a solution in a particular small range,

**The pK$_a$ of the buffer should be as close as possible to the desired pH of the solution.**

---

## Example 15.3 B - Selecting a Buffer

We wish to buffer a solution at pH = 10.07. Which one of the following bases (and conjugate acid salts) would be most useful?

a.  $NH_3$ ($K_b = 1.8 \times 10^{-5}$)
b.  $C_6H_5NH_2$ ($K_b = 4.2 \times 10^{-10}$)
c.  $N_2H_4$ ($K_b = 9.6 \times 10^{-7}$)

## Solution

$$K_a \text{ for } NH_4^+ = K_w/K_b = 1.0 \times 10^{-14} / 1.8 \times 10^{-5} = 5.6 \times 10^{-10}$$
$$\textbf{pK}_a \textbf{ (NH}_4\textbf{) = 9.26}$$

$$K_a \text{ for } C_6H_5NH_3^+ = K_w/K_b = 1.0 \times 10^{-14} / 4.2 \times 10^{-10} = 2.4 \times 10^{-5}$$
$$\textbf{pK}_a \textbf{ (C}_6\textbf{H}_5\textbf{NH}_3^+\textbf{) = 4.62}$$

$$K_a \text{ for } N_2H_5^+ = K_w/K_b = 1.0 \times 10^{-14} / 9.6 \times 10^{-7} = 1.0 \times 10^{-8}$$
$$\textbf{pK}_a \textbf{ (N}_2\textbf{H}_5^+\textbf{) = 7.98}$$

$NH_3$ would be the best choice for a buffer system. (Please keep in mind that a good deal of thought must not only go into the mathematics, but the **chemistry as well**. That is, how will $NH_3$ affect your experiment? So many things to think about!)

Problems 20-21 at the end of this study chapter cover material from this section.

## 15.4 - Titrations and pH Curves

When you finish this section you will be able to calculate the pH at any point along a curve for the following titrations:

- strong acid - strong base
- weak acid - strong base
- weak base - strong acid

As a precursor to the calculations in this chapter, you should be able to define the following terms: titrant, buret, indicator, titration curve, and endpoint.

In order for a titration to be feasible, it must be **complete** and **fast**. In order to be complete, it should have a value of $K \geq 10^7$ or so.

### Strong Acid - Strong Base

You may wish to go through the "CASE-STUDY" near the start of this section in your textbook before attacking the following problem.

Let's consider the titration of a **HCl** solution with **NaOH**. We say that the process is complete, however, we can calculate an equilibrium constant for the reaction.

$$H^+(aq) + OH^-(aq) \rightleftharpoons H_2O(l)$$

The reaction is the reverse of the autoprotolysis of water, therefore

$$K = 1/K_w = 1 \times 10^{14}$$

This reaction is certainly complete! So much so, in fact, that this type of titration is less an equilibrium problem than a stoichiometry one.

## Example 15.4 A - Strong Acid - Strong Base Titration

Calculate the pH after the following **total** volumes of **0.2500 M HCl** have been **added** to **50.00 mL of 0.1500 M NaOH**.

a. 0.00 mL  c. 29.50 mL  e. 30.50 mL
b. 4.00 mL  d. 30.00 mL  f. 40.00 mL

### Solution

a. **0.00 mL:** We have a 0.1500 M NaOH solution. Because it completely dissociates, and OH⁻ ion is a strong base,

$$[OH^-] = 0.1500 \text{ M}, \quad pOH = 0.82, \quad \textbf{pH = 13.18}$$

b. **4.00 mL total:** As discussed previously, the reaction between H⁺ and OH⁻ is complete. Calculating moles of each reactant is always the preferred way to begin.

$$\text{mmol OH}^-_{\text{initial}} = 0.1500 \text{ mmol/mL} \times 50.00 \text{ mL} = \textbf{7.500 mmol}$$
$$\text{mmol H}^+ \textbf{ added} = 0.2500 \text{ mmol/mL} \times 4.00 \text{ mL} = \textbf{1.000 mmol}$$

|  | $H^+(aq)$ | $+ OH^-(aq)$ | $----> H_2O(l)$ |
|---|---|---|---|
| initial (mmol) | 1.000 | 7.500 | excess |
| final (mmol) | $\approx 0$ | 6.500 | excess |

The relatively small amount of $H^+$ ions added were neutralized by $OH^-$ ions. We have 6.500 mmol of $OH^-$ ion excess, which determines the pH.

$$[OH^-] = 6.500 \text{ mmol}/(50.00 + 4.00) \text{ mL} = \textbf{0.1203 M}$$
$$\text{pOH} = 0.92, \quad \textbf{pH} = \textbf{13.08}$$

The pH has not declined much, because we still have excess of $OH^-$ ion.

c.  **29.50 mL total**: Proceeding as above,

$$\text{mmol OH}^-_{\text{initial}} = \textbf{7.500 mmol}$$
$$\text{mmol H}^+ \textbf{ added} = 0.2500 \text{ mmol/mL} \times 29.50 \text{ mL} = \textbf{7.375 mmol}$$

|  | $H^+(aq)$ | $+ OH^-(aq)$ | $----> H_2O(l)$ |
|---|---|---|---|
| initial (mmol) | 7.375 | 7.500 | excess |
| final (mmol) | $\approx 0$ | 0.125 | excess |

$$[OH^-] = 0.125 \text{ mmol}/(50.00 + 29.50) \text{ mL} = \textbf{1.572} \times \textbf{10}^{-3} \textbf{ M}$$
$$\text{pOH} = 2.80, \quad \textbf{pH} = \textbf{11.20}$$

The pH has begun to come down somewhat, but the solution is still quite basic.

d.  **30.00 mL total**: As above,

$$\text{mmol OH}^-_{\text{initial}} = \textbf{7.500 mmol}$$
$$\text{mmol H}^+ \textbf{ added} = 0.2500 \text{ mmol/mL} \times 30.00 \text{ mL} = \textbf{7.500 mmol}$$

|  | $H^+(aq)$ | $+ OH^-(aq)$ | $----> H_2O(l)$ |
|---|---|---|---|
| initial (mmol) | 7.500 | 7.500 | excess |
| final (mmol) | $\approx 0$ | $\approx 0$ | excess |

We have added **exactly** enough $H^+$ ions to neutralize all of the $OH^-$ ions. We have only water, $Na^+$ and $Cl^-$ in the solution. Of these, only **water** has any acid-base properties. This neutral water solution, as always, has $K_w = 1.0 \times 10^{-14}$. The **pH = 7.00.**

**CRITICAL POINT: THE ONLY CASE IN WHICH THE pH AT THE EQUIVALENCE POINT WILL BE EQUAL TO 7.00 WILL BE THAT OF A STRONG ACID - STRONG BASE TITRATION!!!**

Also notice the rather sharp pH change (4.20 units) within 0.500 mL of the equivalence point. This is typical of a strong acid-strong base titration.

e.  **30.50 mL total**: Proceeding as always,

|  | $H^+(aq)$ | $+ OH^-(aq)$ | $----> H_2O(l)$ |
|---|---|---|---|
| initial (mmol) | 7.625 | 7.500 | excess |
| final (mmol) | 0.125 | $\approx 0$ | excess |

We have **passed** the equivalence point (this is called the "post-equivalence point region"). We have excess $H^+$ ion. The solution is now acidic.

$$[H^+] = 0.125 \text{ mmol}/80.50 \text{ mL} = \mathbf{1.55 \times 10^{-3} \text{ M}}$$
$$\mathbf{pH = 2.81}$$

Being just 0.500 mL beyond the equivalence point caused a sharp drop in the pH. Generally, the stronger the species being titrated, the sharper the pH drop.

f.  **40.00 mL total:** We are now well past the equivalence point.

$$H^+(aq) + OH^-(aq) \text{ ----> } H_2O(l)$$

|  | $H^+$ | $OH^-$ | $H_2O$ |
|---|---|---|---|
| initial (mmol) | 10.00 | 7.500 | excess |
| final (mmol) | 2.50 | $\approx 0$ | excess |

$$[H^+] = 2.50 \text{ mmol}/90.00 \text{ mL} = \mathbf{0.0278 \text{ M}}$$
$$\mathbf{pH = 1.56}$$

If you plot the mL of HCl added vs. pH for these data, your plot should look like that in <u>Figure 15.2 in your textbook.</u>

## Weak Acid - Strong Base

When doing calculations for this type of titration it is useful to relate our thinking back to our sections on **buffered solutions (15.2 and 15.3).**

$$HA(aq) + OH^-(aq) \text{ ----> } A^-(aq) + H_2O(l)$$

You will be generating a buffer (weak acid and conjugate base) as a result of the addition of $OH^-$ ions to the solution. You will have a buffered solution until you exceed the buffer capacity of the solution, and then the pH will rise dramatically.

Let's use the titration of formic acid, HCOOH ($K_a = 1.8 \times 10^{-4}$) with NaOH as an illustrative example. **The equilibrium constant for the titration can be calculated.** The reaction

$$HCOOH(aq) + OH^-(aq) \rightleftharpoons COOH^-(aq) + H_2O(l)$$

can be expressed as a **combination** of

1.  $HCOOH(aq) \rightleftharpoons H^+(aq) + COOH^-(aq)$      $K_a = 1.8 \times 10^{-4}$

and

2.  $H^+(aq) + OH^-(aq) \rightleftharpoons H_2O(l)$        $1\backslash K_w = 1.0 \times 10^{14}$

K for the overall reaction = $K_1 \times K_2 = K_a \times 1/K_w = K_a/K_w$

$$\mathbf{K} = 1.8 \times 10^{-4} / 1.0 \times 10^{-14} = \mathbf{1.8 \times 10^{10}}$$

This is why we say that the titration goes completely.

---

# Example 15.4 B - Weak Acid - Strong Base Titration

Calculate the pH after the following **total** volumes of 0.4000 M NaOH are added to 50.00 mL of 0.2000 M HCOOH.

| | | | | | | | |
|---|---|---|---|---|---|---|---|
| a. | 0.00 mL | c. | 12.50 mL | e. | 25.00 mL | g. | 40.00 mL |
| b. | 5.00 mL | d. | 24.50 mL | f. | 25.50 mL | | |

## Solution

a. **0.00 mL:** We have a weak acid solution, 0.2000 M HCOOH. We have done these kinds of problems many times, and will present a **shortened version here**. The equilibrium of interest is

$$HCOOH(aq) \rightleftharpoons H^+(aq) + COOH^-(aq)$$

$$K_a = \frac{[H^+][COOH^-]}{[HCOOH]} \quad \Rightarrow \quad 1.8 \times 10^{-4} = \frac{X^2}{0.2000}$$

$$[H^+] = 6.0 \times 10^{-3}, \quad \textbf{pH = 2.22}$$

In titration problems, if the titration is feasible (large K), our "5% test" will virtually always hold true.

b. **5.00 mL:** As pointed out previously, the titration reaction of interest is

$$\textbf{HCOOH(aq) + OH}^-\textbf{(aq)} \rightleftharpoons \textbf{COOH}^-\textbf{(aq) + H}_2\textbf{O(l)}$$

The reaction is essentially complete ($K = 1.8 \times 10^{10}$). After this reaction is completed, we will have a **buffered solution** containing HCOOH and COOH$^-$ ion. You must then calculate the pH of the buffered solution.

**Stoichiometry Part:**

$$\textbf{mmoles HCOOH}_{initial} = 0.2000 \text{ mmol/mL} \times 50.00 \text{ mL} = \textbf{10.00 mmol}$$
$$\textbf{mmol OH}^- \textbf{ added} = 0.4000 \text{ mmol/mL} \times 5.00 \text{ mL} = \textbf{2.000 mmol}$$

|  | HCOOH(aq) | + OH$^-$(aq) | ----> COOH$^-$(aq) | + H$_2$O(l) |
|---|---|---|---|---|
| initial (mmol) | 10.00 | 2.000 | ≈0 | excess |
| final (mmol) | 8.00 | ≈0 | 2.00 | excess |

**Equilibrium Part:**

Formic acid is a stronger acid than formate ion is a base. Therefore, we can use the acid dissociation equilibrium to solve the problem (but in this so-called "**buffer region**," the base hydrolysis would work as well). (See section 15.2 if you need a review of buffered solutions.)

$$[HCOOH] = 8.00 \text{ mmol}/(50.00 + 5.00) \text{ mL} = \textbf{0.145 M}$$
$$[COOH^-] = 2.00 \text{ mmol}/55.00 \text{ mL} = \textbf{0.0364 M}$$

$$HCOOH(aq) \rightleftharpoons H^+(aq) + COOH^-(aq)$$

$$K_a = \frac{[H^+][COOH^-]}{[HCOOH]} \quad \Rightarrow \quad 1.8 \times 10^{-4} = \frac{[H^+] \, 0.0364}{0.145}$$

$$[H^+] = 7.2 \times 10^{-4}, \quad \textbf{pH = 3.14}$$

The pH does not rise much in the buffer region of a titration.

c. **12.50 mL:** We shall proceed as in part b.

**Stoichiometry Part:**

$$\textbf{mmoles HCOOH}_{initial} = \textbf{10.00 mmol}$$
$$\textbf{mmol OH}^- \textbf{ added} = 0.4000 \text{ mmol/mL} \times 12.50 \text{ mL} = \textbf{5.000 mmol}$$

|  | HCOOH(aq) | + OH$^-$(aq) | ----> COOH$^-$(aq) | + H$_2$O(l) |
|---|---|---|---|---|
| initial (mmol) | 10.00 | 5.000 | ≈0 | excess |
| final (mmol) | 5.00 | ≈0 | 5.00 | excess |

**Equilibrium Part:**

$$[HCOOH] = 5.00 \text{ mmol}/(50.00 + 12.50) \text{ mL} = \mathbf{0.0800 \ M}$$
$$[COOH^-] = 5.00 \text{ mmol}/62.50 \text{ mL} = \mathbf{0.0800 \ M}$$

$$K_a = \frac{[H^+][COOH^-]}{[HCOOH]} \quad \Rightarrow \quad 1.8 \times 10^{-4} = \frac{[H^+] \ 0.0800}{0.0800}$$

$$[H^+] = K_a = 1.8 \times 10^{-4}, \quad \mathbf{pH = 3.74}$$

Recall from section 15.3 that when [HA] = [A⁻], $pK_a$ = pH. This is the situation at this point in our titration. This is called the **titration midpoint**. We are **halfway** to the equivalence point.

The titration midpoint is an especially important experimental point because if we know pH, we can find $pK_a$ for the weak acid.

d.  **24.50 mL:** Proceeding as always,

$$\text{mmoles HCOOH}_{initial} = \mathbf{10.00 \ mmol}$$
$$\text{mmol OH}^- \text{ added} = 0.4000 \text{ mmol/mL} \times 24.50 \text{ mL} = \mathbf{9.80 \ mmol}$$

| | HCOOH(aq) + | OH⁻(aq) ----> | COOH⁻(aq) + | H₂O(l) |
|---|---|---|---|---|
| initial (mmol) | 10.00 | 9.80 | ≈0 | excess |
| final (mmol) | 0.20 | ≈0 | 9.80 | excess |

We are near the limit of our buffer capacity.

$$[HCOOH] = 0.20 \text{ mmol}/(50.00 + 24.50) \text{ mL} = \mathbf{2.67 \times 10^{-3} \ M}$$
$$[COOH^-] = 9.80 \text{ mmol}/74.50 \text{ mL} = \mathbf{0.132 \ M}$$

$$K_a = \frac{[H^+][COOH^-]}{[HCOOH]} \quad \Rightarrow \quad 1.8 \times 10^{-4} = \frac{[H^+] \ 0.132}{2.67 \times 10^{-3}}$$

$$[H^+] = 3.6 \times 10^{-6}, \quad \mathbf{pH = 5.44}$$

e.  **25.00 mL:**

$$\text{mmoles HCOOH}_{initial} = \mathbf{10.00 \ mmol}$$
$$\text{mmol OH}^- \text{ added} = \mathbf{10.00 \ mmol}$$

| | HCOOH(aq) + | OH⁻(aq) ----> | COOH⁻(aq) + | H₂O(l) |
|---|---|---|---|---|
| initial (mmol) | 10.00 | 10.00 | ≈0 | excess |
| final (mmol) | ≈0 | ≈0 | 10.00 | excess |

**We have reached the equivalence point.** We now have a solution that contains a **base** that can undergo hydrolysis,

$$COOH^-(aq) + H_2O(l) \rightleftharpoons HCOOH(aq) + OH^-(aq)$$

$$K_b = K_w/K_a = \frac{1.0 \times 10^{-14}}{1.8 \times 10^{-4}} = \mathbf{5.6 \times 10^{-11}} = \frac{[HCOOH][OH^-]}{[COOH^-]}$$

$$[COOH^-] = 10.00 \text{ mmol}/(50.00 + 25.00) \text{ mL} = \mathbf{0.133 \ M}$$

Solving this weak base problem in the usual way,

$$5.6 \times 10^{-11} = \frac{X^2}{0.133} \quad \Rightarrow \quad X = [OH^-] = \mathbf{2.73 \times 10^{-6} \ M}$$

$$pOH = 5.56, \quad pH = 8.44$$

We have a mildly basic solution, and that is reflected in the pH value.

f.   **25.50 mL:**

$$\textbf{mmoles HCOOH}_{\textbf{initial}} = \textbf{10.00 mmol}$$
$$\textbf{mmol OH}^- \textbf{ added} = \textbf{10.20 mmol}$$

We have neutralized all of the weak acid and have **0.20 mmol of strong base remaining.** This is the **post-equivalence point region.** The pH will be determined by the excess strong base.

$$[OH^-] = 0.20 \text{ mmol}/(50.00 + 25.50) \text{ mL} = \textbf{2.65} \times \textbf{10}^{\textbf{-3}} \textbf{ M}$$
$$pOH = 2.58, \quad \textbf{pH} = \textbf{11.42}$$

Notice that the pH has risen sharply now that we have **surpassed the buffer capacity** of our formic acid/ formate buffered solution.

g.   **40.00 mL:**  You should be able to show that we have **6.00 mmol of excess OH$^-$ ions** and

$$[OH^-] = 6.00 \text{ mmol}/90.00 \text{ mL} = 0.0667 \text{ M}$$
$$pOH = 1.18, \quad \textbf{pH} = \textbf{12.82}$$

Summing up, note that our "pH break" around the equivalence point region was somewhat less sharp than with the strong acid - strong base titration. The smaller the titration equilibrium constant, the less sharp the break at the equivalence point. If you plot a titration curve, it should look similar to <u>Figure 15.3 in your textbook.</u>

---

## <u>Weak Base - Strong Acid</u>

These kinds of titrations are handled using the same strategies as those involving weak acids and strong bases. We have the **buffer region, equivalence point, and post-equivalence point,** as in previous titrations. Think about the dominant reactions in each part of the problem, be careful regarding math, and **check to make sure that your answer makes sense.**

---

## Example 15.4 C - Weak Base - Strong Acid Titration

Calculate the pH at each of the following points in the titration of **50.00 mL of a 0.01000 M** sodium phenolate ($NaOC_6H_5$) solution with **1.000 M HCl solution** ($K_a$ for $HOC_6H_5 = 1.05 \times 10^{-10}$).

a.   initially      b.   midpoint      c.   equivalence point

## <u>Solution</u>

Sodium phenolate is a strong electrolyte. It will completely dissociate to give $Na^+$ and $OC_6H_5^-$.

$$\textbf{K}_\textbf{b} \textbf{ for OC}_6\textbf{H}_5^- = \textbf{K}_\textbf{w}/\textbf{K}_\textbf{a} = \textbf{1.0} \times \textbf{10}^{\textbf{-14}} / \textbf{1.05} \times \textbf{10}^{\textbf{-10}} = \textbf{9.5} \times \textbf{10}^{\textbf{-5}}$$

a.   This weak base undergoes hydrolysis in the usual way,

$$OC_6H_5^-(aq) + H_2O(l) \; \rightleftharpoons \; HOC_6H_5(aq) + OH^-(aq)$$

$$K_b = \frac{[HOC_6H_5][OH^-]}{[OC_6H_5^-]} \quad \Rightarrow \quad 9.5 \times 10^{-5} = \frac{X^2}{0.01000}$$

$$X = [OH^-] = 9.8 \times 10^{-4} \text{ M}, \quad pOH = 3.01, \quad \textbf{pH = 10.99}$$

b.  By definition, the pH at the titration midpoint equals $pK_a$ for the acid. Therefore, **pOH = $pK_b$ for the base.** For our solution,

$$pK_b = -\log K_b = -\log(9.5 \times 10^{-5}) = \textbf{4.02}$$
$$pOH = 4.02, \quad \textbf{pH = 9.98}$$

c.  All of our base has been converted to acid at the equivalence point. Although we know how to find the pH of a very weak acid, *we need to know its concentration* at the equivalence point. After all, a small *dilution* occurred as a result of adding HCl. A key question then is, *"How many mL of HCl were required to reach the equivalence point?"*

$$\textbf{mmoles OC}_6\textbf{H}_5^-{}_{\textbf{initial}} = 0.01000 \text{ mmol/mL} \times 50.00 \text{ mL} = \textbf{0.5000 mmol}$$

$$\textbf{mL HCl required to just neutralize} = 0.5000 \text{ mmol} \times \frac{1 \text{ mL}}{1.000 \text{ mmol}} = \textbf{0.5000 mL}$$

Our **total volume** is now 50.00 + 0.5000 = **50.50 mL**

$$[HOC_6H_5] = 0.5000 \text{ mmol/50.50 mL} = \textbf{9.90} \times \textbf{10}^{-3} \textbf{ M}$$

To solve for the pH of this weak acid,

$$HOC_6H_5(aq) \rightleftharpoons H^+(aq) + OC_6H_5^-(aq)$$

$$K_a = \frac{[H^+][OC_6H_5^-]}{[HOC_6H_5]} \quad \Rightarrow \quad 1.05 \times 10^{-10} = \frac{X^2}{9.90 \times 10^{-3}}$$

$$[H^+] = 1.02 \times 10^{-6}, \quad \textbf{pH = 5.99}$$

Notice that the pH indicates that we have a dilute acidic solution. In the sense that our pH started high and ended low, our answers make sense. However, the initial solution was so dilute that our pH did not change a great deal. As such, this system would be considered marginal for a titration.

---

Problems 22-34, 36-38, and 42-43 at the end of this study chapter cover material from this section.

---

## 15.5 - Acid-Base Indicators

When you finish this section you will be able to select the proper indicator for an acid-base titration.

The purpose of an **indicator** is to allow you to know when you have reached the equivalence point in a titration. Acid-base indicators are normally organic acids (see <u>Figure 15.6 in your textbook</u>). Thus, when you add an indicator to a solution and perform a titration, **you are titrating the indicator along with you substance of interest.** This requires some small extra amount of titrant.

The volume of titrant at which you visually detect the equivalence point is called the **endpoint. The volume at the equivalence point is never exactly equal to the endpoint, but it is usually quite close.**

There are two criteria that are used to choose an indicator for a given titration:

- The $pK_a$ of the indicator should be within ± 1 unit of the pH of the solution at the equivalence point.
- The color change at the endpoint should be clearly distinguishable.

---

## Example 15.5 A - Proper Indicator Choice

A 0.100 M $NH_3$ solution is being titrated with 0.200 M HCl. Using data from Table 15.8 in your textbook, select a suitable indicator for the titration.

### Solution

The reaction of interest is

$$NH_3(aq) + H^+(aq) ----> NH_4^+(aq)$$

The important question is, "What is the pH at the equivalence point?" We can estimate it by noting that since the HCl solution is **twice as concentrated** as the $NH_3$ solution, it will take **1/2 the volume** of HCl solution to neutralize the $NH_3$ solution. That is,

**Volume at the equivalence point** $= V_0 + 1/2V_0 = 3/2V_0$

where $V_0$ is the original volume of $NH_3$ solution.

Because the volume is 3/2 as much, the $NH_4^+$ solution will be 2/3 as concentrated as the original $NH_3$ solution.

$$[NH_4^+] = 0.100 M \times 2/3 = \mathbf{0.067\ M}$$

$$K_{a\ NH_4^+} = K_w/K_{b\ NH_3} = 5.6 \times 10^{-10} = \frac{[H^+][NH_3]}{[NH_4^+]} = \frac{X^2}{0.0667}$$

$$X = [H^+] = 6.1 \times 10^{-6}, \quad \mathbf{pH = 5.21}$$

The indicator should have a $pK_a$ close to 5.2. **Bromcresol green is a suitable choice.** The color change, blue to yellow (base to acid) is easy to distinguish.

---

## Example 15.5 B - Indicator Colors

A solution has a pH of 7.0. What will be the color of the solution with each of the following indicators? (See Table 15.8 in your textbook.)

    a.   phenolphthalein
    b.   methyl orange
    c.   bromthymol blue
    d.   thymol blue

### Solution

    a.   clear
    b.   yellow
    c.   green-blue
    d.   yellow

Problems 30, 35, and 39-41 at the end of this study chapter cover material from this section.

## 15.6 - Solubility Equilibria

When you finish this section you will be able to:

- interconvert between solubility and $K_{sp}$, and
- solve problems relating to the common ion effect.

As with many of the previous sections, solubility equilibria use principles that you have used before. This section deals with **solubility**, the amount of a salt that can be dissolved in water. Your textbook points out that the **solubility of a salt is variable. The solubility product ("$K_{sp}$") is constant** (at a given temperature).

The solubility equilibrium expression is set up as any other. For example, the equilibrium expression for the dissolution of $Ag_2S$ in water is

$$Ag_2S(s) \rightleftharpoons 2Ag^+(aq) + S^{2-}(aq) \qquad K_{sp} = 1.6 \times 10^{-49}$$

$$K_{sp} = [Ag^+]^2[S^{2-}]$$

Remember that the pure solid, $Ag_2S$, is not included in the equilibrium expression.

## Example 15.6 A - Solubility Equilibrium Expressions

Write products and equilibrium expressions for the following dissolution reactions:

a. $Ba(OH)_2(s) \rightleftharpoons$ _____

b. $PbCO_3(s) \rightleftharpoons$ _____

c. $Ag_2CrO_4(s) \rightleftharpoons$ _____

d. $Ca_3(PO_4)_2(s) \rightleftharpoons$ _____

## Solution

a. $Ba(OH)_2(s) \rightleftharpoons Ba^{2+}(aq) + 2OH^-(aq)$
$K_{sp} = [Ba^{2+}][OH^-]^2$

b. $PbCO_3(s) \rightleftharpoons Pb^{2+}(aq) + CO_3^{2-}(aq)$
$K_{sp} = [Pb^{2+}] [CO_3^{2-}]$

c. $Ag_2CrO_4(s) \rightleftharpoons 2Ag^+(aq) + CrO_4^{2-}(aq)$
$K_{sp} = [Ag^+]^2[CrO_4^{2-}]$

d. $Ca_3(PO_4)_2(s) \rightleftharpoons 3Ca^{2+}(aq) + 2PO_4^{3-}(aq)$
$K_{sp} = [Ca^{2+}]^3[PO_4^{3-}]^2$

## Example 15.6 B - $K_{sp}$ From Solubility Data

Silver sulfide ($Ag_2S$) has a **solubility of $3.4 \times 10^{-17}$ M at 25°C. Calculate $K_{sp}$ for $Ag_2S$.**

## Solution

The key here is *how we define solubility*. For the reaction

$$Ag_2S(s) \rightleftharpoons 2Ag^+(aq) + S^{2-}(aq),$$

*one mole of $S^{2-}$* is produced for *every mole of $Ag_2S$* that dissolves. therefore, the solubility equals the concentration of $S^{2-}$ in solution.

$$\text{solubility} = s = [S^{2-}] = 3.4 \times 10^{-17} \text{ M}$$

The stoichiometry of the reaction indicates that *2 moles of $Ag^+$ ion are produced for each mole of $S^{2-}$ ion*. Therefore,

$$[Ag^+] = 2[S^{2-}] = 6.8 \times 10^{-17} \text{ M}$$

$$K_{sp} = [Ag^+]^2[S^{2-}] = (6.8 \times 10^{-17})^2(3.4 \times 10^{-17}) = 1.6 \times 10^{-49}$$

---

The keys to solving solubility problems are to

- properly define solubility, and
- properly use the reaction stoichiometry.

---

## Example 15.6 C - Solubility From $K_{sp}$ Data

Calculate the solubility of each of the following in moles per liter and grams per liter.

a. $NiCO_3$        $K_{sp} = 1.4 \times 10^{-7}$
b. $Ba_3(PO_4)_2$     $K_{sp} = 6 \times 10^{-39}$
c. $PbBr_2$        $K_{sp} = 4.6 \times 10^{-6}$

## Solution

We need to properly write dissolution reactions and equilibrium expressions for each. We must then carefully define solubility for each salt.

a. $NiCO_3(s) \rightleftharpoons Ni^{2+}(aq) + CO_3^{2-}(aq)$

$$s = [Ni^{2+}] = [CO_3^{2-}]$$
$$K_{sp} = [Ni^{2+}][CO_3^{2-}]$$
$$1.4 \times 10^{-7} = (s)(s) = s^2 \quad \Rightarrow \quad s = 3.7 \times 10^{-4} \text{ M}$$

This means that **$3.7 \times 10^{-4}$ mol/L of $NiCO_3$** will go into solution.

$$\text{g/L} = 3.7 \times 10^{-4} \text{ mol/L} \times 118.7 \text{ g/mol} = s = \textbf{0.44 g/L } NiCO_3$$

b. $Ba_3(PO_4)_2(s) \rightleftharpoons 3Ba^{2+}(aq) + 2PO_4^{3-}(aq)$

Solubility = s = **moles of $Ba_3(PO_4)_2$** that go into solution. The stoichiometry of the reaction dictates that

$$[Ba^{2+}] = 3s \quad [PO_4^{3-}] = 2s$$
$$K_{sp} = [Ba^{2+}]^3[PO_4^{3-}]^2$$
$$6 \times 10^{-39} = (3s)^3(2s)^2 = 108s^5$$

$$s^5 = 5.55 \times 10^{-41} \quad \Rightarrow \quad s = 9 \times 10^{-9} \text{ M}$$
$$g/L = 9 \times 10^{-9} \text{ mol/L} \times 601.8 \text{ g/mol} = s = 5 \times 10^{-6} \text{ g/L}$$

c.   $PbBr_2(s) \rightleftharpoons Pb^{2+}(aq) + 2Br^-(aq)$

$[Pb^{2+}] = s \quad [Br^-] = 2s$
$K_{sp} = [Pb^{2+}][Br^-]^2$
$4.6 \times 10^{-6} = (s)(2s)^2 = 4s^3 \quad \Rightarrow \quad s = 1.0 \times 10^{-2} \text{ M}$
$g/L = 1.0 \times 10^{-2} \text{ mol/L} \times 367 \text{ g/mol} = s = 3.67 \text{ g/L}$

---

Your textbook points out that **relative solubilities** among different compounds **cannot** be measured simply by comparing $K_{sp}$ values.  You must take the **composition of the salt** into account, as illustrated by the next example.

---

# Example 15.6 D - Relative Solubilities

Which of the following compounds is the most soluble?

$AgCl$;              $K_{sp} = 1.5 \times 10^{-10}$
$Ag_2CrO_4$;         $K_{sp} = 9.0 \times 10^{-12}$
$Ag_3PO_4$;          $K_{sp} = 1.8 \times 10^{-18}$

## Solution

$$AgCl(s) \rightleftharpoons Ag^+(aq) + Cl^-(aq)$$

$s = [Ag^+] = [Cl^-]$
$K_{sp} = [Ag^+][Cl^-] = s^2 = 1.5 \times 10^{-10} \quad \Rightarrow \quad s = 1.2 \times 10^{-5} \text{ M}$

$$Ag_2CrO_4(s) \rightleftharpoons 2Ag^+(aq) + CrO_4^{2-}(aq)$$

$s = [CrO_4^{2-}] \quad 2s = [Ag^+]$
$K_{sp} = [Ag^+]^2[CrO_4^{2-}] = (2s)^2(s) = 4s^3 = 9.0 \times 10^{-12}$
$s = 1.3 \times 10^{-4} \text{ M}$

$$Ag_3PO_4(s) \rightleftharpoons 3Ag^+(aq) + PO_4^{3-}(aq)$$

$s = [PO_4^{3-}] \quad 3s = [Ag^+]$
$K_{sp} = [Ag^+]^3[PO_4^{3-}] = (3s)^3(s) = 27s^4 = 1.8 \times 10^{-18}$
$s = 1.6 \times 10^{-5} \text{ M}$

$Ag_2CrO_4$ is the most soluble.

---

## Common Ion Effect

We have encountered this before.  Recall that **Le Chatelier's Principle** predicts that **adding a common ion to the solution shifts the equilibrium to the left** in solubility equations.  In other words **adding a common ion** (that doesn't react with other species in the solution) **reduces the solubility.**

---

## Example 15.6 E - Comon Ion Effect

Calculate the solubility of $SrF_2$ ($K_{sp} = 7.9 \times 10^{-10}$) in
   a.   pure water
   b.   0.100 M $Sr(NO_3)_2$
   c.   0.200 M NaF

## Solution

The reaction of interest is

$$SrF_2(s) \rightleftharpoons Sr^{2+}(aq) + 2F^-(aq)$$

$$K_{sp} = [Sr^{2+}][F^-]^2$$

a.   In pure water, solubility of $SrF_2$ equals

   $s = [Sr^{2+}] \quad 2s = [F^-]$
   $7.9 \times 10^{-10} = (s)(2s)^2 = 4s^3 \quad \Rightarrow \quad s = 5.8 \times 10^{-4}$ M

b.   $Sr(NO_3)_2$ is a strong electrolyte.  This means that a 0.100 M solution will supply **0.100 M $Sr^{2+}$ ion** to the solution

$$[Sr^{2+}] = 0.100 + s \approx 0.100 \text{ M}$$

We are assuming that s is negligible relative to 0.100 M.  We can test this with the 5% test.

   $[F^-] = 2s$
   $K_{sp} = 7.9 \times 10^{-10} = (0.100)(2s)^2 = 0.4s^2$
   $s = 4.4 \times 10^{-5}$ M

A solubility of $4.4 \times 10^{-5}$ M is < 5% of 0.100.  The solubility was thus *decreased by a factor of 13* due to the presence of the common ion.

c.   $NaF_2$ is a strong electrolyte.  The dissociation of 0.200 M $NaF_2$ will give

   $[F^-] = 0.400 \text{ M} + 2s \approx \mathbf{0.400}$
   $[Sr^{2+}] = s$
   $K_{sp} = 7.9 \times 10^{-10} = s(0.400)^2 = 0.16s$
   $s = 4.9 \times 10^{-9}$ M

The solubility was markedly *decreased* by the presence of the common $F^-$ ion.

---

Although the rigorous solutions are beyond the scope of your textbook, in general, substances (such as acids or bases) that can combine with one of the ions of the salt will increase the solubility of the salt by removing product, thus forcing the reaction to the right.

---

## Problems 44-61 at the end of this study chapter cover material from this section.

## 15.7 - Precipitation and Qualitative Analysis

When you finish this study section you will be able to:
- determine whether the mixing of two solutions will produce a precipitate, and
- calculate the concentration of each ion in a solution in which a precipitate is produced.

This section asks the musical question, "If two solutions are mixed, will a precipitate form?" Then, "If it forms, what will be the concentration of each ion in solution?"

Your textbook introduces "Q," the **ion product** of the initial ion concentrations. **If $Q \geq K_{sp}$, precipitation will occur**.

## Example 15.7 A - The Ion Product

A 200.0 mL solution of $1.3 \times 10^{-3}$ M $AgNO_3$ is mixed with 100.0 mL of a $4.5 \times 10^{-5}$ M $Na_2S$ solution. Will precipitation occur?

## Solution

$Na^+$ and $NO_3^-$ ions are completely soluble, so we would presume that the precipitate would be $Ag_2S$. The reaction of interest is

$$2Ag^+(aq) + S^{2-}(aq) \rightleftharpoons Ag_2S(s)$$

The ion product is

$$Q = [Ag^+]_o^2[S^{2-}]_o$$

$$[Ag^+]_o = 1.3 \times 10^{-3} \text{ M} \times \frac{200.0 \text{ mL (original volume)}}{(200.0 + 100.0) \text{ mL (total soln volume)}} = 8.7 \times 10^{-4} \text{ M}$$

$$[S^{2-}]_o = 4.5 \times 10^{-5} \text{ M} \times \frac{100.0 \text{ mL}}{300.0 \text{ mL}} = 1.5 \times 10^{-5} \text{ M}$$

$$Q = (8.7 \times 10^{-4})^2(1.5 \times 10^{-5}) = 1.1 \times 10^{-11}$$

$K_{sp} = 1.6 \times 10^{-49}$. $Q > K_{sp}$, so **precipitation will occur.**

Once we determine that precipitation will occur, we are faced with the problem of determining the **equilibrium concentrations** of each of our ions of interest.

The general strategy involves **assuming** that because $K_{sp}$ is so low, **if a precipitate forms, it will do so quantitatively.** We can then use the equilibrium expression involving the solubility of the salt (with a common ion) to solve for the equilibrium concentration of each ion. Let's illustrate this with the next example.

## Example 15.7 B - Solubility from Mixing Solutions

Calculate the equilibrium concentration of each ion in a solution obtained by mixing 50.0 mL of $6.0 \times 10^{-3}$ M $CaCl_2$ with 30 mL of 0.040 M NaF. ($K_{sp}$ for $CaCl_2 = 4.0 \times 10^{-11}$)

## Solution

First, let's verify that precipitation in fact occurs.

$$[Ca^{2+}]_0 = 6.0 \times 10^{-3} \text{ M} \times \frac{50.0 \text{ mL}}{(50.0 + 30.0) \text{ mL}} = 3.7_5 \times 10^{-3} \text{ M}$$

$$[F^-]_0 = 0.040 \text{ M} \times \frac{30.0 \text{ mL}}{80.0 \text{ mL}} = 0.015 \text{ M}$$

$$Ca^{2+}(aq) \text{ /+ } 2F^-(aq) \rightleftharpoons CaF_2(s)$$

$$Q = [Ca^{2+}][F^-]^2 = (3.7_5 \times 10^{-3})(0.015)^2 = 8.4 \times 10^{-7}$$

**$Q > K_{sp}$, so precipitation occurs.**

We can assume a **quantitative ("stoichiometric") reaction between $Ca^{2+}$ and $F^-$ to form $CaF_2$ solid.** We can then see how much of which ion remains as excess in solution.

$$\text{mmol } Ca^{2+}_{\text{initial}} = 6.0 \times 10^{-3} \text{ mmol/mL} \times 50.0 \text{ mL} = \textbf{0.30 mmol}$$
$$\text{mmol } F^-_{\text{initial}} = 0.040 \text{ mmol/mL} \times 30.0 \text{ mL} = \textbf{1.2 mmol}$$

| | $Ca^{2+}(aq)$ + | $2F^-(aq)$ $\rightleftharpoons$ | $CaF_2(s)$ |
|---|---|---|---|
| initial (mmol) | 0.30 | 1.2 | 0 |
| final (mmol) | ≈0 | 0.60 | 0.30 |

$$[Ca^{2+}] = s \quad [F^-] = 0.60 \text{ mmol/80.0 mL} = 7.5 \times 10^{-3} \text{ M} + 2s \approx 7.5 \times 10^{-3} \text{ M}$$

$$K_{sp} = [Ca^{2+}][F^-]^2 \quad \Rightarrow \quad 4.0 \times 10^{-11} = s(7.5 \times 10^{-3})^2$$

$$s = 7.1 \times 10^{-7} \text{ M} = [Ca^{2+}]$$

$$7.5 \times 10^{-3} \text{ M} = [F^-]$$

Remember again that although the solubility of different ions can change depending upon what is in solution, **the solubility product ($K_{sp}$) remains the same** at a given temperature.

---

Problems 62-66 at the end of this study chapter cover material from this section.

---

## 15.8 - Complex Ion Equilibria

When you finish this section you will be able to:

- calculate the concentrations of species in a solution involving complex ions, and
- determine the increase in solubility of an insoluble salt by adding a complex ion.

Your textbook introduces several new terms here. You should be able to define: **complex ion, ligand,** and **formation constant.** The ultimate goal of this section is to demonstrate that **introducing a Lewis base** into a solution **enhances the solubility** of an otherwise insoluble salt.

The key idea here is that to avoid being drowned in a sea of equations, **we must make (and test!) simplifying assumptions where possible!**

Let's do the following example to demonstrate the idea of **complex ion formation.**

## Example 15.8 A - Complex Ion Formation

Calculate the concentrations of $Ag^+$ and $Ag(CN)_2^-$ in a solution prepared by mixing **100.0 mL of $5.0 \times 10^{-3}$ M AgNO$_3$** with **100.0 mL of 2.00 M KCN.**

$$Ag^+(aq) + 2CN^-(aq) \rightleftharpoons Ag(CN)_2^-(aq) \qquad \beta = 1.3 \times 10^{21}$$

### Strategy and Solution

Note that $\beta$ equals the equilibrium constant for the reaction, and it is overwhelmingly large. This means that **virtually all of the $Ag^+$ ion will end up as $Ag(CN)_2^-$ ion.** The other important point to note is that **our ligand, $CN^-$, is present in huge excess.** Therefore,

$$[CN^-] \approx [CN^-]_0$$

The equilibrium expression is

$$\beta = \frac{[Ag(CN)_2^-]}{[Ag^+][CN^-]^2}$$

To get equilibrium concentrations,

$$[CN^-] \approx [CN^-]_0 = 1.00 \text{ M}$$

(after mixing the volume doubles, so the concentration is halved). Virtually all of the silver is present as $Ag(CN)_2^-$. Therefore,

$$[Ag(CN)_2^-] \approx [Ag^+]_0 = 2.5 \times 10^{-3} \text{ M}$$

(Remember the dilution!) Rearranging our overall equilibrium expression,

$$[Ag^+] = \frac{[Ag(CN)_2^-]}{\beta[CN^-]^2} = \frac{2.5 \times 10^{-3}}{1.3 \times 10^{21} (1.00)^2} = 1.9 \times 10^{-24} \text{ M}$$

---

The important question in all of this is, "How does complex ion formation affect solubility?"

Using our first example (15.8 A), we note that most of the silver was complexed as $Ag(CN)_2^-$. That means that as free $Ag^+$ ion is being produced by an insoluble salt (such as AgI, $K_{sp} = 1.5 \times 10^{-16}$), it is being **removed by complexing with $CN^-$ ions.** This is pulling the dissolution of AgCl(s) **to the right.**

**FORMATION OF A COMPLEX ION INCREASES THE SOLUBILITY OF AN "INSOLUBLE" SALT.**

---

## Example 15.8 B - Solubility and Complex Ion Formation

Calculate the solubility of AgI(s) in 1.00 M $CN^-$ ion   ($K_{sp}$ for AgI = $1.5 \times 10^{-16}$).

### Strategy

We have, up until now, said that the solubility of AgI equals

$$s = [Ag^+] = [I^-]$$

However, $Ag^+$ ion forms complexes which enhance the solubility.  A more precise statement is that

$$s = [I^-] = [\text{all Ag species}] = [Ag^+] + [AgCN] + [Ag(CN)_2^-]$$

We know, based on the overall formation constant, $\beta$, that virtually all of the $Ag^+$ ion is present as $Ag(CN)_2^-$. Therefore,

$$s = [I^-] \approx [Ag(CN)_2^-]$$

The overall equilibrium of AgI(s) and $CN^-$(aq) can be represented as

$$AgI(s) + 2CN^-(aq) \rightleftharpoons Ag(CN)_2^- + I^-(aq)$$

$$K = K_{sp}\beta = \frac{[Ag(CN)_2^-][I^-]}{[CN^-]^2} = 1.5 \times 10^{-16} (1.3 \times 10^{21})$$

**$K = 1.95 \times 10^5$**

**$[CN^-] = 2.00 - 2X$**     (original concentration - amount complexed)
**$[Ag(CN)_2^-] = X$**     (amount formed, X = solubility, s)
**$[I^-] = X$**     (X = solubility, s)

$$K = \frac{X^2}{(2.00 - 2X)^2} \quad \Rightarrow \quad K^{1/2} = \frac{X}{2.00 - 2X}$$

Getting rid of the denominator,

$$883 - 883X = X$$

**X = solubility = 0.999 M**

Note how our solubility has increased from $1.2 \times 10^{-8}$ M to 0.999 M due to the presence of the ligand.

---

## Problems 67-68 at the end of this study chapter cover material from this section.

---

**END OF CHAPTEER FIFTEEN**

# SELF-TEST QUESTIONS

1. Qualitatively, what is the effect on pH of adding sodium benzoate to an aqueous solution of benzoic acid? Why does this change occur?

2. Calculate the pH, and the percent dissociation of the acid, in each of the following solutions.

   a. 0.150 M $C_6H_5COOH$ (benzoic acid, $K_a = 6.14 \times 10^{-5}$)
   b. 0.150 M $C_6H_5COOH$ in the presence of 0.350 M $C_6H_5COONa$

3. Calculate the pH, and the percent hydrolysis of the base, in each of the following solutions.

   a. 0.438 M $NH_3$ ($K_b = 1.8 \times 10^{-5}$)
   b. 0.438 M $NH_3$ to which has been added 0.300 M $NH_4Cl$

4. A solution is prepared from 0.150 mol of $CH_3COOH$, 0.0100 mol of $CH_3COO^-$ (from $CH_3COONa$), and enough water to make a total volume of 1 L. Calculate the value of $[H^+]$ at equilibrium. (acetic acid, $K_a = 1.75 \times 10^{-5}$)

5. A solution contains 0.350 mol of $(CH_3)_3N$, 0.050 mol of $(CH_3)_3NH^+$, and enough water to make a total volume of 1 L. Calculate the value of $[OH^-]$. (trimethylamine, $K_b = 6.25 \times 10^{-5}$)

6. Determine the pH in a solution containing 0.085 M $NH_3$ and 0.247 M $NH_4Cl$ ($K_b$ for $NH_3 = 1.8 \times 10^{-5}$).

7. Calculate the pH of the buffer system with 0.10 M $Na_2HPO_4$/0.15 M $KH_2PO_4$.

8. Is a 6 M HCl solution a good buffer? Is it biologically useful?

9. What is the pH of a 200 mL solution containing 21.46 g of benzoic acid and 37.68 g of sodium benzoate,

   a. initially?
   b. after 30.0 mL of 5.0 M HCl have been added?
   c. Has this buffer exceeded its buffer capacity after addition of the acid?

10. A buffer is prepared by adding 20.5 g of $CH_3COOH$ and 17.8 g of $CH_3COONa$ to enough water to make $5.00 \times 10^2$ mL of solution. Calculate the pH of this buffer solution.

11. Calculate the pH of 1 L of a buffer solution containing 0.10 M HCN and 0.12 M $CN^-$.

12. What must be the ratio of acetic acid to sodium acetate to prepare a buffer whose pH = 4.81 ($K_a$ of $CH_3COOH = 1.8 \times 10^{-5}$).

13. What is the pH of the buffer containing 0.35 M $NH_4Cl$ and 0.15 M $NH_3$?

14. Calculate the pH if 0.01 mol of HCl is added to the buffer in problem 11.

15. A solution contains 0.300 moles of acetic acid and 0.200 moles of sodium acetate in a total volume of 500 mL. How much 6.0 M NaOH must be added so that the pH of the solution equals the $pK_a$ of acetic acid?

16. Calculate the pH if 0.02 mol of NaOH is added to the original buffer in problem 11.

17. A solution contains 0.216 moles of a base and 0.614 moles of the conjugate acid in a total volume of 800 mL. The pH of the solution is 9.65. What is the value for $K_b$ of the base?

18. How many grams of sodium acetate must be dissolved in a 0.200 M acetic acid solution to make a 400 mL buffer solution with a pH of 4.56 ? (Assume that the volume of the solution remains constant.)

19. The pH of a bicarbonate-carbonic acid buffer is known to be 8.00. Calculate the ratio of the concentration of carbonic acid to that of the bicarbonate ion. ($K_a$ for $H_2CO_3 = 4.3 \times 10^{-7}$)

20. What would the pH of the solution in problem #18 be if we added 40.0 mL of 0.30 M HCl?

21. We wish to buffer a solution at pH = 4.58. Which of the following acids (and conjugate base salts) would be most useful?

    a.   $CH_3COOH$        ($K_a = 1.8 \times 10^{-5}$)
    b.   $C_6H_5COOH$       ($K_a = 6.14 \times 10^{-5}$)
    c.   $ClCH_2COOH$       ($K_a = 1.36 \times 10^{-3}$)
    d.   $C_6H_5OH$         ($K_a = 1.6 \times 10^{-10}$)

22. Calculate K for the titration of HCl with NaOH.

23. A volume of 17.8 mL of a 0.344 M $H_2SO_4$ solution is required to completely neutralize 20.0 mL of a KOH solution. Calculate the concentration (in molarity) of the KOH solution.

24. If 37.42 mL of 0.1078 M NaOH solution is required to completely neutralize 25.00 mL of an HCl solution, what is the concentration (in molarity) of the acid solution.

25. Calculate the pH after the following **total** volumes of 0.3000 M NaOH have been added to 40.00 mL of 0.6000 M HCl.

    a.   0.00 mL          d.   79.40 mL          g.   90.00 mL
    b.   5.00 mL          e.   80.00 mL
    c.   40.00 mL         f.   80.50 mL

26. A sample of 50.0 mL of a commercial vinegar solution (which contains acetic acid) is titrated with a 1.00 M NaOH solution. What is the concentration of acetic acid in vinegar if 5.75 mL of the base was required for the titration?

27. Calculate the pH after the following **total** volumes of 0.3200 M HCl have been added to 25.00 mL of 0.1600 M NaOH.

    a.   0.00 mL          c.   12.40 mL          e.   12.60 mL
    b.   1.00 mL          d.   12.50 mL          f.   15.00 mL

28. If a 25.00 mL sample of base "X" requires 18.34 mL of 0.100 M HCl to reach the equivalence point, what is the concentration of base "X"? (Assume that the acid-base stochiometry is 1:1.)

29. If 12.5 mL of 0.500 M $H_2SO_4$ exactly neutralizes 50.0 mL of NaOH, what is the concentration of the NaOH solution?

30. Using information found in **Figure 15.7 in your textbook**, select a suitable indicator for the titrations in problems #25 and #27 **other than phenolphthalein**.

31. Calculate the pH at the equivalence point for the titration of 0.10 M HCl with 0.10 M $NH_3$.

32. Calculate the pH at the equivalence point for the titration of 0.01 M $CH_3COOH$ with 0.10 M NaOH.

33. A solution containing 100.0 mL of 0.1350 M $CH_3COOH$ ($K_a = 1.8 \times 10^{-5}$) is being titrated with 0.5400 M NaOH. Calculate the pH

    a.   initially.
    b.   half way to the equivalence point.
    c.   at the equivalence point.
    d.   5 mL past the equivalence point.

34. Calculate K for the titration in problem #33.

35. Pick the best indicator for the titration in problem #33.

36. Calculate the number of milliliters of a 0.153 M sodium hydroxide solution that must be added to a 20.0 mL sample of a solution of hydrochloric acid, whose pH is 0.747, to reach the equivalence point.

37. The equivalence point for a solution of acid "A" occurs after 46.7 mL of base has been added. After 34.2 mL of base had been added (earlier in the titration) the pH was 4.64. What is the $K_a$ for acid "A"?

38. A solution containing 50.00 mL of 0.1800 M $NH_3$ ($K_b = 1.8 \times 10^{-5}$) is being titrated with 0.3600 M HCl. Calculate the pH

    a. initially.
    b. after the addition of 5.00 mL of HCl.
    c. after the addition of a total volume of 12.50 mL of HCl.
    d. after the addition of a total volume of 25.00 mL of HCl.
    e. after the addition of 26.00 mL of HCl.

39. Pick a suitable indicator for the titration in problem #38.

40. Pick a suitable indicator for the titration in problem #37.

41. What indicator would you choose for the titration of a water solution of HCN (concentration unknown) with a 0.100 M solution of NaOH? ($K_a$ for HCN = $6.2 \times 10^{-10}$)

42. Calculate the equilibrium constant for the titration of benzoic acid ($C_6H_5COOH$) with $NH_3$. Is this titration feasible? Why or why not?

43. A solution containing 2.049 g of a weak acid required 43.88 mL of 0.1207 M NaOH to reach the equivalence point. What is the molar mass of the acid?

44. Write products and equilibrium expressions for the following dissolution reactions.
    a.  $PbI_2(s)$ $\rightleftharpoons$
    b.  $Sr_3(PO_4)_2(s)$ $\rightleftharpoons$
    c.  $MnS(s)$ $\rightleftharpoons$

45. Calculate [I$^-$] in an AgI solution with [Ag$^+$] = $1.2 \times 10^{-8}$ M. (AgI has $K_{sp} = 1.5 \times 10^{-16}$.)

46. Manganese sulfide (MnS) has a $K_{sp} = 2.3 \times 10^{-13}$. Calculate the solubility of MnS.

47. Calculate the solubility product ($K_{sp}$) of silver sulfate if its molar solubility is $1.5 \times 10^{-2}$ mol/L.

48. The $K_{sp}$ for manganese(II) carbonate is $1.8 \times 10^{-11}$. Calculate the solubility of $MgCO_3$.

49. Strontium phosphate ($Sr_3(PO_4)_2$) has a $K_{sp} = 1 \times 10^{-31}$. Calculate the solubility of $Sr_3(PO_4)_2$.

50. Calculate the $K_{sp}$ for calcium sulfate if its solubility is 0.67 g/L.

51. The solubility of AgBr is $7.1 \times 10^{-7}$ M. Calculate $K_{sp}$ for AgBr.

52. Calculate the solubility (in g/L) of $Fe(OH)_3$. The $K_{sp}$ for iron(II) hydroxide [$Fe(OH)_3$] is $1.8 \times 10^{-15}$.

53. The solubility of $SrF_2$ is $5.8 \times 10^{-4}$ M. Calculate $K_{sp}$ for $SrF_2$.

54. The solubility of the ionic compound $M_2X_3$, having a molar mass of 288 g, is $3.6 \times 10^{-17}$ g/L. Calculate the $K_{sp}$ of the compound.

55. Calculate the concentration of $Ag^+$ in a saturated solution of $Ag_2CrO_4$ ($K_{sp}$ = $9.0 \times 10^{-12}$).

56. What is the solubility of silver chloride (in g/L) ($K_{sp}$ = $1.6 \times 10^{-10}$) in a $6.5 \times 10^{-3}$ M silver nitrate solution?

57. Calculate the number of grams of ZnS ($K_{sp}$ = $2.5 \times 10^{-22}$) that will dissolve in $3.0 \times 10^2$ mL of 0.050 M $Zn(NO_3)_2$.

58. Calculate the solubility of $Co(OH)_2$ ($K_{sp}$ = $2.5 \times 10^{-16}$) at pH 11.50.

59. Calculate the solubility of $PbCO_3$ ($K_{sp}$ = $1.5 \times 10^{-15}$) in

    a.  pure water
    b.  0.0400 M $Pb(NO_3)_2$

60. If 50.0 mL solution of $2.0 \times 10^{-3}$ M $CaCl_2$ is mixed with 100.0 mL of $5.0 \times 10^{-2}$ M $NaF_2$, will precipitation occur?

61. Determine if a precipitate will form when exactly 200 mL of 0.0040 M $BaCl_2$ is added to 600 mL of 0.0080 M $K_2SO_4$.

62. Calculate the number of moles of $Ag_2CrO_4$ ($K_{sp}$ = $9.0 \times 10^{-12}$) that will dissolve in 1 L of 0.010 M $K_2CrO_4$ solution. What will be the ion concentrations at equilibrium?

63. Calculate the equilibrium concentration of each ion in a solution obtained by mixing 75.0 mL of $2.5 \times 10^{-2}$ M $AgNO_3$ with 25.0 mL of $3.2 \times 10^{-4}$ M KI.

64. Will $BaCO_3$ ($K_{sp}$ = $1.6 \times 10^{-9}$) precipitate if a sample of 20.0 mL of 0.10 M $Ba(NO_3)_2$ is added to 50.0 mL of 0.10 M $Na_2CO_3$?

65. Write equations for the stepwise and overall formation of each of the following complex ions:

    a.  $Cd(NH_3)_4^{2+}$
    b.  $PbBr_3^-$

66. What is the molar solubility of silver chloride in a 1.0 M $NH_3$ solution?

67. Calculate the concentrations of $Ag^+$ and $Ag(NH_3)_2^+$ in a solution with an initial $Ag^+$ concentration of $4.0 \times 10^{-3}$ M and an initial $NH_3$ concentration of 0.500 M. ($K_1$ = $2.1 \times 10^3$, $K_2$ = $8.2 \times 10^3$).

# ANSWERS TO SELF-TEST QUESTIONS

1. The pH is raised due to the common-ion effect.

2. a.  pH = 2.52, 2% dissociated.
   b.  pH = 4.58, less than $2.0 \times 10^{-2}$% dissociated

3. a.  pH = 11.45, 0.6% hydrolyzed
   b.  pH = 9.42, less than $6 \times 10^{-3}$% hydrolyzed

4. $[H^+] = 2.7 \times 10^{-4}$ M

5. $[OH^-] = 4.4 \times 10^{-4}$ M

6. pH = 8.79

7. pH = 7.03

8. Yes, a 6 M HCl solution is a superior buffer in that it resists changes in pH when an acid or base is added. It is not biologically useful, as it is far too acidic to be biologically safe.

9. a.  pH = 4.38
   b.  pH = 3.75
   c.  No, the buffer gas has not exceeded its capacity.

10. pH = 4.54

11. pH = 9.29

12. The ratio must be 0.862 to 1.00.

13. pH = 8.88

14. pH = 9.21

15. A total of 8.33 mL of 6.0 M NaOH must be added.

16. pH = 9.45

17. $K_b = 1.27 \times 10^{-4}$

18. 4.3 g

19. 0.024

20. pH = 4.51

21. $CH_3COOH$ would be the most useful ($pK_a = 4.74$).

22. $K = 1/K_w = 1 \times 10^{14}$

23. 0.610 M KOH

24. 0.1614 M HCl

25. a.  pH = 0.22      d.  pH = 2.90      g.  pH = 12.36
    b.  pH = 0.30      e.  pH = 7.00
    c.  pH = 0.82      f.  pH = 11.10

26. 0.115 M

27. a.  pH = 13.20      c.  pH = 10.93      e.  pH = 2.93
    b.  pH = 13.15      d.  pH = 7.00       f.  pH = 1.67

28. $7.32 \times 10^{-2}$ M

29. 0.25 M

30. Bromthymol blue or thymol blue are suitable choices.

31. 5.13

32. 8.37

33. a.  pH = 2.81      c.  pH = 8.89
    b.  pH = 4.74      d.  pH = 12.32

34. $K = 1.8 \times 10^{9}$

35. Thymol blue or phenolphthalein are suitable choices.  Alizarin is not as good because the color change is harder to tell.

36. 23.4 mL

37. $6.3 \times 10^{-5}$

38. a.  pH = 11.25      c.  pH = 9.25      e.  pH = 2.32
    b.  pH = 9.86       d.  pH = 5.09

39. Bromcresol green is a suitable choice.

40. Methyl orange or bromcresol green

41. Alizarin yellow

42. $K = K_a K_b / K_w = 1.1 \times 10^{5}$.  This titration is not feasible.  The value for K is too low (below $10^{7}$ or so).

43. Molar mass = 386.0 g/mol.

44. a.   $PbI_2(s) \rightleftharpoons Pb^{2+}(aq) + 2I^{-}(aq)$
    b.   $Sr_3(PO_4)_2(s) \rightleftharpoons 3Sr^{2+}(aq) + 3PO_4^{3-}(aq)$
    c.   $MnS(s) \rightleftharpoons Mn^{2+}(aq) + S^{2-}(aq)$

45. $1.25 \times 10^{-8}$ M

46. Solubility = $4.8 \times 10^{-7}$ M

47. $K_{sp} = 1.4 \times 10^{-5}$

48. $4.2 \times 10^{-6}$ M

49. Solubility = $2 \times 10^{-7}$ M

50. $K_{sp} = 2.4 \times 10^{-5}$

51. $K_{sp} = 5.0 \times 10^{-13}$

52. $9.66 \times 10^{-3}$

53. $K_{sp} = 7.8 \times 10^{-10}$ (actual = $7.9 \times 10^{-10}$)

54. $4.0 \times 10^{-93}$

55. $[Ag^+] = 2.6 \times 10^{-4}$ M

56. solubility = $3.6 \times 10^{-6}$ g/L

57. $1.5 \times 10^{-19}$ g ZnS

58. Solubility = $6.3 \times 10^{-12}$ M

59. a.   solubility = $3.9 \times 10^{-8}$ M
    b.   $3.75 \times 10^{-14}$ M

60. Yes, $Q = 7.3 \times 10^{-7}$ ($> 4.0 \times 10^{-11}$, the $K_{sp}$ of $CaF_2$)

61. Yes, $Q = 6.0 \times 10^{-6}$. $BaSO_4$ will precipitate out of solution until $[Ba^{2+}][SO_4^{2-}] < 1.5 \times 10^{-9}$, the $K_{sp}$ of $BaSO_4$.

62. $1.5 \times 10^{-5}$ moles; solubility = $1.5 \times 10^{-5}$ mol/L;  $[Ag^+] = 3.0 \times 10^{-5}$ M;  $[CrO_4^{2-}] = 0.010$ M

63. $[Ag^+] = 1.9 \times 10^{-2}$ M; $[I^-] = 8.0 \times 10^{-15}$ M.

64. Yes, $BaCO_3$ will precipitate out of solution.

65. stepwise:  $Cd^{2+} + NH_3 \rightleftharpoons Cd(NH_3)^{2+}$          $K_1$
    $Cd(NH_3)^{2+} + NH_3 \rightleftharpoons Cd(NH_3)_2 2+$          $K_2$
    $Cd(NH_3)_2^{2+} + NH_3 \rightleftharpoons Cd(NH_3)_3^{2+}$          $K_3$
    $Cd(NH_3)_2^{2+} + NH_3 \rightleftharpoons Cd(NH_3)_4^{2+}$          $K_4$
    overall $Cd^{2+} + 4NH_3 \rightleftharpoons Cd(NH_3)_4^{2+}$          $\beta_4 = K_1 K_2 K_3 K_4$

66. solubility = 0.045 M

67. $[Ag^+] = 9.3 \times 10^{-10}$, $[Ag(NH_3)_2^+] = 4.0 \times 10^{-3}$

# CHAPTER 16

# Spontaneity, Entropy and Free Energy

| Section | Contents | Page in this Study Guide |
|---|---|---|
| 16.1 | Spontaneous Processes and Entropy | 352 |
| 16.2 | Entropy and the Second Law of Thermodynamics | 354 |
| 16.3 | The Effect of Temperature on Spontaneity | 354 |
| 16.4 | Free Energy | 356 |
| 16.5 | Entropy Changes in Chemical Reactions | 357 |
| 16.6 | Free Energy and Chemical Reactions | 359 |
| 16.7 | The Dependence of Free Energy on Pressure | 361 |
| 16.8* | Free Energy and Equilibrium | 362 |

Recall the First Law of Thermodynamics from Chapter 6. It says that energy can neither be created nor destroyed - just changed from one form to another. The law describes energy changes in chemical reactions but does not answer the more fundamental question, "Can we predict **IF** reactions will occur using thermodynamics?" This chapter shows how such predictions can be made.

## 16.1 - Spontaneous Processes and Entropy

When you finish this section you will be able to:

- define entropy,
- choose among alternatives that which has the greatest positional entropy.

Your textbook makes several points regarding **spontaneous** reactions.

- **Spontaneous** means occurring **without outside intervention**.
- Rate of reaction is **irrelevant** to spontaneity. Spontaneous means it will happen, not necessarily quickly, or even in the Earth's lifetime.
- Spontaneous processes **increase the entropy of the universe**. (The entropy of a **system can decrease** if that of the **surroundings increases**.)

⇒ Entropy is a complex mathematical function that describes the number of <u>possible</u> arrangements (**positional probability**) of the states of a substance.
⇒ Each arrangement available to a substance is called a **microstate**.
⇒ Gases (in general) have a much higher positional entropy than liquids or solids.

Let's have some practice with microstates and entropy.

---

*Section 16.9 presents an interesting discussion on free energy and work and will not be covered in this study guide.*

## Example 16.1 A - Microstates and Entropy

You have 3 identical atoms, "A," "B" and "C." They can go back and forth from one "dash" to the other via the "wall" as shown.

$$||$$

$$\underline{AB} \quad \rightleftarrows \quad \underline{C} \quad \text{(for example)}$$

side 1     side 2

a. List all the possible microstates that these three atoms can have on our two dashes. In other words, list all the ways they can arrange themselves.

b. How much more probable is arrangement II than arrangement I?

arrangement I    •••___  ||  ___

arrangement II    ••___  ||  •___

c. Does this make sense in terms of our understanding of entropy?

## Solution

a.                **Microstates**

| | | | | |
|---|---|---|---|---|
| ABC | \|\| | ___(I) | A \|\| | BC |
| AB_ | \|\| | _C_ (II) | B_ \|\| | AC_ |
| AC_ | \|\| | _B_ (II) | C_ \|\| | AB_ |
| BC_ | \|\| | _A_ (II) | ___ \|\| | ABC |

b. Arrangement I can happen **1 way**. Arrangement II can happen 3 ways. Arrangement II is thus 3 times as likely as arrangement I.

c. The "atoms" are **more likely** to spread than be confined to one side of the wall. That agrees with our concept of entropy as positional probability.

## Example 16.1 B - Positional Entropy

Which of the following pairs is likely to have the higher positional entropy per mole at a given temperature?

a. Solid or gaseous phosphorus
b. $CH_4(g)$ or $C_3H_8(g)$
c. KOH(s) or KOH(aq)

## Solution

a. Gaseous phosphorus will have the higher positional entropy. It is not as constrained by intramolecular bonds as solid phosphorus is.

b. All other things being equal, larger molecules containing many single bonds have more positional possibilities than smaller ones. $C_3H_8(g)$ has the higher entropy.

c.  In general (all other things being equal), liquids have **slightly** higher entropies than solids do.  **However,** in this case, the positional probabilities of KOH(aq) are constrained due to hydrogen bonding interactions.  KOH(s) has the higher entropy.

---

# Problems 1-4 at the end of this study chapter cover material from this section.

---

## 16.2 - Entropy and the Second Law of Thermodynamics

The following questions will test your understanding of the "second law."

1.  State the second law of thermodynamics.
2.  State the second law in terms of the system and surroundings.
3.  Under what circumstance can the **entropy of the system decrease** for a spontaneous process?

---

## 16.3 - The Effect of Temperature on Spontaneity

When you finish this section you will be able to:

- state the importance of exothermic reactions to entropy, and
- calculate the change in entropy of the surroundings from the heat of reaction of the solution.

Exothermic reactions **give off energy to the surroundings**.  Therefore, **random motions** of particles in the surroundings **increase**.  When random motions increase, **positional probabilities increase**.

The <u>key point</u> from all this is that **exothermic reactions increase the entropy of the surroundings ($\Delta S_{surr}$).**

The **magnitude** of the increase in $\Delta S_{surr}$ depends on the temperature.  (See the money-related discussion in your textbook.)

Recall from Chapter 6 that we think of **heat flow** in terms of **the system**.

**Exothermic reaction (at constant pressure),   $\Delta H = -$**
**Endothermic reaction (at constant pressure),   $\Delta H = +$**

$$\Delta S_{surr} = \frac{-\Delta H}{T \text{ (in Kelvins)}}$$

---

## Example 16.3 A - $\Delta S_{surr}$ and the Heat of Reaction

Calculate $\Delta S_{surr}$ for each of the following reactions at 25°C and 1 atm.

a.  $C_3H_8(g) + 5O_2(g) \longrightarrow 3CO_2(g) + 4H_2O(g)$        $\Delta H = -2045 \text{ kJ}$
b.  $(NH_4)_2Cr_2O_7(s) \longrightarrow N_2(g) + Cr_2O_3(s) + 4H_2O(g)$    $\Delta H = -315 \text{ kJ}$
c.  $H_2O(l) \longrightarrow H_2O(g)$            $\Delta H = +44 \text{ kJ}$

## Solution

a.  $\Delta S_{surr} = \dfrac{-\Delta H}{T}$

T = (25 + 273) = 298 K

$\Delta S_{surr} = \dfrac{-(-2045 \text{ kJ})}{298 \text{ K}} = 6.86 \text{ kJ/K} = \textbf{6860 J/K}$

There is an increase, because the reaction is exothermic.

b.  $\Delta S_{surr} = \dfrac{-(-315 \text{ kJ})}{298 \text{ K}} = 1.06 \text{ kJ/K} = \textbf{1060 J/K}$

There is an increase here as well because the reaction is exothermic.

c.  $\Delta S_{surr} = \dfrac{-(44 \text{ kJ})}{298 \text{ K}} = -0.14 \text{ kJ/K} = \textbf{-140 J/K}$

Energy must be taken from the surroundings to the system in this reaction. Therefore, $\Delta S_{surr}$ will be negative.

---

The relationship between **entropy change** and reaction spontaneity is summarized in <u>Table 16.3 in your textbook</u>. Consider the information in that table and try the following example.

---

## Example 16.3 B - Reaction Spontaneity

Determine if the values for entropy in each of the following will produce a spontaneous process. Also, which of the following processes is endothermic (from the perspective of the system)?

a.  $\Delta S_{sys} = 30 \text{ J/K}$    $\Delta S_{surr} = 50 \text{ J/K}$
b.  $\Delta S_{sys} = -27 \text{ J/K}$    $\Delta S_{surr} = 40 \text{ J/K}$
c.  $\Delta S_{sys} = 140 \text{ J/K}$    $\Delta S_{surr} = -85 \text{ J/K}$
d.  $\Delta S_{sys} = 60 \text{ J/K}$    $\Delta S_{surr} = -85 \text{ J/K}$

## Solution

$$\Delta S_{univ} = \Delta S_{sys} + \Delta S_{surr}$$

- A reaction will be spontaneous if $\Delta S_{univ} > 0$.
- A reaction will be endothermic if $\Delta S_{surr} < 0$.

a.  $\Delta S_{univ} = 30 + 50 \quad = 80 \text{ J/K}$, **spontaneous, exothermic.**
b.  $\Delta S_{univ} = -27 + 40 \quad = 13 \text{ J/K}$, **spontaneous, exothermic.**
c.  $\Delta S_{univ} = 140 + (-85) = 55 \text{ J/K}$, **spontaneous, endothermic.**
d.  $\Delta S_{univ} = 60 + (-85) \ = -25 \text{ J/K}$, **not spontaneous, endothermic.**

---

Notice also from Table 16.3 that there are circumstances under which **temperature plays the deciding factor in whether a reaction will be spontaneous.**

---

Problems 5-7 at the end of this study chapter cover material from this section.

## 16.4 - Free Energy

When you finish this section you will be able to relate free energy and spontaneity.

**Free energy** is a mathematical term that describes **unequivocally** whether a reaction will be spontaneous. It is experimentally useful because **it reflects** $\Delta S_{univ}$. Your textbook states two important relationships,

$$\Delta G = \Delta H - T\Delta S$$

(When no subscript appears, it is assumed that we are referring to the system.)

$$\Delta S_{univ} = \frac{-\Delta G}{T}$$

The first equation gives us an explicit way of calculating free energy. It further says that there are circumstances under which **temperature will determine whether a reaction is spontaneous.** (See Table 16.5 in your textbook.) Note also that since the temperature in Kelvins will always be greater than zero, that $-\Delta G = +\Delta S_{univ}$, **always.**

The case of the temperature dependence of ice melting, as described in your textbook, is a perfect example of this. You have two opposing entropy factors. On the one hand, the reaction is **endothermic**, which **opposes the process** ($\Delta S_{surr} = -$). On the other hand, melting **increases the positional probability of the system** ($\Delta S_{sys} = +$). The temperature will determine which process will dominate (whether ice melts).

The second equation says the $\Delta G$ must be **negative** ($< 0$) in order for a reaction to proceed.

## Example 16.4 A - Free Energy and Spontaneity

Given the values for $\Delta H$, $\Delta S$ and T, determine whether each of the following sets of data represent spontaneous or nonspontaneous processes.

|     | $\Delta H$ (kJ) | $\Delta S$ (J/K) | T (K) |
| --- | --- | --- | --- |
| a.  | 40  | 300  | 130 |
| b.  | 40  | 300  | 150 |
| c.  | 40  | -300 | 150 |
| d.  | -40 | -300 | 130 |
| e.  | -40 | 300  | 150 |

## Solution

$$\Delta G = \Delta H - T\Delta S$$

For a reaction to be spontaneous, $\Delta G$ must be $< 0$. When you do the calculations, make sure that you either **change $\Delta H$ to joules or $\Delta S$ to kilojoules!**

a.   $\Delta G = 40$ kJ - 0.300 kJ/K (130 K)   $= +1$ kJ, **nonspontaneous**
b.   $\Delta G = 40$ kJ - 0.300 kJ/K (150 K)   $= -5$ kJ, **spontaneous**
c.   $\Delta G = 40$ kJ - (-0.300 kJ/K)(150 K) $= +85$ kJ, **nonspontaneous**

    d.   $\Delta G = -40$ kJ - (-0.300 kJ/K)(130 K) = -1 kJ, **spontaneous**

    e.   $\Delta G = -40$ kJ - (0.300 kJ/K) (150 K) = -85 kJ, **spontaneous**

Note that temperature is NOT important to having a spontaneous reaction **only** when the reaction is exothermic and there is an increase in entropy.

---

## Example 16.4 B - Free Energy and Temperature

You know that the boiling point of water is 373 K. See how this compares to the minimum temperature for reaction that you determine thermodynamically for the phase change:

$$H_2O(l) \longrightarrow H_2O(g)$$

where $\Delta H = $ **44 kJ** and $\Delta S = $ **119 J/K**.

### Solution

The criterion for spontaneity is $\Delta G < 0$. This means that $\Delta H - T\Delta S < 0$.

Adding $T\Delta S$ to both sides, $\Delta H < T\Delta S$.

Dividing both sides by $\Delta S$, $\dfrac{\Delta H}{\Delta S} < T$.

Using the data from our problem, $T > \dfrac{4.4 \times 10^4 \text{ J}}{119 \text{ J/K}}$, T > 370 K, which is close to the actual value.

---

Problems 8-16 at the end of this study chapter cover material from this section.

---

## 16.5 - Entropy Changes in Chemical Reactions

When you finish this section you will be able to:

- predict the sign of entropy changes for a given reaction, and
- calculate $\Delta S$ from thermodynamic data tables.

This section in your textbook begins by reminding us that $\Delta S_{surr}$ is related to **heat flow** from the system.

BUT, $\Delta S_{sys}$ is related to the positional **probabilities** for each of the reactants.

For example,

$$2H_2(g) + O_2(g) \longrightarrow 2H_2O(g)$$

$\Delta S = -89$ kJ (at 25°C). We lose entropy because **3 total moles** of the gases on the left have **more** positional possibilities than **2 moles** of vapor on the right.

- **For a chemical reaction involving only the gas phase, entropy is related to the total number of moles on either side of the equation. A decrease means lower entropy, an increase means higher entropy.**

- For a chemical reaction involving different phases, the production of a gas will (in general) increase the entropy much more than an increase in the number of moles of a liquid or solid.

For example,

$$2HNO_3(aq) + Na_2CO_3(s) ----> 2NaNO_3(aq) + H_2O(l) + CO_2(g)$$

$$\Delta S = +88 \text{ kJ (at } 25°C).$$

---

## Example 16.5 A - The Sign of Entropy Changes

Predict the sign of $\Delta S°$ for each of the following reactions:

  a.  $(NH_4)_2Cr_2O_7(s) ----> Cr_2O_3(s) + 4H_2O(l) + N_2(g)$
  b.  $Mg(OH)_2(s) ----> MgO(s) + H_2O(g)$
  c.  $PCl_5(g) ----> PCl_3(g) + Cl_2(g)$

## Solution

  a.  There are many more moles on the right hand side. In addition, a gas is formed on the right hand side. **The entropy will be positive.**

  b.  A gas is formed on the right hand side. **The entropy will be positive.**

  c.  More moles of gas are present on the right hand side. **The entropy will be positive.**

---

The **Third Law of Thermodynamics** says that **the entropy of a perfect crystal at 0 K is zero.** This means that the absolute entropy of substances can be explicitly measured. (See Appendix 4 in your textbook.)

As was true with $\Delta H°$ (a state function), $\Delta S°$ (also a state function) can be determined as the difference between the sum of the entropy of products minus the sum of the entropy of the reactants,

$$\Delta S°_{reaction} = \Sigma n_p S°_{products} - \Sigma n_r S°_{reactants}$$

---

## Example 16.5 B - Entropy of Reaction

Calculate $\Delta S°$ for each of the following reactions, using data from Appendix 4 in your textbook.

  a.  $N_2O_4(g) ----> 2NO_2(g)$
  b.  $Fe_2O_3(s) + 2Al(s) ----> 2Fe(s) + Al_2O_3(s)$
  c.  $4NH_3(g) + 5O_2(g) ----> 4NO(g) + 6H_2O(g)$

## Solution

$$\Delta S°_{reaction} = \Sigma n_p S°_{products} - \Sigma n_r S°_{reactants}$$

  a.  $\Delta S°_{rxn} = 2S°(NO_2) - S°(N_2O_4)$
       $= 2 \text{ mol } (240 \text{ J/K·mol}) - 1 \text{ mol } (304 \text{ J/K·mol})$
      $\Delta S°_{rxn} = \textbf{176 J/K}$  (The entropy should increase due to an increase in the number of moles of gas.)

b.  $\Delta S°_{rxn}$  = [2S°(Fe) + S°(Al$_2$O$_3$)] - [S°(Fe$_2$O$_3$) + 2S°(Al)]

$\phantom{b.  \Delta S°_{rxn}}$ = [2 mol (27 J/K·mol) + 1 mol (51 J/K·mol)] - [1 mol (90 J/K·mol) + 2 mol (28 J/K·mol)]

$\phantom{b.}$  $\Delta S°_{rxn}$ = **41 J/K**  (This is fairly small because there were no net phase changes.)

c.  $\Delta S°_{rxn}$  = [4S°(NO) + 6S°(H$_2$O)] - [4S°(NH$_3$) + 5S°(O$_2$)]

$\phantom{c.  \Delta S°_{rxn}}$ = [4 mol (211 J/K·mol) + 6 mol (189 J/K·mol)] - [4 mol (193 J/K·mol) + 5 mol (205 J/K·mol)]

$\phantom{c.}$  $\Delta S°_{rxn}$ = **181 J/K**  (We increased the number of moles of gas, which is reflected by the increase in entropy.)

---

Problems 17-21 at the end of this study chapter cover material from this section.

---

## 16.6 - Free Energy and Chemical Reactions

When you finish this section you will be able to calculate the standard free energy of formation and use it to predict spontaneity of chemical reactions.

The **standard free energy change ($\Delta G°$)** is the free energy change that occurs if reactants **in their standard states (1 atm, 25°C)** are converted to products in their standard states.

Your textbook points out that $\Delta G°$ *cannot be measured directly*, but is an important value because it represents a *standard set of conditions* at which to compare properties of reactions (as we will see later).

**Three methods** of calculating $\Delta G°$ are introduced in this section.

1.  $\Delta G° = \Delta H° - T\Delta S°$
2.  **by manipulating known equations, à la Hess' Law problems for $\Delta H°$.**
3.  $\Delta G° = \Sigma n_p \Delta G°products - \Sigma n_r \Delta G°reactants$

The following examples illustrate each of these methods.

---

## Example 16.6 A - Standard Free Energy from Entropy and Enthalpy

Using data for **$\Delta H°$ and $\Delta S°$**, calculate $\Delta G°$ for the following reactions at 25°C and 1 atm.

a.  Cr$_2$O$_3$(s) + 2Al(s) ----> Al$_2$O$_3$(s) + 2Cr(s)

b.  C$_3$H$_8$(g) + 5O$_2$(g) ----> 3CO$_2$(g) + 4H$_2$O(g)

## Solution

$$\Delta G° = \Delta H° - T\Delta S°$$

We need to calculate $\Delta H°$ and $\Delta S°$ for each reaction.

$$\Delta H° = \Sigma n_p \Delta H°_{products} - \Sigma n_r \Delta H°_{reactants}$$
$$\Delta S° = \Sigma n_p \Delta S°_{products} - \Sigma n_r \Delta S°_{reactants}$$

a.  $\Delta H°$ = [1 mol (-1676 kJ/mol) + 2 mol (0 kJ/mol)] - [1 mol (-1128 kJ/mol) + 2 mol (0 kJ/mol)]

$\phantom{a.  \Delta H° = [1 mol (}$↑$\phantom{1676 kJ/mol) + 2 mol (0}$↑$\phantom{kJ/mol)] - [1 mol (-11}$↑$\phantom{28 kJ/mol) + 2 mol (0}$↑

$\phantom{a.  \Delta H° = [}$Al$_2$O$_3$$\phantom{1676 kJ/mol) + 2 m}$Cr$\phantom{l)] - [1 mol (-11}$Cr$_2$O$_3$$\phantom{8 kJ/mol) + 2 mo}$Al

$\phantom{a.  \Delta H° = }$**$\Delta H°$ = -548 kJ**

$\Delta S° = [1 \text{ mol } (51 \text{ J/K·mol}) + 2 \text{ mol } (24 \text{ J/K·mol})] - [1 \text{ mol } (81 \text{ J/K·mol}) + 2 \text{ mol } (28 \text{ J/K·mol})]$

$\quad\quad\quad\quad\quad\uparrow\quad\quad\quad\quad\quad\quad\quad\uparrow\quad\quad\quad\quad\quad\quad\quad\uparrow\quad\quad\quad\quad\quad\quad\quad\uparrow$

$\quad\quad\quad\quad\quad Al_2O_3 \quad\quad\quad\quad\quad Cr \quad\quad\quad\quad\quad Cr_2O_3 \quad\quad\quad\quad\quad Al$

$\Delta S° = -38 \text{ J/K}$

$\Delta G° = \Delta H° - T\Delta S° = -548 \text{ kJ} - 298 \text{ K } (-0.038 \text{ kJ/K})$

$\Delta G° = -537 \text{ kJ},$   **The reaction is spontaneous.**

b.   $\Delta H° = [3 \text{ mol } (-393.5 \text{ kJ/mol}) + 4 \text{ mol } (-242 \text{ kJ/mol})] - [5 \text{ mol } (0 \text{ kJ/mol}) + 1 \text{ mol } (-104 \text{ kJ/mol})]$

$\quad\quad\quad\quad\quad\uparrow\quad\quad\quad\quad\quad\quad\quad\uparrow\quad\quad\quad\quad\quad\quad\quad\uparrow\quad\quad\quad\quad\quad\quad\quad\uparrow$

$\quad\quad\quad\quad\quad CO_2 \quad\quad\quad\quad\quad H_2O \quad\quad\quad\quad\quad O_2 \quad\quad\quad\quad\quad C_3H_8$

$\Delta H° = -2044 \text{ kJ}$

$\Delta S° = [3 \text{ mol } (214 \text{ J/K·mol}) + 4 \text{ mol } (189 \text{ J/K·mol})] - [5 \text{ mol } (205 \text{ J/K·mol}) + 1 \text{ mol } (270 \text{ J/K·mol})]$

$\quad\quad\quad\quad\quad\uparrow\quad\quad\quad\quad\quad\quad\quad\uparrow\quad\quad\quad\quad\quad\quad\quad\uparrow\quad\quad\quad\quad\quad\quad\quad\uparrow$

$\quad\quad\quad\quad\quad CO_2 \quad\quad\quad\quad\quad H_2O \quad\quad\quad\quad\quad O_2 \quad\quad\quad\quad\quad C_3H_8$

$\Delta S° = 103 \text{ J/K}$

$\Delta G° = \Delta H° - T\Delta S° = -2044 \text{ kJ} - 298 \text{ K } (0.103 \text{ kJ/K})$

$\Delta G° = -2075 \text{ kJ},$   **This combustion reaction is highly spontaneous.**

---

# Example 16.6 B - Standard Free Energy by Combining Equations

Given the following data:

(Equation 1)   $S(s) + 3/2O_2(g) \longrightarrow SO_3(g)$          $\Delta G° = -371 \text{ kJ}$

(Equation 2)   $2SO_2(g) + O_2(g) \longrightarrow 2SO_3(g)$          $\Delta G° = -142 \text{ kJ}$

Calculate $\Delta G°$ for the reaction

(Goal Equation)   $S(s) + O_2(g) \longrightarrow SO_2(g)$

Is the reaction spontaneous?

## Solution

We solve this problem just as we would a Hess' Law problem (see Chapter 6). We must manipulate equations 1 and 2 so that we can combine them to get the reaction of interest. (Remember what you do to the equations, must also be done to the $\Delta G°$ values!)

**One mole of S(s)** appears only once on the left hand side of the **goal equation**, and **one mole** on the left hand side of equation 1.

### Equation 1 must remain as it is.

**One mole of SO$_2$(g)** appears on the **right hand side** of the goal equation. **Two moles** of $SO_2(g)$ are present on the **left hand side** of equation 2. We must therefore

**multiply equation 2 by -1/2**

to get one mole of $SO_2(g)$ on the right hand side.  (Remember to multiply $\Delta G°$ by -1/2 as well!)

$$S(s) + 3/2O_2(g) \longrightarrow SO_3(g) \qquad \Delta G° = -371 \text{ kJ}$$
$$SO_3(g) \longrightarrow SO_2(g) + 1/2O_2(g) \qquad \Delta G° = +71 \text{ kJ}$$
$$S(s) + O_2(g) \longrightarrow SO_2(g) \qquad \Delta G° = \textbf{-300 kJ}$$

## Example 16.6 C - Standard Free Energy from "Products - Reactants"

Calculate $\Delta G°$ for the reaction

$$C_3H_8(g) + 5O_2(g) \rightleftharpoons 3CO_2(g) + 4H_2O(g)$$

using $\Delta G°$ data from Appendix 4.  Compare the answer with that from our Example 16.6 A part b.

### Solution

$\Delta G° = \Sigma n_p \Delta G°_{products} - \Sigma n_r \Delta G°_{reactants}$
   $= [3\Delta G°(CO_2) + 4\Delta G°(H_2O)] - [\Delta G°(C_3H_8) + 5\Delta G°(O_2)]$
   $= [3 \text{ mol } (-394 \text{ kJ/mol}) + 4 \text{ mol } (-229 \text{ kJ/mol})] - [1 \text{ mol } (-24 \text{ kJ/mol}) + 5 \text{ mol } (0 \text{ kJ/mol})]$
$\Delta G° = \textbf{-2074 kJ}$

The $\Delta G°$ values agree within about 1%.  Also note that as with $\Delta H°$, $\Delta G°$ for elements in their standard states equal 0.

Problems 22-29 at the end of this study chapter cover material from this section.

## 16.7 - The Dependence of Free Energy on Pressure

When you finish this section you will be able to calculate $\Delta G$ at pressures other than 1 atmosphere.

We have until now assumed standard conditions.  This section deals with **free energy at nonstandard pressures**.

For an ideal gas, enthalpy is not pressure dependent.  **Entropy, however is affected by pressure.**  More positions are possible at lower pressure than higher pressure, therefore

$$S_{low pressure} > S_{high pressure}$$

Your textbook derives the relationship

$$\Delta G = \Delta G° + RT \ln(Q)$$

where **R** = 8.3145 J/K·mol
       **T** = temperature in Kelvins
       **Q** = reaction quotient (the mass action expression relating to **initial** quantities.)

It is important to learn to use **and interpret** the results from this equation. Through a similar equation (to be introduced in the next section) we can relate equilibrium constants to ΔG.

---

## Example 16.7 - Relating Free Energy and Pressure

Calculate ΔG at 700 K for the following reaction:

$$C(s, \text{graphite}) + H_2O(g) \rightleftharpoons CO(g) + H_2(g)$$

Initial pressures are $P_{H_2O}$ = 0.85 atm, $P_{CO}$ = 1.0 × 10⁻⁴ atm, $P_{H_2}$ = 2.0 × 10⁻⁴ atm.

### Solution

Our first task is to determine ΔG°.

ΔG° = [ΔG°(CO) + ΔG°(H₂)] - [ΔG°(C) + ΔG°(H₂O)]

ΔG° = [1 mol (-137 kJ/mol) + 1 mol (0 kJ/mol)] - [1 mol (0 kJ/mol) + 1 mol (-229 kJ/mol)]

**ΔG° = +92 kJ**

This tells us that the reaction is not spontaneous **under standard conditions**. To evaluate at 700 K and the given pressures,

$$Q = \frac{P_{H_2}\, P_{CO}}{P_{H_2O}} = \frac{2.0 \times 10^{-4}\ \text{atm} (1.0 \times 10^{-4}\ \text{atm})}{0.85\ \text{atm}} = \mathbf{2.35 \times 10^{-8}\ atm}$$

ΔG = ΔG° + RT ln(Q) = 9.2 × 10⁴ J + 8.3145 J/K mol(700 K)ln(2.35 × 10⁻⁸)

= 9.2 × 10⁴ J + (-1.022 × 10⁵J)

= -1.02 × 10⁴ J

**ΔG = -10 kJ***

Under conditions of high temperature and low product pressure, the reaction becomes spontaneous.

The final subheading in Section 16.7 of your textbook addresses "The Meaning of ΔG for a Chemical Reaction." It stresses that a spontaneous reaction will not necessarily go to completion. Rather, there may be some intermediate point that reflcets the lowest possible ΔG value for the reaction.

---

Problem 30 at the end of this study chapter covers material from this section.

---

## 16.8 - Free Energy and Equilibrium

When you finish this section you will be able to:

- interpret the direction of a reaction given appropriate data, and
- interconvert between K and ΔG°.

---

*Note the discussion on units and "rounds" in the margin next to <u>Sample Exercise 16.13 in your textbook.</u>

**Equilibrium occurs at the lowest free energy available to the system.** In Chapter 13, we defined equilibrium as occurring when the forward rate of reaction is equal to the reverse rate of reaction. That is still valid. Thermodynamically, your textbook defines equilibrium as occurring when

$$G_{forward\ reaction} = G_{reverse\ reaction}$$

In other words, **$\Delta G = 0$ at equilibrium.**

If $\Delta G < 0$, it means that $G_{reactants} > G_{products}$ and the reaction will **go to the right until** $G_{reactants} = G_{products}$. If $\Delta G > 0$, it means that $G_{reactants} < G_{products}$ and the reaction will **go to the left until** $G_{reactants} = G_{products}$.

<u>Table 16.7 in your textbook</u> mathematically relates **$\Delta G°$ to K.** Remember that if K = 1 (the equilibrium condition), ln K = 0 and $\Delta G = \Delta G°$. At equilibrium, therefore,

$$\Delta G° = \text{-RT ln K}$$

---

# Example 16.8 A - Free Energy and the Direction of Reaction

Using the same reaction as in our Example 16.7,

$$C(s,\ graphite) + H_2O(g) \rightleftharpoons CO(g) + H_2(g)$$

where **T = 700 K** and **$\Delta G° = 92$ kJ**, determine the direction of reaction if the following initial pressure of each gas is

$$P_{H_2O} = 0.67\ \text{atm}, \quad P_{CO} = 0.23\ \text{atm}, \quad P_{H_2} = 0.51\ \text{atm}$$

## Solution

The goal is to calculate $\Delta G$.

$$\Delta G = \Delta G° + RT\ \ln K$$

$$Q = \frac{P_{H_2}\ P_{CO}}{P_{H_2O}} = \frac{0.51\ \text{atm}\ (0.23\ \text{atm})}{0.67\ \text{atm}} = 0.175\ \text{atm}$$

$$\Delta G = 9.2 \times 10^4\ J + 8.3145\ J/K \cdot mol\ (700\ K)\ln(0.175)$$

$$\Delta G = 82000\ J = \textbf{82 kJ}$$

**This reaction will go very far to the left.**

---

# Example 16.8 B - Free Energy and Equilibrium Constant

Given the values for $\Delta G°$ that you calculated in our Example 16.6 A, calculate K for the following reaction (at 25°C):

$$Cr_2O_3(s) + 2Al(s) \rightleftharpoons Al_2O_3(s) + 2Cr(s)$$

## Solution

$$\Delta G° = -537\ kJ$$

$$\Delta G^\circ = \text{-RT ln K}$$

$$\ln K = \frac{\Delta G^\circ}{\text{-RT}}, \text{ or } K = e^{-\Delta G^\circ/RT}$$

$$\ln K = \frac{\text{-}5.37 \times 10^5 \text{ J}}{\text{-}8.3145 \text{ J/K·mol (298 K)}} = 217$$

$K = e^{217} = 2 \times 10^{94}$, this reaction is quite spontaneous.

---

# Example 16.8 C - Summing It All Up

We have previously (Chapter 15) studied the weak base - strong acid titration

$$NH_3(aq) + H^+(aq) \rightleftharpoons NH_4^+(aq)$$

We said that you can look at it as the sum of two reactions (for purposes of calculating K).

1.  $NH_3(aq) + H_2O(l) \rightleftharpoons NH_4^+(aq) + OH^-(aq)$        ($K_b = 1.8 \times 10^{-5}$)
2.  $H^+(aq) + OH^-(aq) \rightleftharpoons H_2O(l)$        ($1/K_w = 1.0 \times 10^{+14}$)

Please do the following:

   a.   Calculate K for the titration reaction using the two equilibrium expressions.
   b.   Calculate $\Delta H^\circ$, $\Delta S^\circ$, $\Delta G^\circ$ and K for the titration reaction using data from Appendix 4 of your textbook.
   c.   How do the calculated K values compare?
   d.   Is the titration spontaneous?

## Solution

   a.   Summing the reactions requires multiplying the equilibrium constants

$$K_{titration} = K_b \times 1/K_w = 1.8 \times 10^9$$

   b.   $\Delta H^\circ$ and $\Delta S^\circ = $ S(products) - S(reactants)

$$\Delta H^\circ = [1 \text{ mol } (\text{-}132 \text{ kJ/mol})] - [1 \text{ mol } (\text{-}80 \text{ kJ/mol}) + 1 \text{ mol } (0 \text{ kJ/mol})]$$
$$\qquad\qquad \uparrow \qquad\qquad\qquad \uparrow \qquad\qquad\qquad \uparrow$$
$$\qquad\quad NH_4^+ \qquad\qquad\quad NH_3 \qquad\qquad\quad H^+$$

$\Delta H^\circ = \text{-}52 \text{ kJ}$

$$\Delta S^\circ = [1 \text{ mol } (113 \text{ J/K·mol})] - [1 \text{ mol}(111 \text{ J/K·mol}) + 1 \text{ mol } (0 \text{ J/K·mol})]$$

$\Delta S^\circ = .2 \text{ J/K}$

$$\Delta G^\circ = \Delta H^\circ - T\Delta S^\circ = \text{-}52000 \text{ J} - 298 \text{ K } (2 \text{ J/K}) = \text{-}52596 \text{ J}$$

$\Delta G^\circ = \text{-}53 \text{ kJ}$

$$\ln K = \frac{\Delta G}{\text{-RT}} = \frac{+ 52596 \text{ J}}{8.3145 \text{ J/K·mol (298 K)}} = 21.23$$

$K = e^{21.23} = 1.7 \times 10^9$

c. The K values are very comparable (within about 5%).

d. The titration is spontaneous (as we know from experience).

---

Problems 31-34 at the end of this study chapter cover material from this section.

---

**END OF CHAPTER SIXTEEN**

# SELF-TEST QUESTIONS

1.  Given 8 molecules in the two-bulb set up described in <u>Table 16.1 in your textbook</u>, calculate the relative probability of finding all 8 molecules in the left-hand bulb. What does this tell you regarding entropy and probability?

2.  Which of the following pairs of substances is likely to have the higher positional entropy?

    a.  $HCl(aq)$ or $HCl(g)$          d.  $H_2O$ (s) or $H_2O(l)$
    b.  $P_4(s)$ or $P_4O_{10}(g)$          e.  $NO_2(g)$ or $N_2O_4(g)$
    c.  $O_2(g)$ or $P_4O_{10}(g)$          f.  $Ar(g)$ at 5 atm or $Ar(g)$ at 0.30 atm.

3.  Predict the sign of the entropy change for each of the following processes.

    a.  Potassium hydroxide pellets are dissolved in water.
    b.  Solid ammonium dichromate is burned to give solid chromium oxide, water vapor and nitrogen gas.
    c.  Saturated calcium acetate is mixed with ethanol to form a gel.

4.  Predict the sign of the entropy change for each of the following reactions:

    a.  $Ag^+(aq) + Cl^-(aq) \rightarrow AgCl(s)$
    b.  $NH_4Cl(s) \rightarrow NH_3(g) + HCl(g)$
    c.  $H_2(g) + Br_2(g) \rightarrow 2HBr(g)$

5.  Calculate $\Delta S_{surr}$ for each of the following reactions at 25°C and 1 atm:

    a.  $Br(l) \text{ ----> } Br(g)$        $\Delta H = +31$ kJ
    b.  $2C_2H_6(g) + 7O_2(g) \text{ ----> } 4CO_2(g) + 6H_2O(g)$        $\Delta H = -2857$ kJ.

6.  A chemical reaction gives a change in entropy of the universe of -48 J/K. Is the process spontaneous? Why or why not?

7.  Which of the following values represent spontaneous processes? Which ones are exothermic (from the point of view of the system)?

    a.  $\Delta S_{sys} = +358$ J/K,  $\Delta S_{surr} = -358$ J/K
    b.  $\Delta S_{sys} = -358$ J/K,  $\Delta S_{surr} = -52$ J/K
    c.  $\Delta S_{sys} = -358$ J/K,  $\Delta S_{surr} = +463$ J/K
    d.  $\Delta S_{sys} = +358$ J/K,  $\Delta S_{surr} = -463$ J/K

8.  Given the following values for $\Delta H$, $\Delta S$ and T, determine whether each of the following sets of data represent spontaneous or nonspontaneous processes.

    |     | $\Delta H$(kJ) | $\Delta S$(J/k) | T(K) |
    |-----|------|------|------|
    | a.  | -16  | 50   | 300  |
    | b.  | 12   | 40   | 300  |
    | c.  | -5   | -20  | 200  |
    | d.  | -5   | 20   | 200  |
    | e.  | -5   | -20  | 500  |

9.  Given the following $\Delta H$ and $\Delta S$ values, determine the temperature at which the reactions would be spontaneous:

    a.  $\Delta H = 10.5$ kJ;    $\Delta S = 30$ J/K
    b.  $\Delta H = 1.8$ kJ;    $\Delta S = 113$ J/K
    c.  $\Delta H = -11.7$ kJ;    $\Delta S = -105$ J/K
    d.  $\Delta H = -126$ kJ;    $\Delta S = 84$ J/K

10. Predict the sign of the entropy change for each of the following processes:

   a.  evaporating a beaker of ethanol at room temperature
   b.  cooling nitrogen gas from 80°C to 20°C
   c.  freezing liquid bromine below its melting point (-7.2°C)

11. The heat of fusion for actinium is 10.50 kJ/mol. The entropy of fusion is 9.6 J/K·mol. **Calculate the melting point of actinium.**

12. At a constant temperature of 298 K, calculate $\Delta S_{sys}$ and $S_{univ}$ for the free expansion of 3.0 L of an ideal gas at 1.0 atm to 11.0 L.

13. The heat of vaporization for protactinium is 481 kJ/mol. The entropy of vaporization is 109 J/K·mol. Calculate the boiling point of protactinium. Compare with the actual value of approximately 4500 K.

14. If the molar heat of vaporization of ethanol is 39.3 kJ/mol and its boiling point is 78.3°C, **calculate $\Delta S$ for the vaporization of 0.50 mol ehtanol.**

15. The normal boiling point of diethyl ether is 308 K. The enthalpy of vaporization is 27.2 kJ/mol. **Calculate $\Delta S$ for the vaporization of 1.0 mol of diethyl ether under these conditions.**

16. The melting point of silicon is 1683 K. The heat of fusion is 46.4 kJ/mol. **Calculate the entropy of vaporization of silicon.**

17. Predict the sign of $\Delta S°$ for each of the following reactions:

   a.  $Sr(g) + 1/2 O_2(g) ----> SrO(c)$  (c = crystalline)
   b.  $2Al(s) + 3F_2(g) ----> 2AlF_3(s)$

18. Using Appendix 4, calculate the standard enthalpy changes for the following reactions at 25°C:

   a.  $CaCO_3(s) \rightarrow CaO(s) + CO_2(g)$
   b.  $N_2(g) + 3H_2(g) \rightarrow 2NH_3(g)$
   c.  $H_2(g) + Cl_2(g) \rightarrow 2HCl(g)$

19. Using data from Appendix 4 in your textbook, calculate $\Delta S°$ for each of the following reactions:

   a.  $CH_4(g) + N_2(g) ----> HCN(g) + NH_3(g)$
   b.  $2Ag_2O(s) ----> 4Ag(s) + O_2(g)$
   c.  $Cd(s) + 1/2 O_2(g) ----> CdO(s)$

20. Using data for $\Delta H°$ and $\Delta S°$ in Appendix 4 in your textbook, calculate $\Delta G°$ (at 25°) for each of the reactions in **problem #19.**

21. Calculate $\Delta G°$ for each of the reactions in **problem #19** using $\Delta G°$ data from Appendix 4 in your textbook. How do these compare with your answers from **problem #20?**

22. Using Appendix 4, calculate $\Delta G°$ for the combustion of ethane ($C_2H_6$):

$$2C_2H_6(g) + 7O_2(g) \rightarrow 4CO_2(g) + 6H_2O(l)$$

23. Calculate $\Delta G°$ for the following reactions at 25°C:

   a.  $2MgO(s) \rightarrow 2Mg(s) + O_2(g)$
   b.  $CH_4(g) + 2O_2(g) \rightarrow CO_2(g) + 2H_2O(l)$

24. Calculate $\Delta G°$ for the following reactions:

    a.   $2Mg(s) + O_2(g) \rightarrow 2MgO(s)$
    b.   $2C_2H_2(g) + 5O_2(g) \rightarrow 4CO_2(g) + 2H_2O(l)$
    c.   $N_2(g) + O_2(g) \rightarrow 2NO(g)$

25. Given the following data:

    a.   $N_2(g) + 2O_2(g) ----> 2NO_2(g)$      $\Delta G° = 104$ kJ
    b.   $2NO(g) + O_2(g) ----> 2NO_2(g)$      $\Delta G° = -70$ kJ

    Calculate $\Delta G°$ for $N_2(g) + O_2(g) ----> 2NO(g)$.

26. Using Appendix 4, calculate $\Delta G°$ for each of the following reactions at 298 K:

    a.   $2Cu_2O(s) + O_2(g) \rightarrow 4CuO(s)$
    b.   $C_2H_5OH(l) \rightarrow C_2H_4(g) + H_2O(g)$

27. Calculate $\Delta G°$ for the following reaction at 25°C (use Appendix 4):

$$2H_2(g) + O_2(g) \rightleftharpoons 2H_2O(l)$$

28. Given the following data:

    a.   $2H_2(g) + C(s) ---->CH_4(g)$      $\Delta G° = -51$ kJ
    b.   $2H_2(g) + O_2(g) ----> 2H_2O(l)$      $\Delta G° = -474$ kJ
    c.   $C(s) +O_2(g) ----> CO_2(g)$      $\Delta G° = -394$ kJ

    Calculate $\Delta G°$ for $CH_4(g) + 2O_2(g) ----> CO_2(g) + 2H_2O(l)$.

29. Calculate $\Delta G$ at 600 K for the following reaction:

$$P_4(g) + 5O_2(g) \rightleftharpoons P_4O_{10}(s),$$

where the initial pressures are $P_{P_4} = 0.52$ atm and $P_{O_2} = 2.1 \times 10^{-3}$ atm.

30. Calculate the equilibrium constant, K, at 25°C for each of the reactions in **problem #19**.

31. The value of the equilibrium constant for given reaction is $K = 6 \times 10^{-23}$. What does that indicate about the spontaneity of the reaction?

32. The value of the equilibrium constant for a given reaction is $K = 8 \times 10^{58}$. What does this tell us regarding the speed of the reaction?

33. We said in chapter 14 that at 25°C, $K_w = 1.0 \times 10^{-14}$. Calculate $K_w$ thermodynamically and compare it to our chapter 14 value for the reaction:

$$H_2O(l) \rightleftharpoons H^+(aq) + OH^-(aq)$$

34. Calculate $K_p$ for the following reaction at 25° C:  $2H_2O(g) \rightleftharpoons 2H_2(g) + O_2(g)$.

# ANSWERS TO SELF-TEST QUESTIONS

1.  Probability $= 1/2^8 = 1/64$. This says that the probability dictates that molecules will be distributed more evenly throughout the bulbs - in other words, a higher degree of entropy.

2.  a.  $HCl(g)$
    b.  $P_4O_{10}(g)$
    c.  $P_4O_{10}(g)$ (though the difference is quite small.)
    d.  $H_2O(l)$
    e.  $N_2O_4(g)$
    f.  $Ar(g)$ at 0.30 atm.

3.  a.  positive
    b.  sharply positive
    c.  negative

4.  a.  $\Delta S$ = negative
    b.  $\Delta S$ = positive
    c.  We cannot accurately predict the sign of $\Delta S$ but we know that the change is very small.

5.  a.  $\Delta S_{surr}$ = -104 J/K
    b.  $\Delta S_{surr}$ = 9590 J/K

6.  The entropy of the universe decreased. The process is therefore **nonspontaneous.**

7.  Only process "c" is spontaneous. Processes "c" is the only exothermic process.

8.  a.  spontaneous
    b.  nonspontaneous
    c.  spontaneous
    d.  spontaneous
    e.  nonspontaneous

9.  a.  spontaneous at 350 K or above
    b.  spontaneous at 16 K or above
    c.  spontaneous at 111 K or below
    d.  spontaneous at all temperatures above -1.5 K

10. a.  $\Delta S > 0$ (positive)
    b.  $\Delta S < 0$ (negative)
    c.  $\Delta S < 0$ (negative)

11. melting point = 1100 K

12. $S_{sys} = S_{univ} = 0.0131$ L·atm/K

13. boiling point = 4410 K.

14. 56.0 J/K

15. $\Delta S$ = 88.3 J/K

16. Entropy vaporization = 27.6 J/K·mol

17. a.  negative
    b.  negative

18. a.  $\Delta S° = 160.5$ J/K
    b.  $\Delta S° = -199$ J/K
    c.  $\Delta S° = 20$ J/K

19. a.  +17 J/K
    b.  +133 J/K
    c.  -100 J/K

20. a.  +159 kJ
    b.  +22 kJ
    c.  -228 kJ

21. a.  +159 kJ
    b.  +22 kJ
    c.  -228 kJ

22. $\Delta G° = -2935$ kJ

23. a.  G° = 1139 kJ
    b.  G° = -818 kJ

24. a.  -1140 kJ
    b.  -2470 kJ
    c.  173.4 kJ

25. $\Delta G° = 174$ kJ

26. a.  $\Delta G° = -216$ kJ
    b.  $\Delta G° = 14.3$ kJ

27. $\Delta G° = -474.4$ kJ

28. $\Delta G° = -817$ kJ

29. $\Delta G° = -2730$ kJ

30. a.  $K = 1.4 \times 10^{-28}$
    b.  $K = 1.4 \times 10^{-4}$
    c.  $K = 9.2 \times 10^{39}$

31. The reaction is not spontaneous.

32. The equilibrium constant says <u>nothing</u> about the rate of reaction.

33. $K = 9.5 \times 10^{-15}$. This is very close to our value of $K_w$ of $1.0 \times 10^{-14}$.

34. $K_p = 5.3 \times 10^{-81}$

# CHAPTER 17

# Electrochemistry

| Section | Contents | Page in this Study Guide |
|---|---|---|
| 17.1 | Galvanic Cells | 371 |
| 17.2 | Standard Reduction Potentials | 372 |
| 17.3 | Cell Potential, Electrical Work and Free Energy | 376 |
| 17.4 | Dependence of Cell Potential on Concentration | 378 |
| 17.5 | Batteries | 381 |
| 17.6 | Corrosion | 381 |
| 17.7* | Electrolysis | 382 |

Your textbook defines **electrochemistry** as **the study of the interchange of chemical and electrical energy.** In this chapter we are concerned with the **use of chemical reactions to generate an electric current** and **the use of electric current to produce chemical reactions.**

## 17.1 - Galvanic Cells

When you finish this section you will be able to define some of the terms that are commonly used with electrochemistry.

We have discussed reduction-oxidation (redox) reactions in section 4.9. The following terms should be reviewed: **reduction, oxidation, reducing agent, oxidizing agent, half-reaction. Review the half-reaction method of balancing redox equations** (section 4.9) and go over some of the problems at the end of that chapter in this study guide and in your textbook. The rest of this discussion assumes that you are able to balance redox equations.

A **galvanic cell** is a device in which **chemical energy is converted to electrical energy.** There is a need to physically separate the oxidizing and reducing agents in galvanic cells so that energy of reaction can be used. Figures 17.2 and 17.3 in your textbook show the important features of a galvanic cell. You should know the function of the following:

- **cathode** - reduction occurs here. Species undergoing reduction ("oxidizing agent") receive electrons from the cathode.
- **anode** - oxidation occurs here. Species undergoing oxidation ("reducing agent") lose electrons here.
- **salt bridge** (or porous disk) - allows exchange of ions to keep electric neutrality while electroactive solutions remain separated.

*Section 17.8 deals with "Commercial Electrolytic Processes." We will not cover this material in this study guide.*

Also, know the following terms:

- **electromotive force (emf)** - the driving force with which electrons are pulled through a wire.
- **volt (V)** - the unit of electrical potential. It equals 1 joule/coulomb.

In constructing galvanic cells, keep in mind the **direction of electron flow**. Species undergoing *reduction receive electrons from the cathode*. Species undergoing *oxidation donate electrons to the anode*. THE DIRECTION OF ELECTRON FLOW IS THEREFORE **FROM THE ANODE TO THE CATHODE.**

---

## Example 17.1 - A Bit of Review

Identify the species in each of the following equations that would receive electrons from the cathode and that which would lose electrons at the anode in each of the following galvanic cells.

    a.  $Mg(s) + 2H^+(aq) \rightarrow Mg^{2+}(aq) + H_2(g)$
    b.  $MnO_4(aq) + 5Fe^{2+}(aq) + 8H^+(aq) \rightarrow Mn^{2+}(aq) + 5Fe^{3+}(aq) + H_2O(l)$

### Solution

It is quite useful to separate balanced reactions into half-reactions. The **reduction** half-reaction **receives** electrons at the **cathode**. The **oxidation** half-reaction **loses** electrons at the **anode**.

    a.  oxidation (anode):  $Mg(s) \rightarrow Mg^{2+}(aq) + 2e^-$
        reduction (cathode):  $2H^+(aq) + 2e^- \rightarrow H_2(g)$

    b.  oxidation (anode):  $Fe^{2+}(aq) \rightarrow Fe^{3+}(aq) + e^-$
        reduction (cathode):  $MnO_4^-(aq) + 8H^+(aq) + 5e^- \rightarrow Mn^{2+}(aq) + 4H_2O(l)$

As in all galvanic cells, the direction of electron flow is from the anode to the cathode.

---

Problems 1-2 at the end of this study chapter cover material from this section.

---

## 17.2 - Standard Reduction Potentials

When you finish this section you will be able to:

- determine the proper cell reaction, and E° value, given two half-reactions, and
- describe the make up of a galvanic cell involving a given chemical reaction.

The electromotive force (emf) of a galvanic cell is a **combination** of the potentials of two half reactions. Because the cathodic potentials are described relative to anodic reactions, we need **one absolute standard** against which all other half-reactions can be compared. The standard is the **standard hydrogen electrode,**

$$2H^+ + 2e^- \rightarrow H_2 \qquad E° = 0.00 \text{ V (exactly, by definition)}$$

This is an arbitrary, but necessary, assignment. Table 17.1 in your textbook has a list of standard **reduction potentials** for many half-reactions. **Standard means 298 K and 1 atmosphere.**

- Galvanic cells require $E°_{cell} > 0$ V.
- One of the tabulated reduction potentials will have to be reversed (to form an oxidation half-reaction) in every E° calculation.

- To determine which reaction is to be reversed, the **sum of the oxidation and reduction half-reactions must be > 0 V in a galvanic cell.**
- When you **reverse** a reaction, **E° gets the opposite sign.**
- When you **multiply** a reaction by a coefficient (for purposes of balancing), **the E° is NOT changed!**

Let's have some practice with calculating $E°_{cell}$ and determining the reduction and oxidation reactions in a galvanic cell.

---

## Example 17.2 A - Cell Emf's

Using data from <u>Table 17.1 in your textbook</u>, calculate the emf values ($E°_{cell}$) for each of the following reactions. State which are galvanic.

a.  $Mg(s) + 2H^+(aq) \rightarrow Mg^{2+}(aq) + H_2(g)$
b.  $Cu^{2+}(aq) + 2Ag(s) \rightarrow Cu(s) + 2Ag^+(aq)$
c.  $2Zn^{2+}(aq) + 4OH^-(aq) \rightarrow 2Zn(s) + O_2(g) + 2H_2O(l)$

### Solution

We need to split each reaction into two half-reactions and look up their reduction potentials. **(Remember that when a half-reaction is an oxidation, E° must change sign!)**

a.  reduction: $2H^+ + 2e^- \rightarrow H_2$      E°    = 0.00 V
    oxidation: $Mg \rightarrow Mg^{2+} + 2e^-$      E°    = +2.37 V
                              $E°_{cell}$    = +2.37 V, galvanic

b.  reduction: $Cu^{2+} + 2e^- \rightarrow Cu$      E°    = +0.34 V
    oxidation: $2Ag \rightarrow 2Ag^+ + 2e^-$      E°    = -0.80 V
                              $E°_{cell}$    = -0.46 V, **not galvanic - will not run in this direction**

c.  reduction: $2Zn^{2+} + 4e^- \rightarrow 2Zn$      E°    = -0.76 V
    oxidation: $4OH^- \rightarrow O_2 + 2H_2O + 4e^-$      E°    = -0.40 V
                              $E°_{cell}$    = -1.16 V, **not galvanic**

Another key point regarding the magnitude and sign of E° for half-reactions is that **THE MORE POSITIVE THE E° VALUE THE MORE LIKELY THE SPECIES IS TO BE REDUCED** (is a stronger oxidizing agent). (The contrary with regard to oxidation is also true.) For example, $Br_2$ is more likely to be reduced than $I_2$, because $Br_2$ has the more positive E°.

---

## Example 17.2 B - Reduction Strength

Place the following in order of increasing strength as oxidizing agents.

$$Fe^{2+}, \quad ClO_2, \quad F_2, \quad AgCl$$

### Solution

From looking at the E° values in <u>Table 17.1 of your textbook</u>, the order is **$Fe^{2+} < AgCl < ClO_2 < F_2$.**

## Example 17.2 C - Composing Galvanic Cells

Given the following half-cells, decide which is the anode and the cathode, balance, write the overall cell reaction and calculate $E°_{cell}$ if the cells are galvanic.

a.   $Ni^{2+} + 2e^- \rightarrow Ni$           $E° = -0.23$ V
    $O_2 + 4H^+ + 4e^- \rightarrow 2H_2O$     $E° = +1.23$ V

b.   $Ce^{4+} + e^- \rightarrow Ce^{3+}$          $E° = +1.70$ V
    $Sn^{2+} + 2e^- \rightarrow Sn$        $E° = -0.14$ V

## Solution

The goal is to see which arrangement of half-reactions will make $E°_{cell} > 0$.

a.   Reversing the $Ni^{2+}$ reduction will create a galvanic cell.  Note that to balance the overall reaction, it must also be multiplied by 2.

| | | |
|---|---|---|
| cathode: | $O_2 + 4H^+ + 4e^- \rightarrow 2H_2O$ | $E° = 1.23$ V |
| anode: | $2Ni \rightarrow 2Ni^{2+} + 4e^-$ | $E° = 0.23$ V |
| **overall:** | $O_2(g) + 4H^+(aq) + 2Ni(s) \rightarrow 2H_2O(l) + 2Ni^{2+}(aq)$ | $E°_{cell} = 1.46$ V |

b.   Reversing the $Sn^{2+}$ half-reaction will create a galvanic cell.  The cerium half-reaction must be multiplied by 2 to balance the overall reaction.

| | | |
|---|---|---|
| cathode: | $2Ce^{4+} + 2e^- \rightarrow 2Ce^{3+}$ | $E° = 1.70$ V |
| anode: | $Sn \rightarrow Sn^{2+} + 2e^-$ | $E° = 0.14$ V |
| **overall:** | $2Ce^{4+}(aq) + Sn(s) \rightarrow Sn^{2+}(aq) + 2Ce^{3+}(aq)$ | $E°_{cell} = 1.84$ V |

---

Immediately <u>after sample exercise 17.1 in your textbook</u>, a method of representing electochemical cells is introduced.  This **line notation** is a shorthand that chemists use to describe cells.  The general line notation formation is:

anode      ||      cathode
side      ||      side

pourous disk or
salt bridge

Each side will have a "phase boundary" separating, for example, a solid electrode from ions in solution, such as $Cu(s)$ and $Cu^{2+}(aq)$.  Such a boundary is denoted by a single vertical line ("|") with the *solid phase on the outside*.  Therefore, for the Copper/Zinc reactions shown in the photograph in <u>Section 17.2 of your textbook</u>,

anode:      $Zn \rightarrow Zn^{2+} + 2e^-$
(oxidation)

cathode:      $Cu^{2+} + 2e^- \rightarrow Cu$
(reduction)

The line notation is

$$Zn(s) \mid Zn^{2+}(aq) \parallel Cu^{2+}(aq) \mid Cu(s)$$

Your textbook points out that on occasion, only ions (rather than solids) will be involved in the redox process. In that case, "inert" electrodes are used. The line notation is given for such an example (17.1 (b)) in your textbook.

---

## Example 17.2 D - Line Notations

Write line notations for each of the following galvanic cells;

a. $Hg^{2+} + Cd \rightarrow Hg + Cd^{2+}$
b. $Pb + 2Cr^{3+} \rightarrow Pb^{2+} + 2Cr^{2+}$
   (a platinum electrode is used at the cathode)
c. $Cu^{2+} + 2Pu^{4+} + 4H_2O \rightarrow Cu + 2PuO_2^+ + 8H^+$
   (a platinum electrode is used at the anode)

## Solutions

A good strategy is to separate the redox equation into its half reactions, so what happens at the anode (oxidation) and cathode (reduction) become apparent.

a. anode: $Cd \rightarrow Cd^{2+} + 2e^-$
   cathode: $Hg^{2+} + 2e^- \rightarrow Hg$

   line notation: $Cd(s) \mid Cd^{2+}(aq) \parallel Hg^{2+}(aq) \mid Hg$

b. anode: $Pb \rightarrow Pb^{2+} + 2e^-$
   cathode: $2Cr^{3+} + 2e^- \rightarrow 2Cr^{2+}$

   The cathodic reaction involves only ions, so we need an inert electrode (Pt) to conduct electricity from one side of the cell to the other. The line notation is

   line notation: $Pb(s) \mid Pb^{2+} \parallel Cr^{3+}, Cr^{2+} \mid Pt$
   $\nwarrow \quad \nearrow$
   order here is not
   important

c. This seems complex, but becomes clearer when you separate the equation into its half-reactions:

   anode: $2Pu^{4+} + 4H_2O \rightarrow 2PuO_2^+ + 8H^+ + 2e^-$
   cathode: $Cu^{2+} + 2e^- \rightarrow Cu$

   Remember that a Pt electrode is used at the anode!

   line notation: $Pt \mid Pu^{4+}, PuO_2^+, H^+ \parallel Cu^{2+} \mid Cu$

We are now most of the way toward completely characterizing a galvanic cell. We have (à la the blue box above Sample Exercise 17.2 in your textbook)

1. calculated cell potentials,
2. described the direction of electron flow, and
3. designated the anode and cathode.
4. learned electrochemical "line-notation"

We have not yet "*5. designated the electrodes in each compartment.*"  Normally, the solid metal in a half-reaction will serve as the electrode.  If a half reaction contains only ions, a nonreacting conductor (usually platinum) will be the electrode.

---

## Example 17.2 E - Describing Galvanic Cells

Describe a galvanic cell based on the two half-reactions below.  Use the four-step analysis listed above.

$$Cu^{2+} + 2e^- \rightarrow Cu \qquad\qquad E° = 0.34 \text{ V}$$
$$Cr_2O_7^{2-} + 14H^+ + 6e^- \rightarrow 2Cr^{3+} + 7H_2O \qquad E° = 1.33 \text{ V}$$

### Solution

Steps 1 and 3:

In order to have a galvanic cell, E° must be > 0.  We must reverse the copper reduction.  We also need to balance the entire cell reaction electrically by multiplying the copper equation by 3.

| | |
|---|---|
| anode:   $3Cu \rightarrow 3Cu^{2+} + 6e^-$ | E° =  -0.34 V |
| cathode:  $Cr_2O_7^{2-} + 14H^+ + 6e^- \rightarrow 2Cr^{3+} + 7H_2O$ | E° = +1.33 V |
| **overall:  $3Cu(s) + Cr_2O_7^{2-}(aq) + 14H+(aq) \rightarrow 3Cu^{2+}(aq) + 2Cr^{3+}(aq) + 7H_2O(l)$** | **E° =  0.99 V** |

**Step 4:**

A **copper electrode** will be in the **anode** side.  A **platinum electrode** will be in the **cathode** side.

**Step 2:**

The electron flow (from anode to cathode) will be from **copper to platinum**.

---

Problems 3-24 at the end of this study chapter cover material from this section.

---

## 17.3 - Cell Potential, Electrical Work and  Free Energy

When you finish this section you will be able to interconvert between free energy and cell voltage.

The first part of this section emphasizes that the **actual work** that can be achieved is **always less than** the **theoretical work** available.  Nonetheless, your textbook presents the relationship between **free energy** (maximum is assumed though not attainable) and **cell potential**.

$$\Delta G = -nFE$$

at standard conditions,

$$\Delta G° = -nFE°$$

where $\Delta G$ = free energy (in Joules)
   **n** = moles of electrons exchanged in the redox reaction
   **F** = the Faraday, **a constant** (96,486 Coulombs per mole of electrons)
   **E** = cell voltage (Joules/Coulomb).

Note that $\Delta G$ and E have **opposite signs.  For a spontaneous process, $\Delta G$ is "-" and E is "+."**

---

## Example 17.3 A - Cell Voltage and Free Energy

Calculate the $\Delta G°$ for the reaction

$$Zn^{2+}(aq) + Cu(s) \rightleftharpoons Zn(s) + Cu^{2+}(aq)$$

Will zinc ions plate out on a copper strip?

### Solution

The two pertinent half-reactions are

$$Zn^{2+} + 2e^- \rightarrow Zn \qquad E° = -0.76 \text{ V}$$
$$Cu \rightarrow Cu^{2+} + 2e^- \qquad E° = -0.34 \text{ V}$$
$$\overline{\qquad\qquad\qquad\qquad\quad E°_{cell} = -1.10 \text{ V}}$$

$$\Delta G° = -nFE° = -2 \text{ mol } e^- (96486 \text{ C/mol } e^-)(-1.10 \text{ J/C})$$
$$\Delta G° = 2.12 \times 10^5 \text{ J} = \textbf{212 kJ}$$

**Zinc ions will not plate out on a copper strip at standard conditions.** (Note that this result is also indicated by the fact that $E°_{cell} < 0$.)

---

## Example 17.3 B - Practice With Cell Voltage and Free Energy

Vanadium (V) can be reduced to vanadium (IV) by reaction with a "Jones reductor," a Zn-Hg amalgam. The reactions of interest are

$$VO_2^+ + 2H^+ + e^- \rightarrow VO_2^+ + H_2O \qquad E° = +1.00 \text{ V}$$
$$Zn^{2+} + 2e^- \rightarrow Zn \qquad E° = -0.76 \text{ V}$$

Calculate E° and $\Delta G°$ for the reaction.

### Solution

For a spontaneous reduction of vanadium, Zn is oxidized.  The reactions are (watch the electron balance!)

$$2VO_2^+ + 4H^+ + 2e^- \rightarrow 2VO^{2+} + 2H_2O \qquad\qquad E° = +1.00 \text{ V}$$
$$\underline{Zn \rightarrow Zn^{2+} + 2e^- \qquad\qquad\qquad\qquad\qquad\qquad E° = +0.76 \text{ V}}$$
$$2VO_2^+(aq) + 4H^+(aq) + Zn(s) \rightarrow 2VO_2^+(aq) + Zn^{2+}(aq) + 2H_2O(l) \quad E°_{cell} = \textbf{1.76 V}$$

$$\Delta G° = -2 \text{ mol } e^- (96,486 \text{ C/mol } e^-)(1.76 \text{ J/C})$$
$$\Delta G° = -3.40 \times 10^5 \text{ J} = \textbf{-340 kJ}$$

Problems 25-29 at the end of this study chapter cover material from this section.

## 17.4 - The Dependence of Cell Potential on Concentration

When you finish this section you will be able to:

- predict the redox processes in concentration cells,
- use the Nernst equation to solve for cell voltage at nonstandard conditions, and
- calculate equilibrium constants from cell voltages.

This section considers the calculation of cell voltages at nonstandard concentrations (ie; not 1 M). Your textbook introduces the **concentration cell**, a cell in which current flows due only to a **difference in concentration** of an ion in two different compartments of a cell. **Le Chatelier's principle** is applicable here. In a cell where there is an equal concentration of metal ion on both sides, $E^{\circ}_{cell} = 0$.

However, if the concentrations are different, a "stress" is put on the system that will be equalized by electron flow to allow reduction and oxidation to occur. (See Figures 17.9 and 17.10 in your textbook.) When the concentrations in the half-cells become equal, $E^{\circ}_{cell} = 0$ and the system is at equilibrium.

## Example 17.4 A - Concentration Cells

A cell has on its left side a 0.20 M $Cu^{2+}$ solution. The right side has a 0.050 M $Cu^{2+}$ solution. The compartments are connected by Cu electrodes and a salt bridge. Designate the cathode, anode, and direction of current.

### Solution

Current will flow in this cell until the concentration of $Cu^{2+}$ is equal in both compartments. This means that the concentration of $Cu^{2+}$ in the **left hand side (0.20 M) must be reduced** by

$$Cu^{2+} + 2e^- \rightarrow Cu$$

The left hand side will be the **cathode**. The right hand side (0.050 M $Cu^{2+}$) will be the **anode**. Current will flow from right to left (anode to cathode).

The point of introducing concentration cells (which, as your textbook points out, produce a very small voltage) is really to illustrate the fact that **nonstandard concentrations produce a cell voltage that is different from that at standard concentrations**.

$$E_{cell}(\text{nonstandard}) \neq E^{\circ}_{cell}$$

The **Nernst equation** is derived in your textbook. We use the equation to **calculate the cell voltage at nonstandard concentrations**.

$$E_{cell} = E^{\circ}_{cell} - \frac{0.0592}{n} \log(Q) \qquad (\text{at } 25°C)$$

where Q has its usual significance as $[\text{products}]_o/[\text{reactants}]_o$.

The mathematics of the Nernst equation are that if **[reactants]$_o$ are higher than [products]$_o$**, log(Q) will be < 0 and **E$_{cell}$ will be > E°$_{cell}$**. This is consistent with Le Chatelier's principle and concentration cells. Also, when a battery is fully discharged, it is **at equilibrium**, and E$_{cell}$ = 0.

## Example 17.4 B - The Nernst Equation

Calculate E$_{cell}$ for a galvanic cell based on the following half-reactions at 25°C.

$$Cd^{2+} + 2e^- \rightarrow Cd \qquad E° = -0.40 \text{ V}$$
$$Pb^{2+} + 2e^- \rightarrow Pb \qquad E° = -0.13 \text{ V}$$

where [Cd$^{2+}$] = 0.010 M and [Pb$^{2+}$] = 0.100 M.

### Solution

In order for the cell to be galvanic, E° must be greater than 0 V. This means that the cadmium must undergo oxidation, giving a net cell reaction

$$Cd(s) + Pb^{2+}(aq) \rightarrow Cd^{2+}(aq) + Pb(s) \qquad E°_{cell} = +0.27 \text{ V}$$

The Nernst equation for this cell is

$$E = E° - \frac{0.0592}{2} \log \left[ \frac{[Cd^{2+}]}{[Pb^{2+}]} \right] = 0.27 - 0.0296 \log \left[ \frac{0.010}{0.100} \right]$$

$$E = 0.27 - 0.0296 \log (0.10) = 0.27 + 0.0296$$
$$\mathbf{E_{cell} = 0.30 \text{ V}}$$

## Example 17.4 C - Nernst Equation

Calculate E$_{cell}$ for a galvanic cell based on the following half-reactions at 25°C.

$$\text{(Equation 1)} \quad FeO_4^{2-} + 8H^+ + 3e^- \rightarrow Fe^{3+} + 4H_2O \qquad E° = +2.20 \text{ V}$$
$$\text{(Equation 2)} \quad O_2 + 4H^+ + 4e^- \rightarrow 2H_2O \qquad E° = +1.23 \text{ V}$$

where [FeO$_4^{2-}$] = 2.0 × 10$^{-3}$ M
[Fe$^{3+}$]   = 1.0 × 10$^{-3}$ M
[O$_2$]   = 1.0 × 10$^{-5}$ atm
pH   = 5.2

### Solution

The ferrate ion ("FeO$_4^{2-}$") has a higher reduction potential so it will be reduced while water is oxidized to oxygen. To balance electrically, equation 1 must be multiplied by 4 and equation 2 must be reversed and multiplied by 3. This gives

$$4FeO_4^{2-} + 32H^+ + 12e^- \rightarrow 4Fe^{3+} + 16H_2O \qquad \qquad E° \quad = +2.20 \text{ V}$$
$$6H_2O \rightarrow 3O_2 + 12H^+ + 12e^- \qquad \qquad\qquad\qquad E° \quad = -1.23 \text{ V}$$

$$\mathbf{4FeO_4^{2-}(aq) + 20\ H^+(aq) \rightarrow 4Fe^{3+}(aq) + 3O_2(g) + 10H_2O(l) \qquad E°_{cell} = +0.97 \text{ V}}$$

$$E = E°_{cell} = \frac{0.059}{n} \log(Q) = 0.97 - \frac{0.059}{12} \log \left[ \frac{[Fe^{3+}]^4 [O_2]^3}{[FeO_4^{2-}]^4[H^+]^{20}} \right]$$

$$E = 0.97 - \frac{0.059}{12} \log \left[ \frac{(1.0 \times 10^{-3})^4 (1.0 \times 10^{-5})^3}{(2.0 \times 10^{-3})^4 (6.31 \times 10^{-6})^{20}} \right]$$

$$\uparrow$$
$$(pH = 5.2)$$

Notice the extreme $H^+$ ion dependence!  If you try to take $6.31 \times 10^{-6}$ to the 20th power, your calculator will flip out.  It is more appropriate to split it up.

$$(6.31 \times 10^{-6})^{20} = (6.31)^{20} \times (10^{-6})^{20} = 1.0 \times 10^{16} \times 10^{-120}$$

$$= 1.0 \times 10^{-104}$$

$$E = 0.97 - 4.9 \times 10^{-3} \log \left[ \frac{1.0 \times 10^{-27}}{1.6 \times 10^{-115}} \right]$$

$$= 0.97 - 4.9 \times 10^{-3} \log (6.3 \times 10^{87}) = 0.97 - 0.43$$

$$E = 0.54 \text{ V}$$

---

Recall that **at equilibrium both E and $\Delta G = 0$.**  Your textbook uses this information along with the relationship between $\Delta G$ and K to derive a formula that **relates E to K.**

$$\log(K) = \frac{nE°}{0.0592} \qquad \text{(at 25°C)}$$

Remember, we are dealing with **equilibrium conditions** in this case.

---

## Example 17.4 D - Equilibrium Constants and Cell Potential

Calculate the equilibrium constant for the reaction in our example 17.2 D,

$$3Cu(s) + Cr_2O_7^{2-}(aq) + 14H^+(aq) \rightarrow 3Cu^{2+}(aq) + 2Cr^{3+}(aq) + 7H_2O(l) \qquad E°_{cell} = 0.99 \text{ V}$$

### Solution

There are 6 moles of electrons transferred in this redox reaction.

$$\log K = \frac{nE°}{0.059} = \frac{6\,(0.99)}{0.059} = 100.67$$

$$K = 10^{100.67} = 10^{0.67} \times 10^{100} = 5 \times 10^{100}$$

---

## Example 17.4 E - Summing It All Up

Consider the reaction

$$Ni^{2+}(aq) + Sn(s) \rightarrow Ni(s) + Sn^{2+}(aq)$$

Calculate the following:  $E°_{cell}$, $\Delta G°$, and K at 25°C.  In addition, determine the minimum **ratio of $[Sn^{2+}]/[Ni^{2+}]$** necessary in order to make the reaction spontaneous as written.

## Solution

a.   The half-reactions are:

$$Ni^{2+} + 2e^- \rightarrow Ni \qquad E° \quad = -0.23 \text{ V}$$
$$Sn \rightarrow Sn^{2+} + 2e^- \qquad E° \quad = +0.14 \text{ V}$$
$$\overline{\phantom{Sn \rightarrow Sn^{2+} + 2e^-} \quad E°_{cell} = -0.09 \text{ V}}$$

The reaction is not spontaneous under standard conditions.

b.   $\Delta G° = -nFE° = -2 \text{ mol } e^- \ (96{,}486 \text{ C/mol } e^-) \ (-0.09 \text{ J/C}) = 1.74 \times 10^4 \text{ J}$
     $\Delta G° = 17 \text{ kJ}$

c.   We can calculate K using either $\log K = nE°/0.059$ or $\ln K = -\Delta G°/RT$.  Let's do it both ways.

$$\log K = \frac{2 \ (-0.09)}{0.059} = -3.05 \quad \Rightarrow \quad K = 9 \times 10^{-4}$$

$$\ln K = \frac{-17400 \text{ J}}{8.3148 \text{ J/K mol } (298 \text{ K})} = -7.022 \quad \Rightarrow \quad K = 9 \times 10^{-4}$$

d.   In order for the reaction to be spontaneous, E > 0.  Using the Nernst equation,

$$E > 0 > E°_{cell} - \frac{0.059}{2} \ \log \ \left[ \frac{[Sn^{2+}]}{[Ni^{2+}]} \right]$$

reducing,

$$0 > -0.09 - 0.0295 \log \left[ \frac{[Sn^{2+}]}{[Ni^{2+}]} \right]$$

$$\frac{0.09}{-0.0295} \ (= -3.05) > \log \left[ \frac{[Sn^{2+}]}{[Ni^{2+}]} \right]$$

$$\frac{[Sn^{2+}]}{[Ni^{2+}]} < 9 \times 10^{-4}$$

**The ratio of $Sn^{2+}$ to $Ni^{2+}$ must be less than $9 \times 10^{-4}$ for this reaction to be spontaneous.**

---

## Problems 30-38 at the end of this study chapter cover material from this section.

---

## 17.5 and 17.6 - Batteries and Corrosion

The following review questions will help you to test your knowledge of the material in these sections.

## Section 17.5

1.   Define battery.
2.   Why is a lead storage battery especially useful in automobiles?
3.   Give the anode, cathode, and overall cell reactions in the lead storage battery.
4.   Calculate E° for the lead storage battery.
5.   Describe the idea behind how the extent of battery discharge can be measured.

6. Why can "jump starting" a car be dangerous?
7. Describe the parts of a typical dry cell battery.
8. Why does an alkaline dry cell last longer than an acid dry cell?
9. Give the anode, cathode and overall cell reactions for the acid and alkaline dry cells.
10. Why is the nickel-cadmium battery rechargeable?
11. Define "fuel cell."
12. Give the anode, cathode and overall cell reactions for the hydrogen-oxygen fuel cell.

## Section 17.6

1. Define corrosion.
2. Why is corrosion so ubiquitous?  (Consider the reduction potentials in Table 17.1.)
3. If most native metals **theoretically** oxidize in air, why don't these metals **actually** oxidize?
4. List the reactions, and calculate the E° for the corrosion of iron.
5. What is the role of water and salt in the rusting of iron?  Why do cars rust more in wet, cold areas of the United States than in dry, warm areas?
6. What is the role of chromium in corrosion protection?
7. How does galvanizing iron help prevent rust?
8. How does magnesium act as **cathodic protection** for iron pipes?

---

# 17.7 - Electrolysis

When you finish this section you will be able to use Faraday's constant and current to relate time to extent of electrolysis.

This section begins with the definition of **electrolysis** as **the process of forcing a current through a cell to produce a chemical change** for which the cell potential is negative.  In order for electrolysis to occur, you must apply an external voltage that is **greater than** the potential of the galvanic cell if you want to force the reaction in the opposite (electrolytic) direction.

---

# Example 17.7 A - Electrolytic Cells

What voltage is necessary to force the following electrolysis reaction to occur?

$$2I^-(aq) + Cu^{2+}(aq) \rightarrow I_2(s) + Cu(s)$$

Which process would occur at the anode? cathode?  Assuming the iodine oxidation takes place at a platinum electrode, what is the direction of electron flow in this cell?

## Solution

$$
\begin{array}{lll}
\text{anode (oxidation):} & 2I^- \rightarrow I_2 + 2e^- & E° = -0.54 \text{ V} \\
\text{cathode (reduction):} & Cu^{2+} + 2e^- \rightarrow Cu & E° = +0.34 \text{ V} \\
\hline
& & E°_{cell} = -0.20 \text{ V}
\end{array}
$$

More than 0.20 V must be externally applied to make this reaction proceed.  The direction of electron flow is always From Anode To CAThode ("FAT CAT"), so it is from the platinum electrode to the copper electrode.

---

If you put a potential on a system which contains one metal ion, and the **potential is above that at which the metal ion will reduce**, you will plate out that metal.  The amount of metal reduced is directly related to the current, in **amps (Coulombs/sec)**, that flows in the system.  In practice, this is not the case because (as pointed out several times in your textbook) **electrochemistry is not a perfectly efficient process**. However, for purposes of problem solving, we will **assume** it is.

**Dimensional analysis** works wonders with electrolysis problems. Your goal is often to find **moles of electrons** of the metal.

$$\text{moles of e-} = \frac{1 \text{ mol e}^-}{\underset{\underset{\text{Faraday}}{\uparrow}}{96,486 \text{ C}}} \times \underset{\underset{\text{amps}}{\uparrow}}{\frac{\text{C}}{\text{s}}} \times \underset{\underset{\text{time}}{\uparrow}}{\text{s}}$$

## Example 17.7 B - Electrolysis

How many grams of copper can be reduced by applying a 3.00 A current for 16.2 min to a solution containing $Cu^{2+}$ ions?

### Strategy

Our goal is to find grams. As always, we want to work through **moles**, and in this case, **moles of electrons**. Working backwards,

**g of Cu** $\Rightarrow$ **moles of Cu** $\Rightarrow$ **moles of e⁻** $\Rightarrow$ **Coulombs** $\Rightarrow$ **current** $\Rightarrow$ **time**

### Solution

$$\text{g Cu} \; \left\| \; \frac{63.54 \text{ g Cu}}{1 \text{ mol Cu}} \times \frac{1 \text{ mol Cu}}{2 \text{ mol e}^-} \times \frac{1 \text{ mol e}^-}{96,486 \text{ C}} \times \frac{3.00 \text{ C}}{\text{s}} \times \frac{60 \text{ s}}{\text{min}} \times 16.2 \text{ min} = \mathbf{0.96 \text{ g Cu}} \right.$$

Electrolysis can be used to separate a mixture of ions, if the reduction potentials are fairly far apart. Remember that **the metal ion with the highest reduction potential is the easiest to reduce.**

## Example 17.7 C - Order of Reduction

Using Table 17.1 in your textbook, predict the order of reduction and which of the following ions will reduce first at the cathode of an electrolytic cell:

$$Ag^+, \quad Zn^{2+}, \quad IO_3^-$$

### Solution

$$Ag^+ + e^- \rightarrow Ag \qquad\qquad E° = +0.80 \text{ V}$$
$$Zn^{2+} + 2e^- \rightarrow Zn \qquad\qquad E° = -0.76 \text{ V}$$
$$IO_3^- + 6H^+ + 5e^- \rightarrow 1/2I_2 + 3H_2O \qquad E° = +1.20 \text{ V}$$

The $IO_3^-$ will reduce first, followed by $Ag^+$ and $Zn^{2+}$ as the voltage is increased.

## Example 17.7 D - Electrolysis of Water

What volumes of $H_2(g)$ and $O_2(g)$ is produced by electrolyzing water at a current of 4.00 A for 12.0 minutes (assume ideal conditions).

## Strategy

The overall reaction is

$$2H_2O(l) \rightarrow 2H_2(g) + O_2(g)$$

In actual practice, the actual ratio is not exactly 2:1, for a variety of reasons including oxygen solubility. We can, however, run through the calculation for practice.

We are asked to find the **volume** of hydrogen and oxygen. The **ideal gas law** relates moles of a gas to volume. Our strategy must therefore be to **find moles of each gas**, then use the ideal gas relationship that **1 mole of a gas occupies 22.4 L** (under ideal conditions) to find volume. Also, for every mole of water, **2 moles of electrons** are exchanged.

## Solution

$$\text{mol } H_2 \left\| \frac{1 \text{ mol } H_2}{2 \text{ mol } e^-} \times \frac{1 \text{ mol } e^-}{96,486 \text{ C}} \times \frac{4.00 \text{ C}}{s} \times \frac{60 \text{ s}}{min} \times 12.0 \text{ min} \right.$$

$$= \textbf{0.0149 mol } H_2 \quad \Rightarrow \quad \textbf{0.00746 mol } O_2 \text{ (2 moles } H_2 \text{ per 1 mole } O_2)$$

$$\text{L } H_2 = \frac{22.4 \text{ L}}{\text{mol}} \times 0.0149 \text{ mol} = \textbf{0.33 L } H_2 \quad \Rightarrow \quad \textbf{0.17 L } O_2$$

---

Problems 39-40 at the end of this study chapter cover material from this section.

---

**END OF CHAPTER SEVENTEEN**

# SELF-TEST QUESTIONS

1. In each of the following half-reactions, give the species being reduced and the number of electrons needed to balance the half-reaction:

   a. $AgBrO_3 + ?e^- \rightarrow Ag + BrO_3^-$
   b. $HCrO_4^- + 7H^+ + ?e^- \rightarrow Cr^{3+} + 4H_2O$
   c. $WO_3 + 6H^+ + ?e^- \rightarrow W + 3H_2O$

2. Identify the species in each of the following reactions that would receive electrons from the cathode, and that would lose electrons at the anode in each of the following galvanic cells:

   a. $Au^{3+}(aq) + Zn(s) \rightleftharpoons Au^+(aq) + Zn^{2+}(aq)$
   b. $3Pu^{6+}(aq) + 2Cr^{3+}(aq) \rightleftharpoons 2Cr^{6+}(aq) + 3Pu^{4+}(aq)$

Use the following reactions and potentials, taken from the Chemical Rubber Company Handbook of Physics and Chemistry, 53rd Edition, for problems 3-8.

| Reaction | Potential (volts) |
|---|---|
| $2SO_4^{2-} + 4H^+ + 2e^- \rightleftharpoons S_2O_6^{2-} + 2H_2O$ | -0.2 |
| $AuBr_4^- + 3e^- \rightleftharpoons Au + 4Br^-$ | +0.858 |
| $O_3 + 2H^+ + 2e^- \rightleftharpoons O_2 + H_2O$ | +2.07 |
| $Ba_2^+ + e^- \rightleftharpoons Ba$ | -2.09 |
| $In^{3+} + e^- \rightleftharpoons In^{2+}$ | -0.49 |

3. Select the strongest oxidizing agent from the reactions above.

4. Select the strongest reducing agent from the reactions above.

5. Arrange the following species in order of increasing strength as oxidizing agents (assume all species are in their standard states):

$$Sn^{4+}, \quad Au^{3+}, \quad Fe^{3+}, \quad Fe^{2+}, \quad Ca^{2+}$$

6. Write a balanced equation, and calculate the value of $E°$, for the reaction of $AuBr_4^-$ with $In^{2+}$.

7. Write a balanced equation, and calculate the value of $E°$, for the reaction of $Ba^{2+}$ with $O_2$ to form $O_3$. Would this reaction be spontaneous? Why or why not?

8. Write the balanced reaction for the galvanic cell that would yield the highest voltage given the reactions listed above.

9. A galvanic cell consists of an Ag electrode in a 1.0 M $AgNO_3$ solution and a Ni electrode in a 1.0 M $Ni(NO_3)_2$ solution. Calculate the standard emf of this electrochemical cell at 25°C (use Table 17.1 in your textbook.)

10. Regarding the following reaction:

$$F_2(g) + 2I^-(aq) \rightarrow 2F^-(aq) + I_2(s)$$

   a. List the species being oxidized.
   b. List the species being reduced.
   c. Calculate $E°$ for this cell. (See Table 17.1 in your textbook.)
   d. Which species receives electrons from the cathode?
   e. Which species donates electrons to the anode?

11. Answer the same questions that were posed in problem #10 for the following reaction:

$$Hg_2^{2+}(aq) + Zn(s) \rightleftharpoons 2Hg(s) + Zn^{2+}(aq)$$

12. Answer the same questions that were posed in problem #10 for the reaction:

$$Fe^{2+}(aq) + Ce^{4+}(aq) \rightleftharpoons Fe^{3+}(aq) + Ce^{3+}(aq)$$

13. Given the following half-reactions:

$$Co^{2+} + 2e^- \rightleftharpoons Co \qquad\qquad E° = -0.277 \text{ V}$$
$$Ce^{4+} + e^- \rightleftharpoons Ce^{3+} \qquad\qquad E° = 1.61 \text{ V}$$

    a.  Write the overall equation for the galvanic cell.
    b.  Calculate E° for the cell.

14. Given the following half-reactions:

$$PtCl_4^- + 4e^- \rightleftharpoons Pt + 4Cl^- \qquad\qquad E° = 0.73 \text{ V}$$
$$Fe^{3+} + 3e^- \rightleftharpoons Fe \qquad\qquad E° = -0.037 \text{ V}$$

    a.  Write the overall equation for the galvanic cell.
    b.  Calculate E° for the cell.

15. Given the following half-reactions:

$$Ag + 2CN^- \rightleftharpoons Ag(CN)_2^- \qquad\qquad E° = +0.31 \text{ V}$$
$$Ag \rightleftharpoons Ag^+ + e^- \qquad\qquad E° = -0.80 \text{ V}$$

Calculate E° for the reaction:

$$Ag^+(aq) + 2CN^-(aq) \rightleftharpoons Ag(CN)_2^-(aq)$$

16. Given the following half-reactions:

$$La^{3+} + 3e^- \rightleftharpoons La \qquad\qquad E° = -2.52 \text{ V}$$
$$Fe^{3+} + e^- \rightleftharpoons Fe^{2+} \qquad\qquad E° = 0.77 \text{ V}$$

    a.  write the overall equation for the galvanic cell.
    b.  calculate E° for the cell.

17. Regarding the galvanic cell you composed for the reaction in problem #14, designate the anode and cathode.

18. Given the following half-reactions:

$$PrO_2^+ + 4H^+ + e^- \rightleftharpoons Pu^{4+} + 2H_2O \qquad\qquad E° = +1.15 \text{ V}$$
$$Cu^{2+} + 2e^- \rightleftharpoons Cu \qquad\qquad E° = +0.34 \text{ V}$$

    a.  Write the overall equation for the galvanic cell.
    b.  Calculate E° for the cell.

19. Regarding the galvanic cell you composed for the reaction in problem #13:

    a.  designate the anode and cathode
    b.  describe the direction of electron flow
    c.  describe the electrodes in each compartment

20. Regarding the galvanic cell you composed for the reaction in problem #18,

    a.   designate the anode and cathode.
    b.   describe the direction of electron flow.
    c.   describe the electrodes in each compartment.

21. Regarding the galvanic cell you composed in problem #16, designate the anode and the cathode.

22. Calculate the standard free-energy change for the following reaction (use Table 17.1 in your textbook):

$$2H^{2+} + Mn \rightleftharpoons Hg_2^{2+} + Mn^{2+}$$

23. Calculate the standard free-energy change for the following reaction:

$$Na^+ + Cu \rightleftharpoons Na + Cu^+$$

24. Calculate the standard free-energy change for the following reaction:

$$Cr^{3+} + ClO_2^- \rightleftharpoons Cr^{2+} + ClO_2$$

25. Using data from <u>Table 17.1 in your textbook</u>, calculate $\Delta G°$ for the reaction:

$$2Al(s) + 6H^+(aq) \rightleftharpoons 2Al^{3+}(aq) + 3H_2(g)$$

Will the reaction be spontaneous (i.e. will hydrogen gas bubble from solution)?

26. Calculate $\Delta G°$ for the following reaction (use <u>Table 17.1 in your textbook</u>):

$$2K^+ + Cu \rightleftharpoons 2K + Cu^{2+}$$

27. Calculate $\Delta G$ for the following reaction:

$$3Hg_2^{2+} + 2Cr \rightleftharpoons 6Hg + 2Cr^{3+}$$

28. Calculate $\Delta$ for the following reaction:

$$Br_2 + Sn \rightleftharpoons 2Br^- + Sn^{2+}$$

29. Calculate $E°$ and $\Delta G°$ for the reaction:

$$Vo_2^{2+}(aq) + 4H^+ + 2Ag(s) + 2Cl^-(aq) \rightleftharpoons U^{4+}(aq) + 2AgCl(s) + 2H_2O(l)$$

Given the following half-reactions:

$$UO_2^{2+} + 4^+ + 2e^- \rightleftharpoons U^{4+} + 2H_2O \qquad E° = +0.334 \text{ V}$$
$$AgCl + e^- \rightleftharpoons Ag + Cl^- \qquad E° = +0.222 \text{ V}$$

30. A cell has on its left side a $1.0 \times 10^{-3}$ M $Zn^{2+}$ solution. The right side has a 0.030 M $Zn^{2+}$ solution. The compartments contain by Zn electrodes and a salt bridge. Designate the cathode, anode and direction of current.

31. Calculate the emf for each of the following half-reactions:

    a.   $Cr_2O_7^{2-}$ (O.350 M) + 14H$^+$(0.0100 M) + 6e$^-$ $\rightleftharpoons$ 2Cr$^{3+}$ ($1 \times 10^{-3}$) + 7H$_2$O     $E° = +1.33$ V
    b.   $AuBr_2^-$ (0.084 M) + e$^-$ $\rightleftharpoons$ Au + 2Br$^-$ (0.1443 M)     $E° = +0.959$ V

32. For a galvanic cell based on the following half-reactions at 25°C:

$$Cl_2 + 2e^- \rightarrow 2Cl^- \qquad\qquad E° = 1.36$$
$$Ni^{2+} + 2e^- \rightarrow Ni \qquad\qquad E° = -0.23$$

Calculate $E_{cell}$ where $[Cl_2] = 0.5$ atm and $[Ni^{2+}] = 1.0$ M.

33. For a galvanic cell based on the following half-reactions at 25°C:

$$Pb^{2+} + 2e^- \rightarrow Pb \qquad\qquad E° = -0.13$$
$$Zn^{2+} + 2e^- \rightarrow Zn \qquad\qquad E° = -0.76$$

Calculate $E_{cell}$ where $[Pb^{2+}] = 1 \times 10^{-2}$ M and $[Zn^{2+}] = 3 \times 10^{-2}$ M.

34. Calculate the cell voltage at 25°C for the equation in problem #13 given the following equilibruim concentrations:

$$[Al^{3+}] = 0.025 \text{ M}, \quad P_{H_2} = 1.0 \text{ atm}, \quad pH = 3.50.$$

35. Calculate the cell voltage for the reaction:

$$Hg_2^{2+}(aq) + Cd(s) \rightleftharpoons 2Hg(s) + Cd^{2+}(aq)$$

where $[Hg_2^{2+}] = 1.0 \times 10^{-3}$ M and $[Cd^{2+}] = 5.0 \times 10^{-3}$ M

$$Cd^{2+} + 2e^- \rightleftharpoons Cd \qquad\qquad Ex = -0.40 \text{ V}$$
$$Hg_2^{2+} = 2e^- \rightleftharpoons 2Hg \qquad\qquad E° = +0.80 \text{ V}$$

36. Calculate the equilibrium constant, K, for the following reaction at 25°C:

$$PbO_2 + 4H^+ + 2Hg + 2Cl^- \rightleftharpoons Pb^{2+} + 2H_2O + Hg_2Cl2 \qquad E°_{cell} = 1.12$$

37. Given the data in problem #35, calculate the euilibrium constant, K, for the reaction:

$$2Hg(s) + Cd^{2+}(aq) \rightleftharpoons Hg_2^{2+} + Cd(s)$$

How does the value of K in this problem relate to the value for the reverse reaction (as presented in problem #35?)

38. Calculate K for the reaction of aluminum with hydrogen given in problem #25.

39. Bismuth can be electrolytically reduced according to the following reaction:

$$BiO^+ + 2H^+ + 3e^- \rightleftharpoons Bi + H_2O$$

How many grams of bismuth can be reduced by applying a 5.60 A current for 28.3 min. to a solution containing $BiO^+$ ions (assume 100% efficeincy.)

40. A sample containing nitric acid was titrated by electrolytically reducing water to form $OH^-$ ion. How many moles of H+ ion were originallly in the sample if the hydroxide required 356.1 sec of generation at $9.07 \times 10^{-3}$ A?

$$H_2O + 2e^- \rightleftharpoons H_2 + 2OH^-$$

# ANSWERS TO SELF-TEST QUESTIONS

1. a.  $Ag^+ + e^- \rightarrow Ag$
   b.  $HCrO_4^- + 7H^+ + 3e^- \rightarrow Cr^{3+} \rightarrow 4H_2O$   $(Cr^{7+} \rightarrow Cr^{3+})$
   c.  $WO_3 + 6H^+ + 6e^- \rightarrow W + 3H_2O$    $(W^{6+} \rightarrow W)$

2.  <u>receives electrons</u>                              <u>loses electrons</u>
    a.  $Au^{3+}$                                          $Zn$
    b.  $Pu^{6+}$                                          $Cr^{3+}$

3.  $Ca^{2+} < Fe^{2+} < Sn^{2+} < Fe^{3+} < Au^{3+}$

4.  $O_3$ is the strongest oxidizing agent.

5.  Ba is the strongest reducing agent.

6.  $AuBr_4^- + 3In^{2+} \rightleftharpoons Au + 4Br^- + 3In^{3+}$          $E^\circ = +1.35$ V

7.  $Ba^{2+} + O_2 + H_2O \rightleftharpoons Ba + O_3 + 2H^+$          $E^\circ = -4.97$ V
    $E^\circ < O$, this reaction is nonspontaneous.

8.  $Ba + O_3 + 2H^+ \rightleftharpoons Ba^{2+} + O_2 + H_2O$

9.  1.03 V

10. a.  $I^-$ is being oxidized.
    b.  $F_2$ is being reduced.
    c.  $E^\circ = +2.33$ V
    d.  $F_2$ receives electrons
    e.  $I^-$ donates electrons to the anode

11. a.  Zn is being oxidized
    b.  $Hg_2^{2+}$ is being reduced
    c.  $E^\circ = +1.56$ V
    d.  $Hg_2^{2+}$ receives electrons
    e.  Zn donates electrons

12. a.  $Fe^{2+}$ is being oxidized
    b.  $Ce^{4+}$ is being reduced
    c.  $E^\circ = +0.93$ V
    d.  $Ce^{4+}$ receives electrons
    e.  $Fe^{2+}$ donates electrons

13. a.  $2Ce^{4+} + Co \rightleftharpoons Co^{2+} + 2Ce^{3+}$
    b.  $E^\circ = 1.89$ V

14. a.  $3PtCl_4^- + 4Fe \rightleftharpoons 3Pt + 12 Cl^- + 4Fe^{3+}$
    b.  $E^\circ = 0.767$ V

15. $E^\circ = 1.11$ V

16. a.  $3Fe^{3+} + La \rightleftharpoons 3Fe^{2+} + La^{3+}$
    b.  $E^\circ = 3.29$ V

17. Oxidation of Fe takes place at the anode.  Reduction of $PtCl_4^-$ takes place at the cathode.

18. $2PuO_2^+ + Cu + 8H^+ \rightleftharpoons 2Pu^{4+} + Cu^{2+} + 4H_2O$   $E° = +0.81$ V

19. a.   Oxidation of Co takes place at the anode.  Reduction of $Ce^{4+}$ takes place at the cathode.
    b.   Electrons flow from the anode to the cathode.
    c.   Copper can act as the anode.  Platinum can act as the cathode.

20. The oxidation of copper takes place at the anode.  The reduction of $PuO_2^+$ takes place at the cathode.  Electrons flow from the anode to the cathode.  Copper can act as the anode.  Platinum can act as the cathode.

21. Oxidation of La takes place at the anode.  Reduction of $Fe^{3+}$ takes place at the cathode.

22. $E° = 2.09$ V

23. $E° = -3.23$ V

24. $E° = -1.454$ V

25. $\Delta G° = -480$ kJ

26. $\Delta G° = 629$ kJ

27. $\Delta G° = -433$ kJ

28. $\Delta G° = -237$ kJ

29. $E° = +0.112$ V,   $\Delta G° = -21.6$ kJ

30. The right side will be the cathode.  The left side will be the anode.  The current will flow from left to right (anode to cathode).

31. a.   E = +1.11 V
    b.   +1.00 V

32. $E_{cell} = 1.58$ V

33. $E_{cell} = 0.62$ V

34. E = +1.30 V

35. E = +1.18 V

36. $K = 2.9 \times 10^{-41}$.  K for problem # 37 = 1/K for probem # 35.

37. $K = 6.9 \times 10^{37}$

38. $K = 1.4 \times 10^{84}$

39. 6.86 g of Bi.

40. $1.67 \times 10^{-5}$ moles of OH.

# CHAPTER 18

# The Representative Elements- Groups 1A Through 4A

| Section | Contents | Page in this Study Guide |
|---|---|---|
| 18.1 | A Survey of the Representative Elements | 391 |
| 18.2 | The Group 1A Elements | 393 |
| 18.3 | Hydrogen | 395 |
| 18.4 | The Group 2A Elements | 395 |
| 18.5 | The Group 3A Elements | 396 |
| 18.6 | The Group 4A Elements | 396 |

## 18.1 - A Survey of the Representative Elements

When you finish this section you will be able to:

- describe the effect of the size of an atom on reactivity, and
- list some methods of preparing elements.

This section begins by reviewing the positions of various types of elements in the periodic table.

- **representative elements** are those in which **s and p orbitals are being filled.**
- **transition elements** are those in which the **d orbitals are being filled.**
- **lanthanides and actinides** (sometimes called "inner-transition metals") are those in which the **4f or 5f orbitals are being filled.**

You should be able to determine the type of element by its position on the periodic table. (Review section 7.11 if you have forgotten!) You may also wish to review electron configurations to confirm your assignments. Try the following review examples.

# Example 18.1 A - Electron Configuration Review

Write the electron configurations (either shorthand or longhand) for each of the following elements:

    a.  yttrium      b.  silver      c.  bismuth      d.  iodine

## Strategy and Solution

We can either use our configuration triangle (Chapter 7), or use the position in the periodic table to establish the electron configuration.

    a.  Yttrium has atomic number 39. It is in **period 5** and **group 3B**. It must therefore be $[Kr]\ 5s^2\ 4d^1$. Checking with the longhand form,

$$Y:\ 1s^2\ 2s^2\ 2p^6\ 3s^2\ 3p^6\ 4s^2\ 3d^{10}\ 4p^6\ 5s^2\ 4d^1$$

    b  Silver has atomic number 47. It is in **period 5** and **group 1B**. It is $[Kr]\ 5s^1\ 4d^{10}$.

$$Ag:\ 1s^2\ 2s^2\ 2p^6\ 3s^2\ 3p^6\ 4s^2\ 3d^{10}\ 4p^6\ 5s^1\ 4d^{10}$$

    c.  Bismuth has atomic number 83. Elements with high atomic numbers get rather difficult to handle by using the configuration triangle. It is easier to use your knowledge of the periodic table. Bismuth is in **period 6** and **group 5A**. The shorthand form is $[Xe]\ 6s^2\ 5d^{10}\ 4f^{14}\ 6p^3$. The longhand form is

$$Bi:\ 1s^2\ 2s^2\ 2p^6\ 3s^2\ 3p^6\ 4s^2\ 3d^{10}\ 4p^6\ 5s^2\ 4d^{10}\ 5p^6\ 6s^2\ 4f^{14}\ 5d^{10}\ 6p^3$$

    d.  Iodine has atomic number 53. It is in **period 5** and **group 7A**. It is $[Kr]\ 5s^2\ 4d^{10}\ 5p^5$.

$$I:\ 1s^2\ 2s^2\ 2p^6\ 3s^2\ 3p^6\ 4s^2\ 3d^{10}\ 4p^6\ 5s^2\ 4d^{10}\ 5p^5$$

# Example 18.1 B - Types of Elements

Classify each of the following elements as **representative** or **transition**:

    a.  iron      b.  sodium      c.  argon      d.  nickel

## Solution

    a.  transition      c.  representative
    b.  representative      d.  transition

The section focuses attention on chemical differences within groups caused by size differences in atoms. In particular **the first member of a group tends to have different properties than the rest of the group because it is so much smaller than the rest of the atoms in the group.**

- Hydrogen is a nonmetal and forms covalent bonds. Lithium is a metal and forms ionic bonds.
- Beryllium oxide is amphoteric. Other group 2A oxides are basic.
- Boron behaves as a nonmetal or semimetal. Other group 3A elements are metal.
- Carbon can form $\pi$ bonds due to effective overlap between 2p orbitals of carbon and carbon, carbon and nitrogen, and carbon and oxygen. Other group 4A elements do not show this behavior.
- Nitrogen can form $\pi$ bonds because of its size, just as carbon can. Other group 5A elements normally do not. (A few phosphorus double bonded compounds have been known to exist.)
- A similar situation exists with oxygen in group 6A.

- Fluorine has a very weak F-F bond due to the close approach causing large electron-electron repulsions of F-F atoms resulting from the small size of the atoms.

The section ends with a discussion of some methods of preparation of elements. Read over that section and try the next example. Read over that section and try the next example.

---

# Example 18.1 C - Preparation of Elements

Write equations for the following preparations:

a. The preparation of gaseous potassium from the reaction of liquid sodium and liquid potassium chloride.
b. The preparation of xenon gas from the reaction of aqueous xenon difluoride and water to form gaseous hydrogen fluoride, oxygen and xenon.
c. The preparation of solid chromium by reducing solid chromium (III) oxide with aluminum metal.

## Solution

a. $Na(l) + KCl(l) \longrightarrow NaCl(l) + K(g)$
b. $XeF_2(aq) + 2H_2O(l) \longrightarrow Xe(g) + 2HF(g) + O_2(g)$
c. $Cr_2O_3(s) + 2Al(s) \longrightarrow 2Cr(s) + Al_2O_3(s)$

---

# 18.2 - The Group 1A Elements

The following questions and examples will help you review the material in this section.

1. What electron configuration is common to each of the elements in this group?
2. Why does hydrogen act so differently from the metals in this group?
3. Your textbook says that although lithium is a **stronger reducing agent**, it reacts **more slowly** with water than potassium or sodium. What does this illustrate?

---

# Example 18.2 A - Preparation of Alkili Metals

Write the reaction for the preparation of each of the following group 1A metals.

a. Liquid sodium by electrolysis of molten sodium chloride.
b. Solid rubidium by the reaction of solid calcium and liquid rubidium chloride.

## Solution

a. Chapter 17 deals with electrolysis. The reaction of interest (done at 600°C) is

$$2NaCl(l) \longrightarrow 2Na(l) + Cl_2(g)$$

b. Potassium, rubidium and cesium have relatively low melting points and can not be made by electrolysis.

$$Ca(s) + 2RbCl(l) \longrightarrow 2Rb(s) + CaCl_2(l)$$

4.   Why do alkali metals react so well with nonmetals?

The next example will give you a further review of reaction writing as well as practice with oxidation states.

## Example 18.2 B - Reactions of Alkali Metals

Write the chemical equations for each of the following reactions.  Also identify the oxidation state of the oxygen in the metal oxide product.

   a.   Lithium with an excess of oxygen.
   b.   Sodium with an excess of oxygen.
   c.   Cesium with oxygen.
   d.   Cesium superoxide with carbon dioxide.

## Solution

   a.   $4Li(s) + O_2(g) \text{----}> 2Li_2O(s)$

Lithium will form a +1 cation and each oxygen will be in the **-2 oxidation state**.

   b.   $2Na(s) + O_2(g) \text{----}> Na_2O_2(s)$

As with lithium, sodium will have the +1 oxidation state.  In this case, oxygen is in the **-1 oxidation state.**  It is known as a **peroxide**.

   c.   $Cs(s) + O_2(g) \text{----}> CsO_2(s)$

Cesium will be present in the +1 oxidation state.  Oxygen will therefore have a **-1/2 oxidation state**.  It is known as a **superoxide**.

   d.   $4CsO_2(s) + 2CO_2(g) \text{----}> 2Cs_2CO_3(s) + 3O_2(g)$

Your textbook points to this reaction as a good source of oxygen in underwater breathing apparatus.  Oxygen is present in the **-2 oxidation state in carbonate** and the **0 oxidation state in molecular oxygen**.

As a wrap-up to this section, study the material in Table 18.6 in your textbook and try the following example.

## Example 18.2 C - Practice with Reactions

Predict the products of the following reactions:

   a.   $2Rb(s) + Cl_2(g) \text{----}> ?$
   b.   $Rb(s) + O_2(g) \text{----}> ?$
   c.   $12K(s) + P_4(s) \text{----}> ?$
   d.   $2Na(s) + H_2O(l) \text{----}> ?$

## Solution

   a.   $2Rb(s) + Cl_2(g) \text{----}> 2RbCl(s)$
   b.   $Rb(s) + O_2(g) \text{----}> RbO_2(s)$
   c.   $12K(s) + P_4(s) \text{----}> 4K_3P(s)$
   d.   $2Na(s) + H_2O(l) \text{----}> 2NaOH(aq) + H_2(g)$

Answers to review questions are at the end of this study chapter.

## 18.3 - Hydrogen

The following questions will help you review the material in this section.

1. Why does hydrogen have such low boiling and melting points?
2. Write the combustion reaction between hydrogen and oxygen to form water.
3. Write the reaction between methane and water to produce hydrogen.
4. Why isn't electrolysis a very practical method of producing hydrogen?
5. List two commercial uses of hydrogen.
6. Define **hydride**, **ionic hydride**, **covalent hydride** and **interstitial hydride**.
7. How is palladium used to help purify hydrogen?
8. Why might interstitial hydrides be used for storage of hydrogen gas?

Answers to review questions are at the end of this study chapter.

## 18.4 - The Group 2A Elements

The following questions and examples will help you review the material in this section.

1. Why are the elements in this group called **alkaline earth** metals?
2. Why does beryllium display very different properties from the rest of the group 2A elements?

### Example 18.4 A - Reactions of Group 2A Elements

Based on the written material presented *before* <u>Table 18.7 in your textbook</u>, predict the products of the following reactions:

a.  $Sr(s) + 2H_2O(l)$ ----> ?
b.  $Ba(s) + H_2(g)$ ----> ?
c.  $3Mg(s) + N_2(g)$ ----> ?
d.  $Be(s) + 2H_2O$ ----> ?

### Solution

a.  $Sr(s) + 2H_2O(l)$ ----> $Sr^{2+}(aq) + 2OH^-(aq) + H_2(g)$
b.  $Ba(s) + H_2(g)$ ----> $BaH_2(s)$
c.  $3Mg(s) + N_2(g)$ ----> $Mg_3N_2(s)$
d.  $Be(s) + 2H_2O$ ----> No observable reaction!

3. Describe the rationale for the **three-center bond** of beryllium.
4. What kind of hybridization does Be exhibit in $BeCl_2$?
5. Why is $BeCl_2$ very reactive toward Lewis bases?
6. List some practical uses of alkaline earth metals.

## Example 18.4 B - Practice with Group 2A Reactions

Based on the information in <u>Table 18.8 in your textbook</u>, predict the **reactants** given the following **products**.

    a.   ? + ? ----> $Ca_3N_2(s)$
    b.   ? + ? ----> $2RaO(s)$
    c.   ? + ? + ? ----> $Be(OH)_4^{2-}(aq) + H_2(g)$
    d.   ? + ? ----> $SrH_2(s)$

## Solution

    a.   **$3Ca(s) + N_2(g)$ ----> $Ca_3N_2(s)$**
    b.   **$2Ra(s) + O_2(g)$ ----> $2RaO(s)$**
    c.   **$Be(s) + 2OH^-(aq) + 2H_2O(l)$ ----> $Be(OH)_4^{2-}(aq) + H_2(g)$**
    d.   **$Sr(s) + H_2(g)$ ----> $SrH_2(s)$**

7.  What is "hard water?"
8.  Describe the **ion exchange method** for softening hard water.

Answers to review questions are at the end of this study chapter.

## 18.5 and 18.6 - The Group 3A and 4A Elements

The following questions and examples will help you review the material in these sections.

### Section 18.5

1.  Why does boron exhibit different chemical behavior than the other elements in this group?
2.  What is the simplest stable borane?
3.  Describe the bonding in boranes.
4.  Why are boranes highly reactive?
5.  Why is aluminum amphoteric?
6.  Why is gallium so useful in thermometers that measure a wide range of temperatures?
7.  What is meant by the **inert pair effect**? Which group 3A metal exhibits the effect?
8.  Why doesn't aluminum foil corrode in moist air?

## Example 18.5 A - Reactions

Predict the products of each of the following reactions:

    a.   $2Al(s) + N_2(g)$ ----> ?
    b.   $2Ga(s) + 2OH^-(aq) + 6H_2O(l)$ ----> ?
    c.   $2In(s) + 6H^+(aq)$ ----> ?
    d.   $4Tl(s) + 3O_2(g)$ ----> ? (low temperatures)

## Solution

a.  $2Al(s) + N_2(g) ----> 2AlN(s)$
b.  $2Ga(s) + 2OH^-(aq) + 6H_2O(l) ----> 2Ga(OH)_4^-(aq) + 3H_2(g)$
c.  $2In(s) + 6H^+(aq) ----> 2In^{3+}(aq) + 3H_2(g)$
d.  $4Tl(s) + 3O_2(g) ----> 2Tl_2O_3(s)$ (low temperatures)

---

# Example 18.6 A - Electron Configuration

Write shorthand electron configurations for each of the following group 4A elements:

a.  carbon        b.  germanium        c.  tin

## Solution

a.  carbon:  $[He] 2s^2 2p^2$
b.  germanium:  $[Ar] 4s^2 3d^{10} 4p^2$
c.  tin:  $[Kr] 5s^2 4d^{10} 5p^2$

Notice that group 4A elements have $ns^2 np^2$ electron configurations.

---

1.  What is the hybridization on the central atom in molecules such as $PbCl_4$ and $GeBr_4$?
2.  Why can't carbon form a compound such as $CCl_6^{2-}$ while tin can form $SnCl_6^{2-}$?
3.  Why can carbon form $\pi$ bonds with other **period 2** nonmetals while other group 4A elements can not?
4.  Why, with regard to bond energies, does the silicon to oxygen bond dominate silicon chemistry rather than a silicon to silicon bond?
5.  Define **allotrope**.
6.  What are the three allotropic forms of carbon?
7.  What is the cause of carbon monoxide poisoning?
8.  What is the major industrial use of silicon?  germanium?
9.  Under what conditions is each of the allotropes of tin stable?
10. What is **tin disease**?
11. What is the common source of lead?
12. Your textbook says that lead may have contributed to the demise of the Roman civilization.  Why?
13. What has been a significant cause of lead poisoning in the twentieth century?
14. Why are there very few known lead (IV) compounds?
15. What is the essential chemistry of concrete?

---

# Example 18.6 B - Reactions of Group 4A Elements

Predict the products of each of the following reactions:

a.  $Sn(s) + 2H^+(aq) ----> ?$
b.  $Ge(s) + O_2(g) ----> ?$
c.  $Pb(s) + Cl_2(g) ----> ?$

## Solution

a.  $Sn(s) + 2H^+(aq) ----> Sn^{2+}(aq) + H_2(g)$

b.  $Ge(s) + O_2(g) \longrightarrow GeO_2(s)$

c.  $Pb(s) + Cl_2(g) \longrightarrow PbCl_2(s)$  (note that Pb is the only group 4A element that forms the dihalide.  Others form the tetrahalide.  $MX_4$.)

---

Answers to review questions are at the end of this study chapter.

---

**END OF CHAPTER EIGHTEEN**

# ANSWERS TO REVIEW QUESTIONS

## Section 18.2

1. The "$ns^1$" configuration is common to each of the elements in this group.
2. Hydrogen is a nonmetal that is very small and can either gain or lose an electron, but does so covalently rather than ionically.
3. This illustrated that reaction rate and thermodynamics are not directly related.
4. They react so well together because they easily lose electrons to form $M^+$ ions.

## Section 18.3

1. Hydrogen has low boiling and melting points because of its low molecular weight and nonpolarity.
2. $2H_2(g) + O_2(g) \longrightarrow 2H_2O(l)$
3. $CH_4(g) + H_2O(g) \longrightarrow CO(g) + 3H_2(g)$
4. It isn't very practical because of the high cost of electricity.
5. Production of ammonia and hydrogenating unsaturated vegetable oil.
6. **hydride:** Binary compounds containing hydrogen.
   **ionic hydride:** Hydrogen combines with group 1A or 2A metals.
   **covalent hydride:** Hydrogen combines with nonmetals.
   **interstitial hydride:** Hydrogen atoms occupy holes ("interstices") in a metal's crystal structure.
7. Hydrogen diffuses through the palladium metal wall (see "interstitial hydride") leaving impurities behind.
8. Interstitial hydrides lose absorbed ("stored") hydrogen when heated.

## Section 18.4

1. They are called **alkaline earth** metals because their oxides form bases in water.
2. Beryllium displays different properties because of its relatively small size and high electronegativity.
3. The rationale comes from the localized electron model where one pair of electrons is assumed to be shared by three atoms.
4. The beryllium is sp hybridized.
5. $BeCl_2$ is so reactive toward Lewis bases because it is electron-deficient.
6. For example, magnesium in flask bulbs and in structural materials.
7. Hard water is that which contains large concentrations of $Ca^{2+}$ and $Mg^{2+}$ ions.
8. In a nutshell, hard water cations become bound to sites on a polymer resin which releases sodium ions as a result, thus softening the water.

## Section 18.5

1. Boron is very small and has a relatively low electronegativity, thus forming mostly covalent compounds.
2. $B_2H_6$ is the simplest stable borane.
3. Boranes have one or more three-center two-electron bonds similar to that in $BeCl_2$. The remaining bonds are normal covalent bonds.
4. As with beryllium, boranes are highly electron-deficient which accounts for their reactivity.
5. Covalent bonding is responsible for the amphoteric nature of aluminum.
6. Gallium is useful in thermometers because, of all the elements, it has the widest temperature range in which it is a liquid.
7. The **inert pair effect** is the tendency of heavier elements to exhibit an oxidation state that is two less positive than its group number ( i.e., group IV heavy elements can have both +4 and +2 oxidation states.)
8. Aluminum foil does not corrode because it forms a protective oxide layer in air.

## Section 18.6

1. The central atom is $sp^3$ hybridized.
2. There are no d electrons available in carbon to allow it to exceed an octet.

3.  Carbon p orbitals have significant overlap with other **similar sized** period 2 elements.  Other group 4A elements cannot have this degree of overlap.
4.  The bond energy of the Si-O bond is substantially higher than that of the Si-Si bond.
5.  An allotrope has the same element but different structures i.e., S, $S_2$ and $S_8$, or $O_2$ and $O_3$.
6.  Graphite and diamond are allotrophic forms of carbon.
7.  Although more details are presented in Chapter 23 (biochemistry), carbon monoxide poisoning can occur when the supply of oxygen is limited and carbon monoxide can interact with iron in your bloodstream.
8.  The major use of silicon is in semiconductors.  The major use of germanium is also for semiconductors.
9.  White tin is stable at ambient temperatures.  Gray tin is stable below 13.2°C.  Brittle tin is stable above 161°C.
10. Tin disease is the conversion of white tin to powdery gray tin in cold temperatures.
11. Galena (PbS) is the common source of lead.
12. High concentrations of lead are toxic.  Significant levels of lead have been found in the bones of the Roman era.
13. Lead in gasoline is a significant contributor to lead poisoning.
14. Lead (IV) reduces to lead (II) in the presence of bromide or iodide ions, add well at temperatures above 100°C.
15. See the <u>Chemical Impact</u> titled "Concrete Learning" in the section on Group 4A elements.

# CHAPTER 19

# The Representative Elements-Groups 5A Through 8A

| Section | Contents | Page in this Study Guide |
|---|---|---|
| 19.1 | The Group 5A Elements | 401 |
| 19.2 | The Chemistry of Nitrogen | 402 |
| 19.3 | The Chemistry of Phosphorus | 404 |
| 19.4 | The Group 6A Elements | 405 |
| 19.5 | The Chemistry of Oxygen | 405 |
| 19.6 | The Chemistry of Sulfur | 405 |
| 19.7 | The Group 7A Elements | 407 |
| 19.8 | The Group 8A Elements | 407 |

Your textbook opens up the chapter by noting that as we go from group 1A to 7A, the elements change from the tendency to donate electrons (groups 1A, 2A, 3A and 4A) to accepting electrons (groups 5A, 6A, and 7A). However, groups 5A and 6A show great variety in their chemistry. This chapter focuses on the most important elements in groups 5A through 8A - nitrogen, phosphorus, oxygen, sulfur and the halogens.

## 19.1 - The Group 5A Elements

The following questions and examples will help you to review the chemistry of group 5A elements.

## Example 19.1 A - Electron Configurations

Give the shorthand configuration for each of the following elements:

a. nitrogen    b. arsenic    c. bismuth

## Solution

a. nitrogen: $[He]\ 2s^2\ 2p^3$
b. arsenic: $[Ar]\ 4s^2\ 3d^{10}\ 4p^3$
c. bismuth: $[Xe]\ 6s^2\ 4f^{14}\ 5d^{10}\ 6p^3$

Notice the $ns^2\ np^3$ configuration on each of these group 5A elements.

1.  Why are there no known ionic compounds containing bismuth (V) or antimony (V)?
2.  Justify the fact that $NH_3$ and $PH_3$ act as Lewis bases.
3.  What is the hybridization of As in $AsCl_3$?
4.  Why can't nitrogen form compounds with 5 ligands?
5.  What is the hybridization of P in $PF_6^-$?
6.  Discuss the structure of compounds such as $PCl_5$.

## Example 19.1 B - Reactions of Group 5A Elements

Predict the products of the following reactions:

a.  $BiF_5(s) ----> ?$  (decomposition)
b.  $P_4O_{10}(s) + 10C(s) ----> ?$
c.  $SbO^+(aq) + Cl^-(aq) ----> ?$

### Solution

a.  $BiF_5(s) ----> BiF_3(s) + F_2(g)$
b.  $P_4O_{10}(s) + 10C(s) ----> 4P(s) + 10CO(g)$
c.  $SbO^+(aq) + Cl^-(aq) ----> SbOCl(s)$

Answers to review questions are at the end of this study chapter.

## 19.2 - The Chemistry of Nitrogen

The following questions and examples will help you review the material in this section.

1.  Why is $N_2$ such a stable molecule?
2.  Why is nitrogen used as an atmosphere for reactants that normally react with water or oxygen?

## Example 19.2 A - Thermodynamics of Nitrogen Compounds

Using data from <u>Appendix 4 in your textbook</u>, calculate $\Delta S°$ and $\Delta G°$ for each of the following reactions.

a.  $N_2O(g) ----> N_2(g) + 1/2O_2(g)$
b.  $NO_2(g) ----> 1/2N_2(g) + O_2(g)$
c.  $NH_3(g) ----> 1/2N_2(g) + 3/2H_2(g)$

### Solution

Recall that entropy and free energy are both **state functions**.

$$\Delta S°_{reaction} = \Sigma S°_{products} - \Sigma S°_{reactants}$$
$$\Delta G°_{reaction} = \Sigma G°_{products} - \Sigma G°_{reactants}$$

For entropy,

    a.   $\Delta S° = [\Delta S°_{N_2} + 1/2\Delta S°_{O_2}] - \Delta S°_{N_2O}$ = [192 J/K·mol + 1/2(205 J/K·mol)] - [220 J/K·mol]

$$\Delta S°_{reaction} = \textbf{74.5 J/K·mol}$$

    b.   $\Delta S° = [1/2\Delta S°_{N_2} + \Delta S°_{O_2}] - \Delta S°_{NO_2}$= [1/2(192 J/K·mol) + 205 J/K·mol] - [240 J/K·mol]

$$\Delta S°_{reaction} = \textbf{61 J/K·mol}$$

    c.   $\Delta S° = [1/2\Delta S°_{N_2} + 3/2\Delta S°_{H_2}] - \Delta S°_{NH_3}$= [1/2(192 J/K·mol) + 3/2(131 J/K·mol)] - [193 J/K·mol]

$$\Delta S°_{reaction} = \textbf{100 J/K·mol}$$

For free energy, recall that $\Delta G°$ of elements in their standard states = 0.

    a.   $\Delta G° = 0 - \Delta G°_{N_2O}$ = **-104 kJ/mol**

    b.   $\Delta G° = 0 - \Delta G°_{NO_2}$ = **-52 kJ/mol**

    c.   $\Delta G° = 0 - \Delta G°_{NH_3}$ = **+16.6 kJ/mol**

Notice that $\Delta G = \Delta H - T\Delta S$ will serve as a double check in each case!

---

3.   Why are nitrogen-based explosives so effective (with regard to thermodynamics and gas volume)?
4.   Why does your textbook say that the thermodynamics and kinetics of the Haber process are in opposition?
5.   What are the conditions under which the Haber process is performed?
6.   Define **nitrogen fixation**.
7.   How does nitrogen fixation occur in an automobile?
8.   List two natural mechanisms of nitrogen fixation.
9.   Why is there such an interest in **nitrogen-fixing bacteria**?
10.  What is **denitrification**?
11.  What is the problem with accumulating excess nitrogen in soil and bodies of water?

## Nitrogen Hydrides

12.  Why does ammonia have a much lower boiling point than water?
13.  What is the structure of hydrazine?
14.  Why is hydrazine a useful chemical in the space program?

---

# Example 19.2 B - Hydrazine as a Propellant

Sample Exercise 19.2 in your textbook gives the following reaction between the rocket fuel monomethylhydrazine and the oxidizer dinitrogen tetroxide.

$$5N_2O_4(g) + 4N_2H_3(CH_3)(g) ----> 12H_2O(g) + 9N_2(g) + 4CO_2(g)$$

How many liters of carbon dioxide are formed at 1.00 atm and 25°C from the reaction of 3.00 grams of $N_2O_4$ with 5.00 g of $N_2H_3(CH_3)$?

## Strategy

This problem combines a limiting reactant calculation with a gas law problem. The general strategy is

- Determine how much product is formed by each reactant.

- Determine how much product is formed by the limiting reactant.
- Convert moles of product to liters of product using the ideal gas law.

## Solution

$$\text{mol } CO_2 \text{ from } N_2O_4 \ \left\|\ 3.00 \text{ g } N_2O_4 \times \frac{1 \text{ mol } N_2O_4}{92.0 \text{ g } N_2O_4} \times \frac{4 \text{ mol } CO_2}{5 \text{ mol } N_2O_4} = 0.0261 \text{ mol } CO_2 \right.$$

$$\text{mol } CO_2 \text{ from } N_2H_3(CH_3) \ \left\|\ 5.00 \text{ g } N_2H_3(CH_3) \times \frac{1 \text{ mol } N_2H_3(CH_3)}{46.0 \text{ g } N_2H_3(CH_3)} \times \frac{4 \text{ mol } CO_2}{4 \text{ mol } N_2H_3(CH_3)} = 0.109 \text{ mol } CO_2 \right.$$

$N_2O_4$ is the limiting reactant and **0.0261 mol $CO_2$ is formed**.

$$V_{CO_2} = \frac{nRT}{P} = \frac{0.0261 \text{ mol } (0.08206 \text{ L·atm/K·mol})(298.2 \text{ K})}{1.00 \text{ atm}}$$

$$= \mathbf{0.639 \text{ L } CO_2 \text{ formed}}$$

---

15. List two uses of hydrazine in addition to that of a rocket propellant.

## Nitrogen Oxides

16. List the formula and determine the oxidation state of each nitrogen oxide in <u>Table 19.2 in your textbook</u>.
17. List several uses of nitrous oxide.
18. What is the role of nitrous oxide in the earth's climate control?
19. How is $N_2O$ prepared in the laboratory?
20. How is nitric oxide prepared?
21. Why does NO turn brown in air?
22. Why does $NO^+$ have a higher bond energy than NO?
23. What happens to $NO_2$ at low temperatures?

## Oxyacids of Nitrogen

24. List some uses of nitric acid.
25. Give the reactions involved in the **Ostwald process**.
26. How can the concentration of nitric acid be increased form 68% to 95%?
27. Why does a nitric acid solution turn yellow upon constant exposure to sunlight?
28. How is nitrous acid prepared?

---

# 19.3 - The Chemistry of Phosphorus

The following questions and the example will help you review the material in this section.

1. List 4 reasons for the differences in chemical properties between nitrogen and phosphorus.
2. What are the three allotropic forms of phosphorus? How are they different from one another?
3. How is red phosphorus obtained from white phosphorus?
4. How is black phosphorus obtained from either red or white phosphorus?
5. What is the oxidation state of phosphorus in phosphides? Give some examples of phosphides.
6. Why are phosphonium salts uncommon?
7. Why is phosphine viewed as an exception to the VSEPR model?
8. Discuss the use of tripolyphosphoric acid in detergents.
9. Your textbook says that $H_3PO_3$ and $H_3PO_2$ are diprotic and monoprotic, respectively. Why aren't they triprotic?

10. Why isn't naturally occurring phosphorous in soil easily usable by plants?
11. What is the oxidation state of phosphorous in $PF_3$? $PF_5$?

## Example 19.3 - Reactions of Phosphorus

Give the products of each of the following reactions (discussed in this section of your textbook):

    a.   $2Na_3P(s) + 6H_2O(l)$ ----> ?
    b.   $P_4O_{10}(s) + 6H_2O(l)$ ----> ?
    c.   $P_4(s) + 3O_2(g; \text{ limited})$ ----> ?
    d.   $P_4(s) + 5O_2(g; \text{ excess})$ ----> ?
    e.   $P_4O_6(s) + 6H_2O(l)$ ----> ?
    f.   $PF_3(s) + 3H_2O(l)$ ----> ?

## Solution

    a.   $2Na_3P(s) + 6H_2O(l)$ ----> $2PH_3(g) + 6Na^+(aq) + 6OH^-(aq)$
    b.   $P_4O_{10}(s) + 6H_2O(l)$ ----> $4H_3PO_4(aq)$
    c.   $P_4(s) + 3O_2(g; \text{ limited})$ ----> $P_4O_6(s)$
    d.   $P_4(s) + 5O_2(g; \text{ excess})$ ----> $P_4O_{10}(s)$
    e.   $P_4O_6(s) + 6H_2O(l)$ ----> $4H_3PO_3(aq)$
    f.   $PF_3(s) + 3H_2O(l)$ ----> $H_3PO_3(aq) + 3HF(aq)$

Answers to review questions are at the end of this chapter.

## 19.4, 19.5, and 19.6 - The Chemistry of the Group 6A Elements Focussing on Oxygen and Sulfur

The following questions and examples will help you review the material in these sections.

## Example 19.4 - Electron Configurations

Write the shorthand configurations for each of the following elements:

    a.  oxygen   b.  selenium   c.  polonium

## Solution

    a.  oxygen:  $[He]\ 2s^2\ 2p^4$
    b.  selenium:  $[Ar]\ 4s^2\ 3d^{10}\ 4p^4$
    c.  polonium:  $[Xe]\ 6s^2\ 4f^{14}\ 5d^{10}\ 6p^4$

Note that all group 6A elements have the configuration $ns^2\ np^4$.

1. What is the most common oxidation state of group 6A elements in ionic compounds?
2. Give some examples of group 6A covalent hydrides.
3. Why do group 6A elements (other than oxygen) form compounds with more than eight electrons around the central atom?

## Example 19.4 - VSEPR Review

Predict the geometry of $TeI_4$.

## <u>Solution</u>

We learned about the VSEPR model in Chapter 8. Our first step is to draw the Lewis structure.

1. # valence electrons in the system = 6 for Te + 4 x 7 for I = 34 electrons.
2. # electrons if happy = 8 for Te + 4 x 8 for I = 40 electrons.
3. # bonds = (40 - 34)/2 = 3 bonds

This is an **exception**! We must therefore draw the Lewis structure and put extra electron pairs around the central atom.

There are 5 effective electron pairs around the central atom. The structure is **based on a trigonal bipyramid**. The lone pair will be in the equatorial plane. This is a **seesaw** structure.

4. Why don't group 6A elements form +6 cations?
5. Why has there been growing interest in the chemistry of selenium?
6. Why might polonium be related to cancer in smokers?
7. What are some of the important uses of oxygen in our world?
8. What percent of the earth's atmosphere is oxygen?
9. How is oxygen isolated from air?
10. How can we demonstrate the paramagnetism of oxygen?
11. What is the VSEPR geometry of ozone?
12. Why is the bond angle in ozone less than 120°?
13. How is ozone prepared in the laboratory?
14. Why might ozone be useful in municipal water purification?
15. What is the importance of the ozone layer of the earth's atmosphere?
16. Name some minerals that contain sulfur.
17. Describe the **Frasch Process**.
18. Why does elemental oxygen exist as $O_2$ whereas sulfur exists as $S_8$?
19. What is the structural difference between rhombic and monoclinic sulfur?
20. What is **plastic sulfur**?
21. Why is $O_2$ more stable than SO?
22. What role might dust play in the conversion of $SO_2$ to $SO_3$?
23. What is the geometry of $SO_3$?
24. Describe the **contact process** for the manufacture of $H_2SO_4$.
25. List at least one compound that sulfur forms with when in each of its various oxidation states.

## Example 19.6 - Reactions of Sulfur

Based on the material in this section, predict the products of each of the following reactions:

    a.   $2S_2O_3^{2-}(aq) + I_2(aq)$ ----> ?
    b.   $C_{12}H_{22}O_{11}(s) + 11H_2SO_4(conc)$ ----> ?
    c.   $SO_2(aq) + H_2O(l)$ ----> ?
    d.   $2SO_2(g) + O_2(g)$ ----> ?

## Solution

    a   $2S_2O_3^{2-}(aq) + I_2(aq)$ ----> $S_4O_6^{2-}(aq) + 2I^-(aq)$
    b.   $C_{12}H_{22}O_{11}(s) + 11H_2SO_4(conc)$ ----> $12C(s) + 11H_2SO_4\ H_2O(l)$
    c.   $SO_2(aq) + H_2O(l)$ ----> $H_2SO_3(aq)$
    d.   $2SO_2(g) + O_2(g)$ ----> $2SO_3(g)$

---

Answers to review questions are at the end of this study chapter.

---

## 19.7 and 19.8 - The Group 7A And 8A Elements

The following questions and examples will help you review the material in these sections.

---

## Example 19.7 A - Electron Configurations

Write the shorthand configurations for each of the following elements:

    a. bromine   b.   iodine

## Solution

    a.   bromine: $[Ar]\ 4s^2\ 3d^{10}\ 4p^5$
    b.   iodine:   $[Kr]\ 5s^2\ 4d^{10}\ 5p^5$

Note that the halogens have the configuration $ns^2\ np^5$.

---

## Example 19.7 B - The Decomposition of Astatine

Your textbook says that the chemistry of astatine "is of no practical importance" because the longest-lived isotope $^{210}At$ has $t_{1/2} = 8.3$ hours. If you have a 1.000 g sample of $^{210}At$, how much will remain after 24.0 hours? (The radioactive decomposition of astatine obeys a **first-order** kinetic rate law.)

## Solution

For a first-order decay, the rate constant,

$$k = 0.693/t_{1/2} = 0.693 / 8.3\ hr = \mathbf{8.35 \times 10^{-2}\ hr^{-1}}$$

$$\ln \frac{A_t}{A_o} = -kt$$

$$\text{where } A_o = 1.000 \text{ g} \qquad k = 8.35 \times 10^{-2} \text{ hr}^{-1}$$
$$A_t = ? \qquad\qquad t = 24.0 \text{ hr}$$

$$\ln \left[\frac{A_t}{1.000}\right] = -8.35 \times 10^{-2} \text{ hr}^{-1} (24.0 \text{ hr})$$

$$\ln \left[\frac{A_t}{1.000}\right] = -2.00$$

Taking the antilog of both sides,

$$\frac{A_t}{1.000} = 0.135$$

**$A_t$ = 0.135 g of astatine remaining at time t**

You can see that after just one day, the majority of a sample of astatine will decompose.

---

1.  What kind of bonds do halogens tend to form with metals? Nonmetals in lower oxidation states?
2.  What about bonding with metals in higher oxidation states?
3.  Why does HF have such a high boiling point relative to other hydrogen halides?
4.  Why can't the relative strength of hydrogen halides as acids be assessed in water?
5.  What is the order of acid strength of hydrogen halides in acetic acid?
6.  What is the rationale for HF being the least acidic hydrogen halide?
7.  Why is hydrochloric acid such an important industrial chemical?
8.  Why must $HClO_4$ be handled so carefully?
9.  Define "**disproportionation reaction**."
10. List some uses of chlorate salts.
11. Why is $OF_2$ called oxygen difluoride rather than difluorine oxide?
12. Discuss the structure of **Teflon**.
13. What are **Freons**?
14. What is PVC?
15. What are the pertanent reactions which make the darkening of sunglasses reversible?
16. What is the importance of copper ion in the "automatic sunglasses" process?

---

## Example 19.7 C - Reactions of Halogens

Predict the products of the following reactions:

a.  $AgBr(s) + 2S_2O_3^{2-}(aq) \longrightarrow ?$ (See the Chemical Impact on photography)

b.  $H_2(g) + Cl_2(g) \xrightarrow{\text{U.V. light}} ?$

c.  $SiO_2(s) + 4HF(aq) \longrightarrow ?$

d.  $2F_2(g) + H_2O(l) \xrightarrow{\text{NaOH}} ?$

e.  $Cl_2(aq) + H_2O(l) \rightleftharpoons ?$

## Solution

a.   $AgBr(s) + 2S_2O_3^{2-}(aq) \longrightarrow Ag(S_2O_3)_2^{3-}(aq) + Br^-(aq)$

b.   $H_2(g) + Cl_2(g) \xrightarrow{\text{U.V. light}} 2HCl(g)$

c.   $SiO_2(s) + 4HF(aq) \longrightarrow SiF_4(g) + 2H_2O(l)$

d.   $2F_2(g) + H_2O(l) \xrightarrow{\text{NaOH}} 2HF(aq) + OF_2(g)$

e.   $Cl_2(aq) + H_2O(l) \rightleftharpoons HOCl(aq) + H^+(aq) + Cl^-(aq)$

---

17.  What is the major source of helium on earth?
18.  What are the major uses of helium? of neon? argon?
19.  What are some xenon compounds that have been prepared? Some krypton compounds?

---

Answers to review questions are at the end of this study chapter.

---

**END OF CHAPTER NINETEEN**

# ANSWERS TO REVIEW QUESTIONS

## Section 19.1

1. Too much energy is required to remove all 5 electrons from the neutral compound.
2. $NH_3$ and $PH_3$ have lone pairs that they can donate.
3. The hybridization is $sp^3$.
4. Nitrogen has no available d orbitals, so it cannot exceed the octet.
5. $PCl_5$ is among a group of compounds that is of mixed hybridization, $sp^3$ (for $PCl_4^+$) and $d^2sp^3$ (for $PCl_6^-$).
6. $PCl_5$ is an exception to the octet rule. It has trigonal bipyramidal hybridization.

## Section 19.2

1. $N_2$ is so stable because it has a very high bond strength.
2. Nitrogen is used as an "inert atmosphere" because it is so unreactive.
3. Nitrogen based explosives are effective because the decomposition to $N_2$ (gas) is a thermodynamically favorable process.
4. In the Haber process, the equilibrium lies in the direction of $NH_3$ at room temperature. Because the reaction is exothermic, increasing the temperature to increase reaction rate decrease the equilibrium constant.
5. High pressure and moderately high temperature.
6. The process of transforming $N_2$ to other nitrogen containing compounds.
7. Nitrogen from the air reacts with oxygen to form NO and then $NO_2$.
8. Lighting and nitrogen-fixing bacteria.
9. Such bacteria produce ammonia in the soil at standard conditions.
10. Denitrification is the return of nitrogen-containing compounds to the atmosphere as $N_2$ gas.
11. Algae and other undesirable organisms may accumulate because of the presence of too much fixed nitrogen.
12. Water contains two polar bonds and two lone pairs. It can exhibit much more substantial hydrogen bonding than ammonia.
13. See Figure 19.8 in your textbook.
14. Hydrazine is and excellent reducing agent that produces a great deal of energy when reacted with oxygen.
15. It is used as a blowing agent in plastics and is also used in the production of agricultural pesticides.
16. See Table 19.2 in your textbook.
17. It is used by dentists as an anesthetic. It is also used as a propellant in aerosol cans of whipped cream.
18. It strongly absorbs infrared radiation, thus helping with climate control.
19. It is prepared by thermal decomposition of ammonium nitrate.
20. NO is prepared by the reaction of copper with nitric acid.
21. NO is oxidized in air to brown $NO_2$.
22. The M.O. bond order for $NO^+$ is 3, and for NO is 2.5. Therefore, $NO^+$ has a higher bond energy than NO.
23. $NO_2$ can dimerize to form $N_2O_4$.
24. It is used in the manufacture of fertilizer and explosives, among other things.
25. a.  $4NH_3(g) + 5O_2(g) ----> 4NO(g) + 6H_2O(g)$.
    b.  $2NO(g) + O_2(g) ----> 2NO_2(g)$.
    c.  $3NO_2(g) + H_2O(l) ----> 2HNO_3(aq) + NO(g)$.
26. The concentration is increased by treatment with sulfuric acid, a dehydrating agent.
27. $HNO_3$ decomposes in sunlight to give $NO_2(g)$, a brown gas which colors the solution yellow.
28. It is prepared by acidifying the nitrite ion, which is made by mixing NO, $NO_2$ and $OH^-$ ion.

## Section 19.3

1. a.  Nitrogen can form stronger bonds.
   b.  Nitrogen is more electronegativity.
   c.  Phosphorous is larger.
   d.  Phosphorous has available d orbitals.
2. White, red and black phosphorous differ in structure as outlined in Figure 19.12 in your textbook.
3. Red phosphorous is formed by heating white phosphorous in the absence of air.
4. Black phosphorous is made by heating white or red phosphorous at high pressures.

5.  The oxidation state is -3. Examples are $Na_3P$ and $Ca_3P_2$.
6.  Phosphonium salts are uncommon because phosphine is a very weak base.
7.  The bond angle is far smaller (94°) than expected for a tetrahedral structure with one lone pair.
8.  The sodium salt of tripolyphosphoric acid is used in detergents because anion forms complexes with $Ca^{2+}$ and $Mg^{2+}$, which interfere with detergent action.
9.  They are not triprotic because hydrogens attached to phosphorous are not acidic.
10. Phosphorous in soil is often present in insoluble (and thus unusable) minerals.
11. The oxidation state is +3 $PF_3$ and +5 in $PF_5$.

## Section 19.4

1.  The most common oxidation state is -2.
2.  Examples are $H_2S$ and $H_2O$.
3.  Other than oxygen, all group 6A elements have available d orbitals.
4.  It takes far too much energy to remove 6 electrons from these elements.
5.  It is possible that selenium has a role in cancer protection.
6.  Polonium is found in tobacco and is an α emitter.
7.  In general, we need oxygen for respiration and for combustion.
8.  Oxygen makes up 21% of the earth's atmosphere.
9.  Nitrogen distillation isolates oxygen.
10. We can pour liquid oxygen between poles of a magnet.
11. The VSEPR structure is trigonal planar.
12. The lone pair requires more room than bonded pairs.
13. Ozone is prepared by passing an electric charge through pure oxygen gas.
14. Ozone is a strong oxidant.
15. Ozone absorbs ultraviolet radiation.
16. Galena, cinnabar and pyrite, among others.
17. The Frasch process is used to obtain sulfur form underground deposits. Water is pumped in to melt sulfur, which is recovered by air pressure.
18. Oxygen can form  bonds, thus stabilizing $O_2$. Sulfur can only form s bonds, thus stabilizing the larger form, $S_8$.
19. The difference has to do with ring stacking.
20. Plastic sulfur results from heating and cooling $S_8$. Plastic sulfur has rubber-like properties.
21. $O_2$ has stronger  bonding than SO.
22. Dust may act as a catalyst for the conversion.
23. $SO_3$ has a trigonal planar geometry.
24. This reaction between $SO_3$ and $O_2$ is described in the chemical impact section of Chapter 3.
25. See Table 19.5 in your textbook.

## Section 19.7 and 19.8

1.  Halogens form polar covalent bonds with nonmetals and ionic compounds with metals in lower oxidation states.
2.  Bonds with metals in higher oxidation states tend to be polar covalent.
3.  HF can hydrogen bond (due to the small size and high electronegativity of fluorine.)
4.  Hydrogen halides (except HF) completely dissociate in water, so they appear "equally strong."
5.  HI > HBr > HCl >> HF
6.  The H-F bond strength is higher than the other H-X acids.
7.  HCl is used for, among other things, cleaning steel.
8.  $HClO_4$ is an incredibly strong oxidizing agent which reacts explosively with many organic compounds.
9.  A disproportionation reaction occurs when an element is both oxidized and reduced in the same reaction.
10. Among the uses are as weed killers and as oxidizers in fireworks.
11. Fluorine is the anion. IUPAC nomenclature rules dictate that it gets the "ide" ending.
12. Teflon is a polymer of perfluorethylene. The structure leads to high thermal stability and high resistance to chemical attack.
13. Freons are compounds with the general formula $CCl_xF_{4-x}$. The compounds are used because of their stability.

15. $Ag^+ + Cl^-$ ----> $Ag + Cl$ (electron transfer)
    $Cl + Cu^+$ ----> $Cu^{2+} + Cl^-$
    $Cu^{2+}$ ----> $Cu^+ + Ag^+$

16. The copper ion allows for the *reversibility* of the process (as opposed to the irreversibility of the photography process).

17. The major sources are natural gas deposits.

18. Helium is used as a coolant, among other things, neon and argon are used in lighting.

19. Examples are $XePtF_6$, $XeF_4$, $KrF_4$ and $KrF_4$.

# CHAPTER 20

# Transition Metals and Coordination Chemistry

| Section | Contents | Page in this Study Guide |
|---------|----------|--------------------------|
| 20.1 | The Transition Metals - A Survey | 413 |
| 20.2 | The First-Row Transition Metals | 415 |
| 20.3 | Coordination Compounds | 416 |
| 20.4* | Isomerism | 419 |
| 20.6 | The Crystal Field Model | 419 |
| 20.7 | The Molecular Orbital Model | 421 |
| 20.8 | The Biological Importance of Coordination Complexes | 421 |
| 20.9 | Metallurgy and Steel Production | 422 |

## 20.1 - The Transition Metals - A Survey

When you finish this section you will be able to:

- write electron configurations for first-row transition metal ions, and
- analyze trends in periodic properties of transition metals.

This section begins by noting that transition metal chemistry shows **much greater consistency** than that of the representative elements as we go across the periodic table. This is because the filling of transition metals involves d- or f-block electrons, which are core electrons. These do not participate as easily in bonding as s and p electrons. Read the information from below <u>Figure 20.1</u> and above and including <u>Table 20.2 in your textbook</u> then try the following example.

---

*Section 20.5 deals with the Localized Electron model. The model is not a good predictor of properties in complex ions. We will not deal with this material.*

## Example 20.1 A - Properties of Transition Metals

Answer the following questions regarding properties of first-row transition metals.

1. List the properties that they have in common.
2. What is the trend with regard to
   a. density?
   b. electrical conductivity?
   c. atomic radius?
   d. common oxidation states?

## Solution

1. Properties in common: metallic luster, high electrical conductivity, high thermal conductivity.
2. a. The density shows a steady increase with the exception of zinc.
   b. In general, electrical conductivity increases.
   c. Atomic radius decreases then increases after iron. The key argument here is the balance between increased nuclear charge and electron-electron repulsion. In any case the radii are not drastically different across the period.
   d. Vanadium, chromium, and manganese have many common oxidation states, due to the availability of s and d electrons. Atoms such as nickel, copper, and zinc only have available s electrons (it would require too much energy to strip d electrons away), so these elements have fewer oxidation states available.

---

You learned about electron configurations in Chapter 7. In this section your textbook points out that where exceptions to filling rules occur (such as chromium, which is $4s^1 3d^5$) it is because the **4s and 3d levels are close together** and electron-electron repulsions can be minimized by this unusual configuration.

However, your textbook also points out that **IN IONS** the energy of the 3d level is significantly lower than that of the 4s level. This means that the 3d energy level may be occupied while the 4s will be unoccupied for the first-row transition metal ions.

---

## Example 20.1 B - Electron Configurations of Ions

Write the shorthand electron configurations for each of the following:

   a. V    b. $V^{2+}$    c. $V^{3+}$    d. $V^{5+}$

## Solution

   a. V: $[Ar] 4s^2 3d^3$    c. $V^{3+}$: $[Ar] 3d^2$
   b. $V^{2+}$: $[Ar] 3d^3$    d. $V^{5+}$: $[Ar]$

---

The discussion that closes out this section deals with the 4d and 5d transition series. The **lanthanide contraction** is the slight shrinkage in atomic size that occurs across these periods due to extra nuclear-electron charge attraction as 4f or 5f electrons are added. The lanthanide contraction also accounts for the similarity in properties between 4d and 5d elements in a particular group.

---

Problems 1-2 at the end of this study chapter cover material from this section.

## 20.2 - The First-Row Transition Metals

The following questions will help you review your knowledge of the materials in this section.

1.  What elements are being discussed?  What do they have in common?
2.  What is the most common oxidation state for scandium in its compounds?
3.  Why are scandium compounds generally colorless and diamagnetic?
4.  How is scandium prepared?
5.  What is the major industrial use of scandium?
6.  What properties make titanium a useful structural material?
7.  Why is titanium used for reaction vessels in the chemical industry?
8.  List some uses of titanium (IV) oxide.
9.  What are the main natural sources of titanium?
10. Write the reactions for the purification of $TiO_2$ from its ores.
11. What are the major industrial uses of vanadium?
12. How is pure vanadium prepared?
13. What is vanadium steel?
14. Why do $V^{5+}$ and $V^{4+}$ exist in solution as $VO_2^+$ and $VO^{2+}$?

## Example 20.2 - A Bit of Detective Work

An aqueous solution of ammonium vanadate ($NH_4VO_3$) is yellow.  It is added to a "Jones Reductor," which is a Zn-Hg amalgam.  The solution was swirled over the reductor.

1.  An aliquot of the solution was removed.  It was **blue**.
2.  The remaining solution was swirled over the reductor.
3.  Another aliquot was removed.  It was **green**.
4.  The remaining solution was swirled over the reductor.
5.  A **final** aliquot was removed.  It was **violet**.
6.  Upon standing in air, this **final** aliquot turned **green**.

Describe, in terms of the oxidation states of vanadium, the chemistry that occurred.

## Solution

a.  The original solution (yellow) is vanadium (V) (as $VO_2^+$)·
b.  When swirled, the vanadium was reduced to $V^{4+}$ (as $VO^{2+}$), which is blue.
c.  When further swirled, $VO^{2+}$ was reduced to $V^{3+}$, which is green.
d.  A final mixing reduced $V^{3+}$ to $V^{2+}$ (violet).
e.  Upon sitting, $V^{2+}$ was air oxidized to $V^{3+}$ (green).

15. How is chromium prepared?
16. What are the common oxidation states of chromium?
17. Write the half-reaction for the reduction of the dichromate ion in acidic solution.
18. What are the structural differences between chromium (VI) in acidic and basic solutions?
19. What is "cleaning solution" composed of?  What is it used for?
20. What are the most common uses of manganese?

21. What are "manganese nodules?"
22. What are the common oxidation states of manganese?
23. Why is $MnO_4^-$ such a powerful oxidizing agent?
24. What are the important oxidation states of iron?
25. Why are iron (II) solutions light green?
26. Why are iron (III) solutions yellow?
27. Why does $Fe(H_2O)_6^{3+}$ behave as an acid in aqueous solution?
28. What are some commercial uses of cobalt?
29. What species causes the rose color of cobalt in aqueous solution?
30. Why is nickel used for plating active metals?
31. Why is nickel almost always found in the +2 oxidation state when in compounds?
32. Why is copper so useful?
33. What is the major use of copper?
34. What species causes the characteristic blue color of aqueous copper solutions?
35. What is the main industrial use of zinc?

---

# 20.3 - Coordination Compounds

When you finish this section you will be able to interchange between the formula and name of coordination compounds.

This section begins with several important definitions:

1. **coordination compound:** It consists of a **complex ion** (a transition metal with attached ligands) and **counter ions** (anions and cations as needed to balance charge).
2. **ligand:** a neutral molecule or ion **having a lone pair** that can be donated to the central metal ion.
3. **monodentate ligand:** Donates one pair of electrons.
4. **bidentate ligand:** Donates two pairs of electrons.
5. **multidentate ligand:** Donates more than two pairs of electrons.

Table 20.13 in your textbook gives examples of each kind of ligand.

**Nomenclature** of coordination compounds can get sticky, so let's do a variety of problems together. **Memorize** the rules above Table 20.14 in your textbook. Let's try to name $[Cr(H_2O)_4Cl_2]Cl$.

The coordination compound is composed of a **complex ion**, $[Cr(H_2O)_4Cl_2]^+$, which is a cation, and a **counter anion**, $Cl^-$. According to the rules of nomenclature, the cation (complex ion in this case) will be named first.

**Naming the cation:** The ligands are named before the metal ion. The two different ligands are **water** ("aqua") and **chlorine** ("chloro"). The ligands are placed **alphabetically**, so **aqua goes before chloro**. This is done before adding any prefixes to the ligands.

<u>aqua</u>   <u>chloro</u> _____   _____

There are **four** waters ("tetra") and **two** chlorines ("di") so we have

<u>tetra</u>aqua <u>di</u>chloro _____   _____

We must determine the oxidation state of the chromium ion. Water is a neutral Lewis base and each chlorine has a -1 charge. The chromium is therefore in the +3 oxidation state.

tetraaqua dichloro <u>chromium (III)</u> _____

The complex ion is treated as one word,

tetraaquadichlorochromium (III) _____

**Naming the anion:** The anion is chloride. This finishes the naming of the entire compound. Remember the cation and anion are named separately.

**tetraaquadichlorochromium (III) chloride**

Wasn't that fun?

_____

# Example 20.3 A - Naming Coordination Compounds

Name each of the following compounds:

  a.   $K_2[Ni(CN)_4]$
  b.   $K_4[Fe(CN)_6]$
  c.   $(NH_4)_2[Fe(H_2O)Cl_5]$
  d.   $[Co(NH_3)_2(en)_2]Cl_2$
  e.   $[Ag(NH_3)_2]Cl$

## Solution

a.   The complex ion is an **anion**. You know this because potassium, the counter ion, must be a cation. There-fore, potassium is listed first. We **don't** say **di**potassium because we can explicitly determine the number of potassiums necessary based on the complex ion charge.

          **potassium**      _____

The complex ion has four cyanide ions ("tetracyano").

          **potassium tetracyano** _____

The nickel is in the **+2 oxidation state** because 4 CN = -4 and 2 K = +2. We put "ate" on the end of a complex anion.

          **potassium tetracyanonickelate (II)** _____

b.   The complex ion is again an anion. The cation is named first.

          **potassium** _____

There are 6 cyanides in the complex cation ("hexacyano").

          **potassium hexacyano** _____

The iron ("ferrate") has an oxidation state of +2. (Can you see why?) The compound name is

          **potassium hexacyanoferrate (II)** _____

c.   The counter ion, ammonium, is a cation. It is named first.

          **ammonia** _____

The complex ion has one water ("aqua") and five chlorines ("pentachloro"). The iron ("ferrate") is in the +3 oxidation state ($NH_4$ = +1 x 2 = +2 and $Cl_5$ = -5).

          **ammonia aquapentachloroferrate (III)** _____

d.    The counter ion, chloride, is an anion.  It is named last.

_____ **chloride**

The complex ion has 2 ammonia ligands ("diammine") and 2 ethylenediammines ("bis(ethylenediamine)").  The cobalt is in the +2 oxidation state (en and $NH_3$ are neutral).

**diamminebis(ethylenediamine)cobalt (II) chloride**

e.    Again, the counter ion is an anion, chloride.

_____ **chloride**

The complex ion has two $NH_3$ ("diammine") and silver in the +1 oxidation state.

**diamminesilver (I) chloride**

---

Let's try the reverse procedure.

---

# Example 20.3 B - Writing Coordination Compound Formulas

Write the formulas for each of the following compounds.

a.    potassium pentacyanocobaltate (II)
b.    tris(ethylenediamine)nickel (II) sulfate
c.    potassium dicarbonyltricyanocobaltate (I)

## Solution

a.    pentacyano    = $(CN)_5$  (-1 per CN x 5 = -5)
      cobaltate (II)  = $Co$ (oxidation state = +2)

The complex anion is $Co(CN)_5$.  The total charge = -3.  We must therefore have 3 potassium ions to balance electrically.

$K_3Co(CN)_5$

b.    tris(ethylenediamine) = $(en)_3$  (neutral Lewis base)
      nickel (II)                = $Ni$  (oxidation state = +2)
      sulfate                   = $SO_4$  (-2 charge)

$[Ni(en)_3]SO_4$

c.    dicarbonyl  = $(CO)_2$  (neutral Lewis base)
      tricyano     = $(CN)_3$  (-1 per CN x 3 = -3)
      cobaltate (I)  = $Co$ (oxidation state = +1)

The complex anion is $Co(CO)_2(CN)_3$.  The total charge = -2.  We must therefore have 2 potassium ions to balance electrically.

$K_2[Co(CN)_3(CO)_2]$

Problems 3-9 at the end of this study chapter cover material from this section.

## 20.4 - Isomerism

When you finish this section you will be able to define the basic terms and identify examples of isomerism.

There are several new definitions introduced in this section:

1. **isomerism**: Same chemical formula, different compounds.
2. **structural isomerism**: Same atoms, different bond arrangement.
   a. **coordination isomerism**: complex ion and counter ion interchange members.
   b. **linkage isomerism**: point of attachment of ligand to metal is different.
3. **stereoisomerism**: Same bond arrangement, different spatial arrangement.
   a. **geometrical isomers**: atoms or groups of atoms assume different positions around the central atom. (cis- or trans- isomerism).
   b. **optical isomers**: rotates plane polarized light in different directions. Such molecules have **chiral** centers.

Your textbook gives some examples of each kind of isomerism. After reading the section, try the following example.

## Example 20.4 - Recognizing Isomers

Identify the type of isomerism exhibited by each pair of substances.

   a.   trans-$[RuCl_2(H_2O)_4]^+$ and cis-$[RuCl_2(H_2O)_4]^+$

   b.   $[Co(en)_2(NH_3)Cl]^{2+}$ exists in a form that can rotate plane polarized light to the left and one that can rotate it to the right.

   c.   $[Co(NH_3)_5(NO_2)]^{2+}$ and $[Co(NH_3)_5(ONO)]^{2+}$

## Solution

   a.   geometrical isomers
   b.   optical isomers
   c.   linkage isomers

## 20.6 - The Crystal Field Model

When you finish this section you will be able to:

   •   state the underlying assumptions behind the crystal field model, and
   •   use the model to assess magnetism and color of complex ions.

The purpose of the crystal field model is to attempt to **explain the magnetism and colors of complex ions**. The underlying assumptions of the model are:

1.  That ligands are "**negative point charges.**"

2.  That the bonding between ligands and the central atom is **totally ionic**. That is, electrostatic interactions keep ligands bonded to the central atom. (There are no electrons shared or donated by the ligands.)

Regarding **octahedral complex ions**, your textbook discusses two types of field splitting between $t_{2g}$ and $e_g$ states for the d orbitals on the metal. "$\Delta$" is the splitting energy.

**low-spin** = strong field = large energy difference ("$\Delta$") = **minimum** number of unpaired electrons. (See Figure 20.23 in your textbook.)

**high spin** = weak field = small $\Delta$ = **maximum** number of unpaired electrons.
In octahedral complex ions, two things determine the type of "spin state" (low or high).

- The **spectrochemical series** (given just below Sample Exercise 20.4 in your textbook).
- The **charge on the metal ion** (a higher charge causes stronger field-splitting).

---

# Example 20.6 A - Crystal Field Splitting

The complex ion $Mn(H_2O)_6^{3+}$ is observed to have a high spin state. How many unpaired electrons are in the complex?

## Solution

Mn has an $[Ar]\ 4s^2\ 3d^5$ configuration. The $Mn^{3+}$ is $4s^0\ 3d^4$. Having a **high spin state** means that water induces **weak field** splitting in the complex ion. That implies a small $\Delta$.

$$e_g\quad \uparrow\qquad \underline{\phantom{x}}$$
$$t_{2g}\quad \uparrow\qquad \uparrow\qquad \uparrow$$

There are 4 unpaired electrons in the complex.

---

# Example 20.6 B - Determining Splittting from Unpaired Electrons

The complex ion $Co(NH_3)_6^{3+}$ is observed to have no unpaired electrons. Is it a low or high spin complex?

## Solution

Cobalt (III) is a $3d^6$ ion. The two options are

$$e_g\quad \underline{\phantom{x}}\qquad \underline{\phantom{x}}$$
$$t_{2g}\quad \uparrow\downarrow\qquad \uparrow\downarrow\qquad \uparrow\downarrow$$

**large E, low spin**

OR

$$e_g\quad \uparrow\qquad \uparrow$$
$$t_{2g}\quad \uparrow\downarrow\qquad \uparrow\qquad \uparrow$$

**small E, high spin**

The high spin complex would result in 4 unpaired electrons. The low spin complex would result in none. This complex is therefore a **low spin complex**.

## Color of the Complex

Your textbook points out that it takes energy to promote electrons between the $t_{2g}$ and $e_g$ states.  The **higher the energy**, the **shorter the wavelength** of absorbed color.  That is, higher energy means blue and violet will be absorbed (you will see a reddish compound).  Lower energy means red will be absorbed (you will see a bluish compound).

If you look at Table 20.17 in your textbook, you can see that as the number of $NH_3$ groups diminishes, the color goes towards violet, indicating more red is being absorbed (a lower $\Delta$ value).

## Example 20.6 C - Colors and Field Splitting

A solution of $[Cu(en)_2]^{2+}$ is green.  The color of a $[CuBr_4]^{2-}$ solution is violet.  What does this tell you about the relative crystal field splitting energies?  Which ligand causes the greater splitting, **en** or **Br⁻**?

## Solution

Because the $[CuBr_4]^{2-}$ solution is violet, it means that lower energy radiation is being absorbed (so that violet is emitted).  This indicates a **relatively small splitting energy**.  The $[Cu(en)_2]^{2+}$ solution transmits green, indicating that it absorbs blue, a relatively high energy.  This means that the **splitting energy is relatively large**.

The ethylenediamine ligand **causes greater splitting than the bromide ligand** (all other things being equal).

Problems 10-18 at the end of this study chapter cover material from this section.

## 20.7 - The Molecular Orbital Model

The following questions will help you review the material in this section.  Remember, crystal field theory is too good to be true, and molecular orbital theory is too true to be good.

1.  What are the weaknesses of the crystal field model?
2.  What are the advantages of the molecular orbital (M.O.) model in describing the bonding and properties of complex ions?  What are the disadvantages?
3.  Discuss the two important considerations in predicting the extent of orbital overlap.
4.  What orbitals combine to give $s_s$, $s_p$, and $s_d$ orbitals in the octahedral M. O. in Figure 20.31 of your textbook?
5.  Why are $e_g{}^*$ orbitals antibonding?
6.  What is the effect on M. O. splitting of having an electronegative ligand?
7.  If the M. O. model accounts for the properties of complex ions, why did your textbook bother to introduce the crystal field model at all?

## 20.8 - The Biological Importance of Coordination Complexes

The following questions will help you review the material in this section.

1.  What are some biological uses of coordination complexes?

2. Why are coordination complexes ideal for biological applications?
3. What are the principal sources of energy in mammals?
4. What is the **respiratory chain**?
5. What are **cytochromes**? What are their two main parts?
6. What is the structure of chlorophyll?
7. How can $Fe^{2+}$ have room to attach to myoglobin?
8. What is the mechanism for the oxidation of $Fe^{2+}$ in heme?
9. Describe the transport of $O_2$ in blood.
10. What biochemical mechanism leads to sickle cell anemia?
11. What does Le Chatelier's principle have to do with altitude sickness?
12. Describe how CO and CN⁻ interact with iron to cause death.
13. Define **respiratory inhibitor**.

---

## 20.9 - Metallurgy and Steel Production

The following questions and the example will help you to review material in this section.

1. Why are almost all metals found in ores?
2. Your textbook says that "most metals are unsuitable for use in the pure state." Why?
3. Define **alloy**.
4. Define **metallurgy**.
5. List each of the steps involved in preparing a metal for use.
6. What is "gangue"?
7. Why are silicate minerals not often used as a metal source?

---

## Example 20.9 - Minerals and Oxidation States

Determine the oxidation state of each of the elements in the following minerals (see <u>Table 20.19 in your textbook</u>).

    a.  galena      b.  bauxite     c.  siderite

## Solution

a. Galena is **PbS**. Recalling your rules for assigning oxidation states (see Section 4.9),

$$Pb = +2 \quad S = -2$$

b. Bauxite is **$Al_2O_3$**. The oxygen, being very electronegative and in group VI, gets the assignment -2. This forces the aluminum to be +3 to balance electrically.

$$Al = +3 \quad O = -2$$

c. Siderite is **$FeCO_3$**. The carbonate anion should be familiar to you as $CO_3{}^{2-}$. This means that O = -2 and C = +4. Iron must be +2 to balance electrically.

$$Fe = +2 \quad C = +4 \quad O = -2$$

8. Describe the floatation process.
9. What is the function of roasting?
10. Why is it both useful and necessary to collect $SO_2$ gas from the roasting process?
11. Describe the smelting process.
12. Describe the process of zone refining.
13. Why does zone refining purify metals (discuss the crystal lattice)?
14. What are some of the problems with pyrometallurgy?
15. Define leaching.
16. Why does carbon dioxide lower the pH of a solution (in the Bayer process)? How does this help to isolate aluminum?
17. Why is water not often useful as a leaching agent?
18. Why is iron recovered from pyrite not suitable for use in steel?
19. How are iron ore particles separated from gangue?
20. How does a blast furnace work?
21. Write the reactions for the reduction of iron oxide in a blast furnace.
22. What is slag? How is it formed?
23. What is pig iron composed of?
24. What is the advantage of a direct reduction furnace?
25. What is steel?
26. What is the basic chemical difference between the production of iron and steel?
27. Describe the "open hearth" process of steel making.
28. What is the advantage of the "basic oxygen" process of steel making?
29. Describe the "electric arc" method of steel making.
30. What are the different crystal forms of iron? How are they different?
31. Why is "tempering" useful?

---

## END OF CHAPTER TWENTY

# SELF-TEST QUESTIONS

1.  Write electron configurations for each of the following metals.

    a.  Mn          b.  Pd          c.  Zr          d.  Zn          e.  Rh

2.  Write electron configurations for each of the following ions.

    a.  $Cr^{6+}$          b.  $Cr^{2+}$          c.  $Fe^{6+}$          d.  $Fe^{3+}$          e.  $Mn^{2+}$

3.  What is the coordination number on the metal in each of the following complex ions?

    a.  $[CuF_4]^{2-}$          c.  $[Fe(en)(H_2O)_4]^{2+}$
    b.  $[CuF_6]^{3-}$          d.  $[Fe(en)_3]^{2+}$

4.  Gadolinium has the electron configuration $[Xe]\ 6s^2\ 5d^1\ 4f^7$. Why is this configuration more energetically favorable than $[Xe]\ 6s^2\ 5d^0\ 4f^8$?

5.  The final step in the preparation of cobalt is the reduction of $Co_3O_4$ with aluminum:

    $$3Co_3O_4(s) + 8Al(s) \longrightarrow 9Co(s) + 4Al_2O_3(s)$$

    How much cobalt can be obtained from 125.0 g of $Co_3O_4$?

6.  Iron in aqueous solution can undergo a series of reactions with ethylenediamine ("en"),

    $$[Fe(H_2O)_6]^{2+} + en \longrightarrow [Fe(en)(H_2O)_4]^{2+} + 2H_2O$$
    $$[Fe(en)(H_2O)_4]^{2+} + en \longrightarrow [Fe(en)_2(H_2O)_2]^{2+} + 2H_2O$$
    $$[Fe(en)_2(H_2O)_2]^{2+} + en \longrightarrow [Fe(en)_3]^{2+} + 2H_2O$$

    a.  Name each complex ion.
    b.  What is the overall geometry of the complexes?
    c.  What is the coordination number of the iron?
    d.  Will the crystal field splitting increase or decrease as these reactions proceed?

7.  What is the oxidation state of the metal ion in each of the following complex ions?

    a.  $[Mn(CO)_5]^+$          b.  $Pt(CO)Cl_2$          c.  $[Co(CN)_5OH]^{3-}$

8.  Name each of the complex ions in problem #7.

9.  Give the formula for each of the following coordination compounds.

    a.  potassium tetrahydroxonickelate (II)
    b.  tetraaquamanganese (II) sulfate
    c.  tris(ethylenediamine)cobalt (III) chloride

10. How many unpaired electrons will be present in each of the following complex ions?

    a.  $[Fe(NH_3)_6]^{2+}$ (high spin)          b.  $[Co(NH_3)_6]^{3+}$ (low spin)

11. Determine the spin of $[Ru(H_2O)_6]^{3+}$ if it has one unpaired electron.

12. Based on the spectrochemical series, predict whether each of the following will be high or low spin.

    a.  $[Fe(CN)_6]^{4-}$          b.  $[CoF_6]^{3-}$          c.  $[MnCl_6]^{4-}$

13. How many unpaired electrons will be present in each of the complex ions in problem #12?

14. Arrange the following colors in order of increasing wavelength.

red,   violet,   green,   blue,   yellow

15. Arrange the colors in problem #14 in terms of increasing energy.

16. Two compounds are synthesized. One is red. The other is green. Which compound has the larger value for $\Delta$? Why?

17. An electron is excited from the $t_{2g}$ to the $e_g$ state by radiation of wavelength 620 nm.

    a.    What color light did the electron absorb?
    b.    What is the energy of the absorbed light?

18. The complex ion $[Cu(H_2O)_6]^{2+}$ has an absorption maximum at around 800 nm. When four ammines replace water, giving $[Cu(NH_3)_4(H_2O)_2]^{2+}$, the absorption maximum shifts to around 600 nm. What do these results mean in terms of the relative field splittings of $NH_3$ and $H_2O$?

# ANSWERS TO SELF-TEST QUESTIONS

1.  a.   Mn: $[Ar] 4s^2 3d^5$
    b.   Pd: $[Kr] 5s^0 4d^{10}$
    c.   Zr: $[Kr] 5s^2 4d^2$
    d.   Zn: $[Ar] 4s^2 3d^{10}$
    e.   Rh: $[Kr] 5s^1 3d^8$

2.  a.   $Cr^{6+}$: $[Ar]$
    b.   $Cr^{2+}$: $[Ar] 3d^4$
    c.   $Fe^{6+}$: $[Ar] 3d^2$
    d.   $Fe^{3+}$: $[Ar] 3d^5$
    e.   $Mn^{2+}$: $[Ar] 3d^5$

3.  a.   4          c.   6
    b.   6          d.   6

4.  Although electron configurations are hard to nail down unequivocally this deep in the periodic table, it would seem that the 4f and 5d energy levels are close enough together to allow electron cross over.  The half-filled 4f helps minimize the electron-electron repulsion that would occur if the configuration were $5d^0 4f^8$.

5.  91.78 g cobalt

6.  a.   $[Fe(H_2O)_6]^{2+}$ = hexaaquairon (II) ion
        $[Fe(en)(H_2O)_4]^{2+}$ = tetraaquaethylenediamineiron (II) ion
        $[Fe(en)_2(H_2O)_2]^{2+}$ = diaquabis(ethylenediamine)iron (II) ion
        $[Fe(en)_3]^{2+}$ = tris(ethylenediamine)iron (II) ion
    b.   All complexes are octahedral.
    c.   The iron has a coordination number of 6.
    d.   According to the spectrochemical series, the splitting will increase (larger D).

7.  a.   +1              b.   +2              c.   +3

8.  a.   pentacarbonylmanganese (I) ion
    b.   carbonyldichloroplatinum (II)
    c.   pentacyanohydroxycobaltate (III) ion

9.  a.   $K_2[Ni(OH)_4]$     b.   $[Mn(H_2O)_4]SO_4$     c.   $[Co(en)_3]Cl_3$

10. a.   4              b.   0

11. $Ru^{3+}$ is $4d^5$.  One unpaired electron corresponds to **low spin** (strong field case).

12. a.   low spin       b.   high spin       c.   high spin

13. a. 0                b.   4               c.   5

14. (lowest) violet, blue, green, yellow, red (highest)

15. (lowest) red, yellow, green blue, violet (highest)

16. The red compound has the higher D value because it absorbs blue light.  The green compound absorbs red and yellow light, which is of lower energy.

17. a. red light        b.   $E = hc/l = 3.2 \times 10^{-19}$ J/photon

18. An absorption maximum shift to 600 nm means D is larger (600 nm represents greater radiation than 800 nm).  Therefore, a stronger field splitting is occurring.  This is consistent with the spectrochemical series.

# CHAPTER 21

# The Nucleus - A Chemist's View

| Section | Contents | Page in this Study Guide |
|---|---|---|
| 21.1 | Nuclear Stability and Radioactive Decay | 427 |
| 21.2 | The Kinetics of Radioactive Decay | 430 |
| 21.3 | Nuclear Transformations | 431 |
| 21.4 | Detection and Uses of Radioactivity | 432 |
| 21.5 | Thermodynamic Stability of the Nucleus | 433 |
| 21.6 | Nuclear Fission and Nuclear Fusion | 435 |
| 21.7 | Effects of Radiation | 435 |

In our study of chemistry, we have been dealing so far with the movement of electrons within and between atoms. We have been relatively unconcerned about nuclear structure - until now. This chapter deals with **nuclear transformations**. An important goal is to relate **radioactive decay** to nuclear energy and to the age of terrestrial objects.

## 21.1 - Nuclear Stability and Radioactive Decay

When you finish this section you will be able to:

- define each mechanism of nuclear decay,
- list the products, given the mechanism of nuclear decay, and
- determine each member of a decay series, given appropriate information.

**Radioactive decay** is the process by which a nucleus decomposes to form a different nucleus along with additional particles.

The key to determining nuclear decay products (from a general chemistry point of view) is to remember that *the sum of the protons after the decay must equal that before the decay. The same is true with the number of neutrons.* For example, neptunium undergoes decay as follows:

$$^{237}_{93}\text{Np} \longrightarrow {}^{4}_{2}\text{He} + {}^{233}_{91}\text{Pa}$$

Notice that the **sum** of protons and neutrons is the same on both sides of the equation as is the number of protons.

Your textbook discusses **six types of radioactive processes**. Each involves isotopes of various atoms. Note that for *individual atoms*, we prefer the term **nuclide**. These are, briefly,

- **α-particle production** results in the release of $^4_2$He.

$$\text{example:} \quad ^{240}_{94}\text{Pu} \longrightarrow ^4_2\text{He} + ^{236}_{92}\text{U}$$

- **β-particle production** results in the release of $^0_{-1}$e.

$$\text{example:} \quad ^{194}_{76}\text{Os} \longrightarrow ^{194}_{77}\text{Ir} + ^0_{-1}\text{e}$$

- **γ-ray production** (High energy radiation released as other nuclear transformations occur.)

- **positron production** results in the release of $^0_1$e.

$$\text{example:} \quad ^{17}_9\text{F} \longrightarrow ^{17}_8\text{O} + ^0_1\text{e}$$

- **electron capture** is accomplished when the nucleus captures an inner orbital electron.

$$\text{example:} \quad ^{229}_{92}\text{U} + ^0_{-1}\text{e} \longrightarrow ^{229}_{91}\text{Pa}$$

- **spontaneous fission** is the decomposition of the nucleus into two relatively large fractions.

The following example tests your knowledge of the various nuclear transformation processes.

---

## Example 21.1 A - Nuclear Decay

Write equations for each of the following processes:

a. $^{241}_{96}$ Cf undergoes electron capture.

b. $^{241}_{95}$ Am produces an α particle.

c. $^{121}_{54}$ Xe produces a positron.

d. $^{138}_{53}$ I produces a β particle.

### Solution

Remember that the **sum** of protons and neutrons, as well as the **number** of protons is **consistent** on both sides of the equation. The key then is to calculate these values, then determine the element that has the values associated with it.

a. $^{241}_{96}$ Cf + $^0_{-1}$e $\longrightarrow$ $^{241}_{95}$ Am

b. $^{241}_{95}$ Am $\longrightarrow$ $^{237}_{93}$ Np + $^4_2$He

c. $^{121}_{54}$ Xe $\longrightarrow$ $^{121}_{53}$ I + $^0_1$e

d. $^{138}_{53}$ I $\longrightarrow$ $^{138}_{54}$ Xe + $^0_{-1}$e

## Example 21.1 B - Practice with Nuclear Decay

Fill in the missing particle in each of the following equations:

a.   $^{164}_{67}$ Ho + $^{0}_{-1}$e ----> ?

b.   $^{158}_{66}$ ? ----> $^{0}_{1}$ Dy + e

c.   $^{242}_{94}$ Pu ----> ? + $^{238}_{92}$U

## Solution

a.   $^{164}_{67}$ Ho + $^{0}_{-1}$e ----> $^{164}_{66}$ **Dy**

b.   $^{158}_{67}$ **Ho** ----> $^{158}_{66}$ Dy + $^{0}_{1}$ e

c.   $^{242}_{94}$ Pu ----> $^{4}_{2}$ **He** + $^{238}_{92}$ U

There are a limited number of ways that a nucleus can decay.  It may release many particles before finally becoming stable.

A **decay series** is **a number of sequential decays** by an unstable nuclide.  The series continues until a stable nuclide is formed.  For example,

$$^{241}_{95}\text{Am} ----> \,^{237}_{93}\text{Np} + \,^{4}_{2}\text{He} ----> \,^{233}_{91}\text{Pa} + \,^{4}_{2}\text{He}$$

Each step is consistent with the previous one in terms of proton and neutron totals.  Keep in mind that when a particle (such as α) is released, it is lost to the system, so it should not enter into the calculation.

## Example 21.1 C - Decay Series

$^{247}_{97}$ Bk undergoes decay to $^{208}_{82}$ Pb in the following order:

$$\alpha, \alpha, \beta, \alpha, \alpha, \beta, \alpha, \beta, \alpha, \alpha, \alpha, \alpha, \beta, \beta, \alpha$$

Write equations for the first **six** steps.

## Solution

1.   $^{247}_{97}$ **Bk** ----> $^{243}_{95}$**Am** + $^{4}_{2}$ He

2.   $^{243}_{95}$ Am ----> $^{239}_{93}$ **Np** + $^{4}_{2}$He

3.   $^{239}_{93}$Np ----> $^{239}_{94}$ **Pu** + $^{0}_{-1}$e

4.   $^{239}_{94}$Pu ----> $^{235}_{92}$ **U** + $^{4}_{2}$He

5.   $^{235}_{92}$U  ----> $^{231}_{90}$ **Th** + $^{4}_{2}$He

6.   $^{231}_{90}$Th ----> $^{231}_{91}$ **Pa** + $^{0}_{-1}$e

Problems 1-6 at the end of this study chapter cover material from this section.

## 21.2 - The Kinetics of Radioactive Decay

When you finish this section you will be able to interconvert between the half-life of a nuclide, and the amount of that nuclide remaining at time "t."

The decay of nuclides follows a **first-order rate law**.  That means that the kinetics are governed by **the same equations** that were introduced for first-order kinetics in section 12.3.  These are

$$\ln\left[\frac{N}{N_o}\right] = -kt$$

where $N_o$ = the original number (mass) of nuclides
  $N$ = the number remaining at time t
  $k$ = the first-order rate constant
  $t$ = the time

Also,

$$t_{1/2} = \frac{0.693}{k}$$

where $t_{1/2}$ is the half-life of the nuclide.

The following examples are intended to show how you can apply these equations directly to radioactive decay data.

## Example 21.2 A - Half-Life and Concentration

The half-life of $^{239}_{94}$Pu is $2.411 \times 10^4$ years.  How many years will elapse before 99.9% of a given sample decomposes?

### Strategy

We have no specific amounts, however, we do know that (from a **fractional** point of view) **0.999 of our original 1.000** decomposes, leaving **0.001** remaining.  We can thus establish the **ratio** $N/N_o$ as **0.001/1.000**.  We can find k from $t_{1/2}$.

### Solution

$$k = \frac{0.693}{t_{1/2}} = \frac{0.693}{2.411 \times 10^4 \text{ y}} = 2.87 \times 10^{-5}/\text{y}$$

$$\ln\left[\frac{N}{N_o}\right] = -kt$$

$$\ln(0.001) = -2.87 \times 10^{-5}/\text{y (t)}$$

$$t = 2.4 \times 10^5 \text{ yr}$$

## Does the Answer Make Sense?

Our total time, $2.4 \times 10^5$ y, is about **10 half-lives**. This means that we should have about $1/2^{10}$, (or 0.001) of our original material remaining. This is in fact the case, so the answer makes sense.

---

## Example 21.2 B - Practice with Half-Life and Concentration

The half-life of protactinium-217 is $4.9 \times 10^{-3}$ s. How much of a 3.50 mg sample of $^{217}_{91}$Pa will remain after 1.000 sec?

## Solution

This can either be solved using $t_{1/2}$ or k. Let's use $t_{1/2}$ in this problem (we used k in the previous one). The key question is "How many half-lives have passed in 1.000 sec?"

$$\# \text{ half-lives} = \frac{1 \text{ half-life}}{4.9 \times 19^{-3} \text{ s}} \times 1.000 \text{ s} = 204$$

The fraction of the original sample remaining will be $1/2^{204} = 3.9 \times 10^{-62}$. The final amount will be 3.50 $(3.9 \times 10^{-62}) = \textbf{1.4} \times \textbf{10}^{-61}$ **g**. Because of the short half-life, essentially none of the original nuclide remains after one second.

---

# Problems 8-12 at the end of this study chapter cover material from this section.

---

## 21.3 - Nuclear Transformations

The following problem is intended to give you practice with the particles involved in nuclear transformations. Remember that in this case nuclides are not spontaneously decaying, but are instead being bombarded with particles (either neutrons or other nuclides) that **cause a heavier element to form**.

---

## Example 21.3 - Nuclear Transformations

Fill in the missing particles in each of the following nuclear transformations:

   a.   $? + {}^{4}_{2}\text{He} \longrightarrow {}^{243}_{97}\text{Bk} + 2{}^{1}_{0}\text{n}$

   b.   ${}^{253}_{99}\text{Es} + {}^{4}_{2}\text{He} \longrightarrow ? + {}^{1}_{0}\text{n}$

   c.   ${}^{250}_{98}\text{Cf} + {}^{11}_{5}\text{B} \longrightarrow {}^{257}_{103}\text{Lr} + ?$

## Solution

   a.   $^{241}_{95}\textbf{Am} + {}^{4}_{2}\text{He} \longrightarrow {}^{243}_{97}\text{Bk} + 2{}^{1}_{0}\text{n}$

   b.   ${}^{253}_{99}\text{Es} + {}^{4}_{2}\text{He} \longrightarrow {}^{256}_{101}\textbf{Md} + {}^{1}_{0}\text{n}$

   c.   ${}^{250}_{98}\text{Cf} + {}^{11}_{5}\text{B} \longrightarrow {}^{257}_{103}\text{Lr} + \textbf{4}{}^{1}_{0}\textbf{n}$

## 21.4 - Detection and Uses of Radioactivity

When you finish this section you will be able to solve problems involving radioactive dating of materials.

The **key assumptions** when using radioactivity to date objects using **radiocarbon dating** are

- the ratio of $^{14}C/^{12}C$ has been constant over the years, and
- living systems have that **same constant ratio** until they die.

The mathematical basis of the technique is that once the living thing dies, the **ratio of $^{14}C/^{12}C$ diminishes** with a **half-life of 5730 years**. This means that after 17,190 years (3 half-lives), the $^{14}C/^{12}C$ ratio (be it amount, or "disintegrations per second" or whatever) will be $1/2^3 = 1/8$ what it was before.

## Example 21.4 A - Radiocarbon Dating

A sample of bone taken from an archeological dig was determined by radiocarbon dating to be 12,000 years old. If we assume that a constant atmospheric $^{14}C/^{12}C$ **ratio has 13.6 disintegrations per minute per gram of carbon, how many disintegrations per minute per gram** did our sample give off ($t_{1/2}$ for $^{14}C$ = 5730 years).

## Strategy and Solution

We know the age of our solution and we know the half-life. We can therefore determine, by using out first-order decay formula, the ratio "$N/N_o$" which in this case, represents the ratios of disintegrations per minute per gram.

$$k = 0.693/t_{1/2} = 0.693/5730 \text{ y} = 1.21 \times 10^{-4}/\text{y}$$

$$\ln(N/N_o) = -kt \quad \Rightarrow \quad -1.21 \times 10^{-4}/\text{y} \, (12000 \text{ y}) = -1.45$$

$$N/N_o = e^{-1.45} = \mathbf{0.234}$$

$$N_o = 13.6 \text{ disintegrations, therefore, } N = 0.234 \, (13.6)$$

$$\mathbf{N = 3.2 \text{ disintegrations per minute per gram}}$$

In determining the age of exceptionally old objects (billions of years), your textbook discusses the nuclear transformation

$$^{238}_{92}\text{U} \text{ ----> } ^{206}_{82}\text{Pb} \qquad t_{1/2} = 4.5 \times 10^9 \text{ y}$$

<u>Sample Exercise 21.6 in your textbook</u> deals with the age of a rock based on comparing **atoms** of $^{238}_{92}\text{U}$ and $^{206}_{82}\textbf{Pb}$. There is a 1:1 stoichiometry in this relationship. Let's take this one step further and compare **grams** of U and Pb. The relationship in this case is **not** 1:1, but rather **238 g U/206 g Pb.**

## Example 21.4 B - Dating Via Uranium

A rock contains 0.141 g of $^{206}_{82}$ Pb for every 1.000 g $^{238}_{92}$ U. How old is the rock ($t_{1/2}$ $^{238}_{92}$ U = $4.5 \times 10^9$ y, and you are to assume that the intermediate nuclides decay instantaneously).

### Strategy and Solution

The goal is really to find the ratio $N/N_o$ for uranium decay. The amount of lead found is a **direct reflection** of the amount of uranium that has decayed.

$$^{238}_{92} \text{ U decayed} = 0.141 \text{ g Pb} \times \frac{238 \text{ g U}}{206 \text{ g Pb}} = \textbf{0.163 g } ^{238}_{92} \text{ U}$$

The original amount of $^{238}_{92}$ U = g found + g decayed = 1.000 + 0.163 = **1.163 g**.

$$k = \frac{0.693}{t_{1/2}} = \frac{0.693}{4.5 \times 10^9 \text{ y}} = \textbf{1.54} \times \textbf{10}^{\textbf{-10}}\textbf{/y}$$

$$\ln(N/N_o) = -kt \quad \Rightarrow \quad \ln(1.000/1.163) = -1.54 \times 10^{-10}\text{/y (t)}$$

$$t = \textbf{9.8} \times \textbf{10}^\textbf{8} \textbf{ years old}$$

Problems 13-15 at the end of this study chapter cover material from this section.

## 21.5 - Thermodynamic Stability of the Nucleus

When you finish this section you will be able to use the **mass defect** of an atom **to calculate its binding energy**.

A nucleus is a bundle of protons and neutrons bound together. The nucleus is more energetically stable than the individual array of protons and neutrons. This is known because the **sum of the masses** of protons and neutrons in a nucleus is **less than** the sum of the masses of individual nucleons. The difference in the mass is called the **mass defect ("$\Delta m$")**. The mass has been converted to energy, as predicted by

$$E = mc^2, \text{ or } \Delta E = \Delta mc^2$$

"$\Delta E$" is called the binding energy. For example,

$$^{187}_{77} \textbf{Ir} = 187.958830 \text{ g/mol}$$

1 mole of 77 (protons + electrons) + 110 neutrons =

| | | |
|---|---|---|
| 77 × 1.007825 g per proton or electron = | 77.602525 g | |
| 110 × 1.008665 g per neutron           = | 110.95315  g | |
| | 188.555675 g | |

The **mass defect, $\Delta m$** = 188.555675 g - 187.958830 g = **0.596845 g/mol**. That is the mass that has been converted to energy. To find out how much, remember that

$$1 \text{ Joule} = 1 \text{ kg m}^2\text{/s}^2$$
$$\text{speed of light} = c = 3.00 \times 10^8 \text{ m/s}$$

$$\Delta E = \Delta mc^2 = \frac{-0.596845 \text{ g}}{\text{mol}} \times \frac{1 \text{ kg}}{1000 \text{ g}} \times \left[\frac{3.00 \times 10^8 \text{ m}^2}{\text{s}}\right] = 5.37 \times 10^{13} \text{ kgm}^2/\text{s}^2 \text{ mol}$$

$$\Delta E = -5.37 \times 10^{13} \text{ J/mol}$$    (The "-" indicated energy is released (exothermic process) when the nucleus is formed.)

In order to calculate $\Delta E$ per nucleus, we must divide by Avogadro's number.

$$\Delta E = \frac{-5.37 \times 10^{13} \text{ J}}{\text{mol}} \times \frac{1 \text{ mol}}{6.02 \times 10^{23} \text{ nuclei}} = -8.92 \times 10^{-11} \text{ J/nucleus}$$

In order to convert to J/nucleon, divide by 187 nucleons in the $^{187}_{72}$ Ir nucleus.

$$\Delta E = \frac{-8.92 \times 10^{-11} \text{ J}}{\text{nucleus}} \times \frac{1 \text{ nucleus}}{187 \text{ nucleons}} = -4.77 \times 10^{-13} \text{ J/nucleon}$$

Finally the binding energy per nucleon in "million electron volts" (MeV),

$$\Delta E = \frac{-4.77 \times 10^{-13} \text{ J}}{\text{nucleon}} \times \frac{1 \text{ MeV}}{1.60 \times 10^{-13} \text{ J}} = -2.98 \text{ MeV/nucleon}$$

This means that 2.98 MeV **per nucleon** would be released if the $^{187}_{77}$Ir nucleus were formed from individual protons and neutrons (with electrons surrounding the nucleus).

---

## Example 21.5 - Mass Defect and Binding Energy

Determine the binding energy in **J/mol** and **MeV/nucleon** for $^{101}_{46}$Pd (atomic mass = 100.908287 g/mol).

### Solution

Mass of individual nucleons,

$$\begin{array}{lr} 46 \times 1.007825 \text{ g/proton or electron} & = 46.35995 \text{ g} \\ + \ 55 \times 1.008665 \text{ g/neutron} & = 55.476575 \text{ g} \\ \hline & 101.836525 \text{ g/mol} \end{array}$$

mass defect ($\Delta m$) = 101.836525 g - 100.908287 g = **0.928238 g**

$$\Delta E = \Delta mc^2 = -9.28238 \times 10^{-4} \text{ kg } (3.00 \times 10^8 \text{ m·s})^2$$
(minus sign because mass is <u>lost</u> in forming the nuclide)

$$\Delta E = -8.35 \times 10^{13} \text{ J/mol}$$

$$\frac{\text{MeV}}{\text{nucleon}} \ \bigg|\bigg| \ \frac{-8.35 \times 10^{13} \text{ J}}{\text{mol}} \times \frac{1 \text{ MeV}}{1.60 \times 10^{-13} \text{ J}} \times \frac{1 \text{ mol}}{6.02 \times 10^{23} \text{ nuclides}} \times \frac{1 \text{ nucleide}}{101 \text{ nucleons}}$$

$$\Delta E = -8.59 \text{ MeV/nucleon}$$

---

Problems 16-18 at the end of this study chapter cover material from this section.

## 21.6 and 21.7 - Nuclear Fission and Nuclear Fusion and Effects of Radiation

The following questions will help you review the material in these two sections of your textbook.

### Section 12.6

1. Define **fusion**.
2. Define **fission**.
3. Is $^{235}_{92}U + ^{1}_{0}n \rightarrow ^{137}_{52}Te + ^{97}_{40}Zr + 2\ ^{1}_{0}n$ an example of fission or fusion?
4. How does a **chain reaction** work?
5. Describe the process of fission in a nuclear warhead (also, define **subcritical**, **critical** and **supercritical**).
6. Outline how nuclear reactors work. Focus on the functions of the **reactor core**, **moderator**, **control rods**, and **cooling system**.
7. What are the consequences for failure of each of the systems in a nuclear reactor?
8. What is the principle behind **a breeder reactor**?
9. What are the advantages and disadvantages of using liquid sodium as a coolant in breeder reactors (see the Chemical Impact on "Nuclear Power..." in your textbook).
10. What is cold fusion? Why are there doubts about the process?
11. Why is fusion a process worth exploring?
12. What is the difficulty with doing fusion reactions on Earth?
13. Discuss possible methods of nuclear waste disposal (as outlined in the Chemical Impact at the end of this section in your textbook).

### Section 21.7

1. Why is any energy source potentially dangerous?
2. Why does exposure to radioactivity seem less hazardous than it really can be (discuss the exposure "per event").
3. Define **somatic damage** and **genetic damage**.
4. What variables affect the degree of damage that radiation can cause? Discuss each one.
5. Your textbook says that our exposure to natural radiation is much greater than to man-made sources. Why are we so worried about a relatively small exposure level?
6. Describe the **linear model** and **threshold model** of radiation damage.

# SELF-TEST QUESTIONS

1. Write equations for each of the following processes:

    a. $^{73}_{31}$Ga produces a β particle.

    b. $^{68}_{31}$Ga undergoes electron capture.

    c. $^{192}_{78}$Ga produces an α particle.

2. Write equations for each of the following processes:

    a. $^{207}_{87}$Fr produces an α particle.

    b. $^{234}_{90}$Th produces a β particle.

    c. $^{62}_{29}$Cu produces a positron.

3. Fill in the missing particle in each of the following equations:

    a. $^{129}_{51}$Sb ---->_____ + $^{0}_{-1}$e

    b. _____ + $^{0}_{-1}$e ----> $^{7}_{3}$Li

    c. $^{205}_{83}$Bi ----> $^{205}_{82}$Pb + _____

    d. $^{206}_{87}$Fr ----> _____ + $^{4}_{2}$He

4. Fill in the missing particle in each of the following equations:

    a. _____ + $^{0}_{-1}$e ----> $^{212}_{86}$Rn

    b. _____ ----> $^{208}_{85}$At + $^{4}_{2}$He

    c. $^{226}_{90}$Th ----> _____ + $^{4}_{2}$He

    d. $^{186}_{77}$Ir ----> _____ + $^{0}_{1}$e

5. The radioactive isotope $^{242}_{96}$Cm undergoes decay to $^{206}_{82}$Pb in the following order:

$$\alpha, \alpha, \alpha, \alpha, \alpha, \alpha, \alpha, \beta, \beta, \alpha, \beta, \beta, \alpha.$$

    Write equations for the first <u>six</u> steps.

6. This sixth step in the previous problem should leave you with $^{218}_{84}$Po. Continue writing equations in the series until you reach $^{206}_{82}$Pb.

7. Calculate the rate constant for each of the following values of half-life:

    a. $^{182}_{72}$Hf, $t_{1/2} = 9 \times 10^6$ yr

    b. $^{228}_{91}$Pa, $t_{1/2} = 26$ hr

    c. $^{225}_{88}$Ra, $t_{1/2} = 14.8$ dy

    d. $^{181}_{78}$Pt, $t_{1/2} = 51$ s

8. The half-life of $^{161}_{65}$Tb is 6.9 days. How many grams of an original 3.000 g sample will remain after two weeks?

9.  The half-life of $^{129}_{53}I$ is $1.7 \times 10^7$ yr. How many grams of original 50.00 g sample will remain after $5.0 \times 10^8$ yr?

10. How long will it take for 98.6% of a sample of $^{188}_{79}Au$ to decompose ($t_{1/2}$ = 8 min.)?

11. How long will it take for 99.99% of a sample of $^{194}_{84}Po$ to decompose ($t_{1/2}$ = 8 min.)?

12. How many half-lives have passed if 87.5% of a substance has decomposed? How many if 99.999% has decomposed?

13. Let us assume a constant $^{14}C/^{12}C$ ratio of 13.6 disintegrations per minute per gram of living matter. A sample of a petrified tree was found to give 1.2 disintegrations per minute per gram. How old is the tree? ($t_{1/2}$ = $^{14}C$ = 5730 years.)

14. The radioactive isotope $^{237}_{90}Th$ has a rate constant, k = $4.91 \times 10^{-11}$ yr$^{-1}$. Is this nuclide useful for determining the age of bone samples? Why or why not?

15. A rock contains 0.688 g of $^{206}_{82}Pb$ for every 1.00g of $^{238}_{92}U$. How old is the rock ($t_{1/2}$ $^{238}_{92}U$ = $4.5 \times 10^9$ yr; assume intermediate nuclide decay instantaneously to the stable Pb nuclide.)

16. Calculate the mass defect in grams for one mole of each of the following:

    a.  $^{235}_{92}U$, atomic mass = 235.0439 g/mol

    b.  $^{127}_{53}I$, atomic mass = 126.9004 g/mol

    c.  $^{75}_{33}As$, atomic mass = 74.9216 g/mol

17. Determine the binding energy in J/mol and MeV/nucleon for $^{66}_{30}Zn$ (atomic mass = 65.9260 g/mol).

18. Determine the binding energy in J/mol and MeV/nucleon for $^{150}_{60}Nd$ (atomic mass = 149.9207 g/mol).

# ANSWERS TO SELF-TEST QUESTIONS

1.  a. $^{73}_{31}$Ga ----> $^{69}_{29}$Cu + $^{4}_{2}$He

    b. $^{68}_{31}$Ga + $^{0}_{-1}$e ----> $^{68}_{30}$Zn

    c. $^{192}_{78}$Pt ----> $^{188}_{76}$Os + $^{4}_{2}$He

2.  a. $^{207}_{87}$Fr ----> $^{203}_{85}$At + $^{4}_{2}$He

    b. $^{234}_{90}$Th ----> $^{234}_{91}$Pa + $^{0}_{-1}$e

    c. $^{62}_{29}$Cu ----> $^{62}_{30}$Zn + $^{0}_{1}$e

3.  a. $^{129}_{51}$Sb ----> $^{129}_{52}$Te + $^{0}_{-1}$e

    b. $^{7}_{4}$Be + $^{0}_{-1}$e ----> $^{7}_{3}$Li

    c. $^{205}_{83}$Bi ----> $^{205}_{82}$Pb + $^{0}_{-1}$e

    d. $^{206}_{87}$Fr ----> $^{202}_{85}$At + $^{4}_{2}$He

4.  a. $^{212}_{87}$Fr + $^{0}_{-1}$e ----> $^{212}_{86}$Rn

    b. $^{212}_{87}$Fr ----> $^{208}_{85}$At + $^{4}_{2}$He

    c. $^{226}_{90}$Th ----> $^{222}_{88}$Ra + $^{4}_{2}$He

    d. $^{186}_{77}$Ir ----> $^{186}_{76}$Os + $^{0}_{1}$e

5.  a. $^{242}_{96}$Cm ----> $^{238}_{94}$Pu + $^{4}_{2}$He

    b. $^{238}_{94}$Pu ----> $^{234}_{92}$U + $^{4}_{2}$He

    c. $^{234}_{92}$U ----> $^{230}_{90}$Th + $^{4}_{2}$He

    d. $^{230}_{90}$Th ----> $^{226}_{88}$Rn + $^{4}_{2}$He

    e. $^{226}_{88}$Ra ----> $^{222}_{86}$Rn + $^{4}_{2}$He

    f. $^{222}_{86}$Rn ----> $^{218}_{84}$Po + $^{4}_{2}$He

6.  a. $^{218}_{84}$Po ----> $^{214}_{82}$Pb + $^{4}_{2}$He

    b. $^{214}_{82}$Pb ----> $^{214}_{83}$Bi + $^{0}_{-1}$e

    c. $^{214}_{83}$Bi ----> $^{214}_{84}$Po + $^{0}_{-1}$e

    d. $^{214}_{84}$Po ----> $^{210}_{82}$Pb + $^{4}_{2}$He

    e. $^{210}_{82}$Pb ----> $^{210}_{83}$Bi + $^{0}_{-1}$e

    f. $^{210}_{83}$Bi ----> $^{210}_{84}$Po + $^{0}_{-1}$e

    g. $^{210}_{84}$Po ----> $^{206}_{82}$Pb + $^{4}_{2}$He

7.  a. $k = 7.7 \times 10^{-8}$/yr (use an extra figure, because this conversion is an intermediate calculation-you will use k further.)
    b. $k = 0.0267$/hr
    c. $k = 0.04683$/dy
    d. $k = 0.0136$/s

8.   0.74 g will remain.

9.   $7.0 \times 10^{-8}$ g will remain.

10.  49.17 min, rounds to 50 min.

11.  8.3 s, rounds to 8 s.

12.  a.   3 half-lives.          b.   19.93 half-lives

13.  The tree is 20,000 years old.

14.  No, it is not.  If we assume that bones may be as old as $2 \times 10^6$ yr, this is far less than one half-life, and therefore probably undetectable.

15.  The rock is $3.8 \times 10^9$ yrs old.

16.  a.   mass defect = 1.915095 g          c.   mass defect = 1.700555 g
     b.   mass defect = 0.155535 g

17.  $\Delta E$ = binding energy = $-5.59 \times 10^{13}$ J/mol and -8.79 MeV/nucleon

18.  $\Delta E = 1.20 \times 10^{14}$ J/mol and 18.28 MeV/nucleon.

# CHAPTER 22

# Organic Chemistry

| Section | Contents | Page in this Study Guide |
|---|---|---|
| 22.1 | Alkanes - Saturated Hydrocarbons | 440 |
| 22.2 | Alkenes and Alkynes | 444 |
| 22.3 | Aromatic Hydrocarbons | 444 |
| 22.4 | The Petrochemical Industry | 448 |
| 22.5 | Hydrocarbon Derivatives | 449 |
| 22.6 | Polymers | 451 |
| 22.7 | The Polymer Industry | 451 |

**Organic chemistry is the study of compounds that contain carbon.** Such compounds are ubiquitous because carbon **forms strong bonds** to hydrogen, oxygen and nitrogen, among others. They also have the **unique** ability to form **chains and rings** with other carbon atoms. This chapter serves as a simple introduction to the tens of thousands of known organic compounds.

## 22.1 - Alkanes - Saturated Hydrocarbons

When you finish this section you will be able to:

- draw isomers of simple alkanes, and
- name isomers of alkanes.

**Alkanes** are a group of **saturated hydrocarbons.** **Saturated** means that the carbon is bound to **four atoms** (each by a single bond). Each carbon is **sp$^3$** hybridized. Alkanes have the general formula $C_nH_{2n+2}$. Table 22.1 in your textbook gives names, formulas and some properties of straight-chain ("n") alkanes.

# Example 22.1 A - Alkanes

Give the formula and name for the straight-chain alkanes with n = 6, and n = 8.  Compare their boiling and melting points.  Justify the difference.

## Solution

$$\text{For n = 6, } C_nH_{2n+2} = C_6H_{14} = \textbf{hexane}$$
$$\text{For n = 8, } C_nH_{2n+2} = C_8H_{18} = \textbf{octane}$$

The boiling point of octane is 58° higher than hexane.  The melting point is 38° higher.  Hexane is considerably heavier and longer.  The London forces are more extensive, thus more energy is required to break intermolecular bonds.

---

**Isomers** are compounds with the **same formula** but **different structures**.  Let's draw some of the isomers for heptane, $C_7H_{16}$.  The key is to make sure that if the main chain lengths are the same, the **type** or **location** of groups on the chain are **unique**.

Given the straight chain,

```
    H H H H H H H
    | | | | | | |
  H-C-C-C-C-C-C-C-H        H3C-CH2-CH2-CH2-CH2-CH2-CH3     C7H16
    | | | | | | |
    H H H H H H H
```

An example of an isomer is

$$H_3C\text{-}CH\text{-}CH_2\text{-}CH_2\text{-}CH_2\text{-}CH_3$$
$$|$$
$$CH_3$$

- The formula is still $C_7H_{16}$.
- The longest chain has **6 carbons, a** <u>hexane</u>.
- The numbering of carbons puts the "functional group" closest to the "#1" carbon.  Therefore the **methyl** group is attached to the **"#2" carbon**.
- The name of this isomer of heptane is **2-methylhexane**.
- Carefully examine the nomenclature rules listed just before <u>Table 22.2 in your textbook</u>.

Another isomer is

$$H_3C\text{-}CH_2\text{-}CH\text{-}CH_3$$
$$|$$
$$H_3C\text{-}CH\text{-}CH_3$$

The longest chain looks like 4 carbons.  It is actually 5 (eliminating hydrogens for clarity):

$$C_5\text{-}C_4\text{-}C_3\text{-}C$$
$$|\qquad\text{This is 2,3-methylpentane}$$
$$C\text{-}C_2\text{-}C_1$$

Remember that what you **draw on paper** is only a **shorthand representation** of a 3 dimensional structure.  Look very hard for the longest chain.

## Example 22.1 B - Isomers

There are **9 isomers** of heptane, $C_7H_{16}$. We have drawn and named **3**. **Draw and name the other 6**. (You may eliminate hydrogens for clarity, if you wish.)

## Solution

|   Structure | Name |
|---|---|

4.  C-C-C-C-C
         |   |
         C   C          2,4-dimethylpentane

5.  C-C-C-C-C-C
         |
         C                              3-methylhexane

6.     C
       |
    C-C-C-C
     |
    C-C                                                          2,2-ethylmethylbutane

7.     C-C
       |
    C-C-C-C
       |
       C                                          2,2,3-trimethylbutane

8.       C
         |
    C-C-C-C-C
         |
         C                              3,3-dimethylpentane

9.       C-C
         |
    C-C-C-C-C                                                    3-ethylpentane

Let's try naming some compounds and drawing some structures.

## Example 22.1 C - Nomenclature

Give IUPAC names for the following structures:

a.  $CH_3CH_2CHCH_2CHCH_2CH_3$
             |      |
            $CH_3$   $CH_2CH_2CH_3$

b.  $CH_3CHCHCH_2CH_3$
          |
         Cl-Cl

$$CH_3$$
$$|$$
c.   $CH_3CH_2CHCH_2CHCHCH_3$
   $|$         $|$
   $CH_3$    $CH_2CH_2CH_3$

## Solution

a.   The longest chain has 8 members, an octane.

$$C_1\text{-}C\text{-}C\text{-}C\text{-}C\text{-}C$$
$$|\quad|$$
$$C\quad C\text{-}C\text{-}C_8$$

The number would start on the **left hand carbon** because the methyl group is closest to that side (in the #3 position).  The name of the structure is **3-methyl-5-ethyloctane**.

b.   **2,3-dichloropentane**

c.   The longest chain has 8 members, an octane.

$$C$$
$$|$$
$$C\text{-}C\text{-}C\text{-}C\text{-}C\text{-}C\text{-}C$$
$$|\quad|$$
$$C\quad C\text{-}C\text{-}C$$

The closest group to a #1 carbon is the methyl group in the 3 position.  The name of the structure is **3-methyl-5-isopropyloctane**.

---

# Example 22.1 D - Drawing Structures

Draw structures for the following compounds:

a.   3,4-dimethyl-4-ethylnonane
b.   1-chloro-2-bromobutane

## Solution

$$CH_3$$
$$|$$
a.   $CH_3\text{-}CH_2\text{-}CH\text{-}C\text{-}CH_2\text{-}CH_2\text{-}CH_2\text{-}CH_2\text{-}CH_3$
        $|\quad|$
        $H_3C\quad CH_2CH_3$

b.   $CH_2\text{-}CH\text{-}CH_2\text{-}CH_3$
      $|\quad|$
      $Cl\quad Br$

---

Cyclic alkanes have the general formula $C_nH_{2n}$ (for example, $C_6H_{12}$ is cyclohexane).  Note the use of the **short-hand notation** for each of the cyclic compounds in your textbook.  Notice also that as with straight chain alkanes, numbering is done so that **substituents are attached to carbons with the lowest numbers possible**.

---

## Example 22.1 E - Cyclic Alkanes

Name the following compounds:

a.

c.

b.

### Solution

a.   chlorocyclopropane
b.   1,2-diethylcyclohexane

c.   1-bromo-1-methylcyclohexane

---

## Problems 1-3 at the end of this study chapter cover material from this section.

---

## 22.2 and 22.3 - Alkenes and Alkynes and Aromatic Hydrocarbons

When you finish these sections you will be able to name and draw structures for:

- simple alkenes and alkynes, and
- simple benzene derivatives.

Straight chain **alkenes** have the general formula $C_nH_{2n}$. They are characterized by the presence of **at least one carbon-carbon double bond**. The bond is formed by sharing p-orbitals. The rules for naming alkenes are the same as for alkanes, with the following exceptions:

- **-ane** is changed to **-ene** (ie; hex**ane** becomes hex**ene**).
- The position of the **double bond** has **highest priority** in terms of nomenclature. For example

$$\underset{H}{\overset{C}{|}}C=C\text{-}C\text{-}\underset{C}{\overset{|}{C}} \text{ is } \textbf{4-methyl-trans-2-pentene}$$

NOT 2-methyl-trans-4-pentene.
- **cis** and **trans** isomers exist because rotation around a carbon-carbon double bond is **restricted** due to p-p orbital interaction between carbons.

---

# Example 22.2 A - Naming Alkenes

Name the following alkenes:

a.

$$H_3C \quad\quad CH_3$$
$$\diagdown\quad\diagup$$
$$C=C$$
$$\diagup\quad\diagdown$$
$$Cl \quad\quad Cl$$

b.

$$Cl \quad\quad CH_3$$
$$\diagdown\quad\diagup$$
$$C=C$$
$$\diagup\quad\diagdown$$
$$H_3C \quad\quad Cl$$

c.

$$CH_2CH_2CH_3$$
$$\diagup$$
$$CH_3CH_2CH=C$$
$$\diagdown$$
$$CH_2CH_2CH_3$$

d.   $CH_3CH_2CH_2CH_2CH_2C=CH_2$
$$|$$
$$CH_3$$

## Solution

a.   The longest straight chain is a butene.  In this case it is **2-butene**.  The chlorines are **cis** to each other and in the **2 and 3 positions**.

### 2,3-dichloro-cis-2-butene

b.   Everything is the same except the chlorines are **trans** to one another.

### 2,3-dichloro-trans-2-butene

c.   The longest straight chain is a 7-membered, or **heptene** chain.  The double bond is in the **3** position.  The attached group is a **propyl** group.

### 4-propyl-3-heptene

d.   The longest chain here is a **heptene**.  The double bond is in the **1** position.  The methyl group is in the **2** position.

### 2-methyl-1-heptene

# Example 22.2 B - Drawing Alkenes

Draw the following alkenes:

a.   3-methyl-1-hexene
b.   1-chloro-4-ethyl-3-hexene

c.   4-methyl-cis-2-pentene

## Solution

a.   $CH_2=CHCHCH_2CH_2CH_3$
$$| $$
$$CH_3$$

b.   $CH_2CH_2CH=CCH_2CH_3$
$$|\qquad\quad|$$
$$Cl\qquad\quad C_2H_5$$

c.

---

**Alkynes** are molecules that contain triple bonds (1s, 2 bonds involving 2 carbons).  The nomenclature follows the same strategy as always, except the compound ends in **"yne"**.

---

## Example 22.2 C - Naming Alkynes

Name the following:

a.   $CH{\equiv}C(CH_2)_6CH_3$

b.

## Solution

a.   1-nonyne          b.  3,3-dimethyl-1-butyne

---

## Example 22.2 D - Just for Fun

Name this compound:

## Solution

The three double bonds are in the **1,3 and 5** positions.  This is a **7-membered ring**.

**1,3,5-cycloheptatriene**  (also called **Tropilidene**)

---

The main idea in Section 22.3 is that **electrons** in benzene and benzene-related compounds are **delocalized** (can move freely around the molecule.  This makes benzene family compounds unreactive to addition, but reactive instead to **substitution** (where hydrogen atoms are replaced by other atoms).

The system of naming benzene-related compounds is shown in <u>Figure 22.12 in your textbook</u>.  Notice that the numbering system is similar to that with alkanes, alkenes and alkynes.  Note as well that when there is a substituent in the "1" position, the substituent in the "2" position is called "ortho."  The "3" position is "meta" and the "4" position is "para."

---

# Example 22.3 A - Naming Aromatic Compounds

Name the following compounds ($NO_2$ = "nitro"):

a.

c.

b.

## Solution

   a.   1-nitro-2,4,5-trichlorobenzene
   b.   hexabromobenzene
   c.   1,3,5-trimethyl-2,4-dinitrobenzene

---

# Example 22.3 B - Drawing Aromatic Compounds

Draw the following compounds:

   a.   1-nitro-2,3,6-triiodobenzene
   b.   1-ethyl-2-methylbenzene ("2-ethyltoluene")
   c.   1,4-bis(dibromomethyl)benzene

## Solution

a.                                          b.

c.    **"bis"** implies that the dibromomethyl **appears twice**, once in the 1 and once in the 4 position.

---

Problems 4-7 at the end of this study chapter cover material from this section.

---

## 22.4 - The Petrochemical Industry

The following examples will help you to review the material in this section.

---

Your textbook discusses a number of terms that are used in the petroleum industry.  Read the remainder of this section, then try the following exercise.

---

## Example 22.4 C - Petrochemical Terms

Match the term with its meaning.

| Term | | Meaning |
|------|---|---------|
| 1. pyrolysis | a. | hydrocarbons are broken down by steam |
| 2. oligomerization | b. | the combination of alkenes to form a larger molecule |
| 3. catalytic reforming | c. | the portion of petroleum used to make petrochemicals |
| 4. cracking | d. | cracking at high temperatures |
| 5. feedstock | e. | conversion of alkanes to aromatics |
| 6. steam cracking | f. | large molecules are broken down by breaking C-C bonds |

## Solution

1.  d      2.  b      3.  e      4.  f      5.  c      6.  a

---

## 22.5 - Hydrocarbon Derivatives

When you finish this study section you will be able to name compounds containing various functional groups.

**Hydrocarbon derivatives** are molecules that have substituents ("**functional groups**") that contain some atoms which are **not carbon or hydrogen.** Your textbook discusses the properties of several functional groups in this section. You should know the properties of

- alcohols
- aldehydes and ketones
- carboxylic acids and esters
- amines

The functional groups are summarized in Table 22.5 in your textbook. Let's try some naming and drawing exercises (remembering that when **benzene** is an attached group rather than the main focus of the molecule, it is called a "**phenyl**" group).

---

## Example 22.5 A - Functional Groups

Name the following compounds:

a.

c.

b.

(See Table 22.7 in your textbook.)

d.

## Solution

a.   4-propylphenol

c.   2,2,3-trimethylcyclohexane

b.   2-chloroaniline
     (or "o-chloroaniline")

d.   4-chloro-2-butanone

## Example 22.5 B - More Functional Groups

Draw the following compounds:

a.   1,2-pentanediol
b.   3-fluoro-benzoic acid
c.   1,2-cyclopentanedicarboxylic acid

## Solution

a.   $CH_2CH_2CH_2CH_2CH_2$
　　　|　　　　　　|
　　　OH　　　　　OH

c.

b.

---

## Example 22.5 C - Revenge of the Functional Groups

List all of the functional groups in each of the following molecules:

a.

c.

b.

$CH_3CH_2CH_2CH_2-\overset{\overset{\textstyle H}{|}}{N}-CH_2\,CH_3$

d.

## Solution

    a.   alcohol, carboxylic acid
    b.   secondary amine (2 hydrogens have been substituted for)
    c.   tertiary amine (3 hydrogens have been substituted for)
    d.   secondary amine, aldehyde

---

Problems 8-12 at the end of this study chapter cover material from this section.

---

## 22.6 - Polymers

The following questions will help you test your knowledge of the material in this section.

1. Define **polymer**.
2. What are polymers made from?
3. How can you vary properties when making polymers?
4. What is Teflon made of?
5. Why is Teflon so widely used?
6. Define **addition polymerization**.
7. What is a free radical?
8. Define **condensation polymerization**.
9. Why is nylon called a copolymer?
10. Use equations to show how water is a product when nylon is formed.
11. Why is Dacron a "polyester?"

---

Problems 13-18 at the end of this study chapter cover material from this section.

---

## 22.7 - The Polymer Industry

The following questions will help you to review the material in this section.

1. Why does your textbook say that "serendipity" was behind the first known polymer synthesis?
2. Define **thermoset polymer** and **thermoplastic polymer**.
3. List the chemical reaction for the formation of nylon.
4. How does "drawing out" change the structure of nylon?
5. Define **crosslinking**.
6. What is **vulcanization**? How does it help to strengthen rubber?
7. What are the two major forms of polyethylene?
8. What are the differences in properties of HDPE and LDPE?
9. What is the function of $Al(C_2H_5)_3 + TiCl_4$?
10. Why is chromium oxide and aluminosilicate often used in place of the Ziegler-Natta catalyst?
11. List some properties of polymers that are affected by molecular weight and chain length.
12. Define **isotactic, syndiotactic** and **atactic chains**.
13. How is an isotactic polypropylene chain made?
14. What monomer is the basis for polystyrene?
15. Why can't pure polystyrene be used more universally?
16. What additives help make polystyrene-based polymers more useful?
17. Draw the structure of PVC.
18. Define **plasticizer**.

## Example 22.7 - The Polymer Industry

Fill in the blanks after reviewing the material in <u>Table 22.7 in your textbook</u>.

a. _____ lowers flammability.

b. _____ softens the polymer.

c. _____ prevents fungal attack.

d. _____ prevents damage by sunlight.

e. _____ increases resistance to thermal decomposition.

## <u>Solution</u>

a.  flame retardant          d.  ultraviolet absorber
b.  plasticizer              e.  heat stabilizer
c.  biocide

## SELF-TEST QUESTIONS

1.  Name the following compounds using IUPAC nomenclature.

    a.   $CH_3-\overset{\overset{\displaystyle CH_3}{|}}{\underset{\underset{\displaystyle CH_3}{|}}{C}}-CH_2-CH_2-CH_2-CH_2-CH_3$

    b.   $CH_3-CH_2-\overset{}{\underset{\underset{\underset{\underset{\displaystyle CH_3}{|}}{CH_2}}{|}}{\underset{\displaystyle CH_2}{|}}{CH}}-CH_2-\overset{\overset{\overset{\displaystyle CH_3}{|}}{\underset{\displaystyle CH_2}{|}}}{CH}-CH_3$

    c.   $CH_3-CH-CH_3$
         $CH_3-\overset{}{\underset{\underset{\displaystyle CH_3}{|}}{\overset{\displaystyle |}{C}}}-CH_3$

    d.   $CH_3-\overset{\overset{\displaystyle CH_3}{|}}{C}-CH_3$

2.  Write structures for the following systematic names.

    a.   1-ethyl-3-propylcyclohexane
    b.   1,1,2-trichloroethane

3.  Are saturated hydrocarbons (alkanes) soluble in water?

4.  Match the following structures with their correct systematic name.

    a.   $CH_2=CHCHCH_3$          b.   $CH_3CHC\ CH$          c.   $CH_3CH_2\underset{\underset{\displaystyle CH_2}{||}}{C}CH_3$
         $\underset{\displaystyle CH_3}{|}$                     $\underset{\displaystyle CH_3}{|}$

    1.  3-methyl-1-butyne      2.  3-methyl-butene      3.  2-methyl-1-butene

5.  What products would form after hydrogenation of 2-methyl-1-butene?  After halogenation ($Cl_2$)?

6.  Why, in general, are alkenes more reactive in addition reactions than alkanes?

7.  What are the products of the following reactions?

    a.

       +    $Br_2$ / $FeBr_2$   $--\rightarrow$    ?

b.

+ Br$_2$ / FeBr$_2$ ----► ?

c.

+ CH$_3$C-Cl / AlCl$_3$ ----► ?

with O double bonded above the carbonyl: $\overset{O}{\overset{\|}{CH_3C-Cl}}$

d.

+ H$_2$SO$_4$ (fuming) ----► ?

8. Name the functional group(s) in each of the following compounds:

a. CH$_3$-O-CH$_2$-CH$_2$OH
      (1)        (2)

b. CH$_3$CH
     |
     O
    (1)

c. CH$_3$CH$_2$C-O-CH$_2$CHCH$_3$
          |        |
          O     NH$_2$
        (1)   (2)

d.
      CH$_3$  O
       |    ||
  CH$_3$-C- - -C-OH
       |
      CH$_3$ (1)

9. Name the following compounds or give their structure:

a.
OH

c. CH$_3$CH$_2$CCH$_2$CH$_2$CH$_3$
       with O double bonded: $\overset{O}{\overset{\|}{\quad}}$

b. 3-chlorobenzaldehyde

d. CH$_3$CH$_2$CHCH$_2$CHCH$_3$
         |      |
        OH   Cl

10. Arrange the molecules in order from lowest to highest boiling point:

$$CH_3CH_2CH_2OH \qquad CH_3CH_2\overset{\displaystyle O}{\overset{\displaystyle \|}{C}}H \qquad CH_3\overset{\displaystyle O}{\overset{\displaystyle \|}{C}}CH_3$$

11. Name the reactants in each equation below. Give the structure of the products that would form.

a. $CH_3\underset{\underset{\displaystyle CH_3}{|}}{CH}-\overset{\displaystyle O}{\overset{\displaystyle \|}{C}}-OH$  +  [benzene ring with $CH_2OH$ substituent]  ------>

b. $CH_2ClCOOH + CH_3-\underset{\underset{\displaystyle OH}{|}}{CH}-CH_3$ ---->

c. $CH_3\underset{\underset{\displaystyle CH_3}{|}}{\overset{\overset{\displaystyle CH_3}{|}}{C}}-CH_2CH_2OH \xrightarrow{\text{KMnO}_4(aq)}$

12. Label the following amines as 1°, 2°, or 3°.

a. $N(CH_2CH_3)_3$

c. $CH_3CH_2NHCH_3$

b. $(CH_3)_2N\underset{\underset{\displaystyle CH_3}{|}}{\overset{\overset{\displaystyle CH_3}{|}}{C}}H$

d. $NH_2CH_2\underset{\underset{\displaystyle CH_3}{|}}{\overset{\overset{\displaystyle CH_3}{|}}{C}}H$

13. Define or explain the following terms:

    a.  dimer          c.  copolymer      e.  polymer
    b.  free radical    d.  homopolymer

14. Distinguish between addition polymerization and condensation polymerization.

15. Write the cis- and trans- chair conformations of 1,2-dichlorocyclohexane.

16. Arrange the following alkenes from most stable to least stable. (hint, stability is directly related to the substitution of the double bond).

a. $H-\overset{\overset{\displaystyle H}{|}}{C}=\overset{\overset{\displaystyle H}{|}}{C}-H$

c. $H-\overset{\overset{\displaystyle CH_3}{|}}{C}=\overset{\overset{\displaystyle H}{|}}{C}-H$

b. $CH_3CH_2-\overset{\overset{\displaystyle H}{|}}{C}=\overset{\overset{\displaystyle CH_2CH_2CH_3}{|}}{C}-H$

d. $CH_3-\overset{\overset{\displaystyle CH_3}{|}}{C}=\overset{\overset{\displaystyle CH_2CH_3}{|}}{C}-CH_3$

17. Alcohols are capable of forming strong hydrogen bonds to each other making them polar. Why is ethyl alcohol greatly soluble in water while heptyl alcohol is almost insoluble in water?

18. Arrange the following amines from highest to lowest boiling point. Give an explanation of your answer.

    a.  $(CH_3-CH_2)_2N-H$      b.  $CH_3-CH_2-N(CH_3)_2$   c.  $CH_3(CH_2)_3-NH_2$

## ANSWERS TO SELF-TEST QUESTIONS

1.  a.  2,2-dimethylheptane        c.  2,2,3-trimethylbutane
    b.  5-ethyl-3-methyloctane     d.  tert-butylcyclopentane

2.  a.         b.  Cl-CH-CH$_2$
                                        |    |
                                        Cl   Cl

3.  Alkanes are almost totally insoluble in water.  This is due to their non-polar nature (water is polar) and their inability to form hydrogen bonds.

4.  a.  (2)      b.  (1)      c.  (3)

5.  Hydrogenation:

$$CH_2=C\text{-}CH_2\text{-}CH_3 + H_2 \longrightarrow CH_3\text{-}CH\text{-}CH_2\text{-}CH_3 \quad (\text{2-methylbutane})$$
$$\qquad\quad | \qquad\qquad\qquad\qquad\qquad | $$
$$\qquad\quad CH_3 \qquad\qquad\qquad\qquad\quad CH_3$$

Halogenation:

$$\qquad\qquad\qquad\qquad Cl \quad Cl$$
$$\qquad\qquad\qquad\qquad | \quad\; |$$
$$CH_2=C\text{-}CH_2\text{-}CH_3 + Cl_2 \longrightarrow CH_2\text{-}C\text{-}CH_2\text{-}CH_3 \quad (\text{1,2-dichloro-2-methylbutane})$$
$$\qquad\quad | \qquad\qquad\qquad\qquad\qquad |$$
$$\qquad\quad CH_3 \qquad\qquad\qquad\qquad\; CH_3$$

6.  Alkenes have a carbon-carbon double bond consisting of a C-C σ bond and weaker C-C π.  Alkanes consist of C-C σ bonds.  Thus the presence of the weaker  bond and its exposed electrons make alkenes more susceptible to addition reaction than alkanes.

7.  a.   +  HBr        c.  [benzene ring with]
                                                O
                                                ||
                                                C-CH$_3$
                                                +  HCl

    (bromobenzene)                              (methylphenylketone)

    b.  No Reaction.  (No Lewis acid catalyst.)    d.  SO$_3$H
                                                       [benzene ring]  +  H$_2$O

                                                       (benzenesulfonic acid)

8.  a.  (1) ether, (2) 1° alcohol    c.  (1) ester, (2) 1° amine
    b.  (1) aldehyde                 d.  (1) carboxylic acid

9.  a.   phenol
    b.

c.   3-hexanone
d.   5-chloro-3-hexanol

$$H-C\overset{O}{\parallel}$$

10. $CH_3CH_2\overset{\overset{O}{\parallel}}{C}H$ < $CH_3\overset{\overset{O}{\parallel}}{C}CH_3$ < $CH_3CH_2CH_2OH$

11. a.   2-methylpropanoic acid + benzyl alcohol ---->

$$CH_3-\underset{\underset{CH_3}{|}}{CH}-\overset{\overset{O}{\parallel}}{C}-O-CH_2-\bigcirc + H_2O$$

b.   chloroethanoic acid + 2-propanol or isopropyl alcohol ---->

$$CH_2ClC-O-\underset{\underset{CH_3}{|}}{\overset{\overset{O}{|}}{C}}H-CH_3 + H_2O$$

c.   3,3-dimethyl-11-butanol $\xrightarrow{\text{KMnO}_4\text{(aq)}}$

$$CH_3-\underset{\underset{CH_3}{|}}{\overset{\overset{CH_3}{|}}{C}}-CH_2COH$$

12. a.   3°        b.   2°        c.   3°        d.   1°

13. a.   Two identical monomers joining together to form a molecule.
    b.   A species with an unpaired electron.
    c.   More than one type of monomer combining to form the chain in a polymer.
    d.   Identical monomers combining to form the chain in a polymer.
    e.   A large chainlike molecule formed from small monomer units.

14. **Addition polymerization** involves the formation of the polymer by free radical mechanism. The only product formed is the polymer, and the polymerization stops when two radicals react to form a bond without producing any other radicals.

**Condensation polymerization** produces a product other than the polymer itself. The side products are most commonly small molecules such as water or alcohols.

15.

cis                              trans

16. d > b > c > a

17. Heptyl alcohol has a seven carbon chain while ethyl alcohol has a two carbon chain. Due to this fact, the non-polar characteristics in the seven carbon chain dominate over the polar characteristics making it insoluble in water.

18. c > a > b. Due to hydrogen bonding. A 1° amine has a greater amount of hydrogen bonding when compared to a 2° amine. A 3° amine has little hydrogen bonding when compared to a 1° and 2° amine.

# ANSWERS TO REVIEW QUESTIONS

## Section 22.4

1. See the <u>paragraphs regarding the work of Christian Schoenbein</u> at the beginning of this section of <u>your textbook</u>.
2. A thermoset polymer cannot be softened or dissolved again. A thermoplastic polymer can be molded again.
3. See <u>Figure 22.19 in your textbook</u>.
4. "Drawing out" forces the polymer chains to line up, which leads to increased hydrogen bonding and stronger fibers.
5. Crosslinking is the formation of covalent bonds between adjacent polymer chains.
6. Vulcanization is the addition of sulfur to and subsequent heating of rubber. Covalent bonds to sulfur link chains with a resulting increase in strength.
7. Low-density polyethylene (LDPE) and high-density polyethylene (HDPE) are the two major forms.
8. LDPE is less dense and is used in manufacturing transparent films. HDPE is more dense and is used in bottles and similar blow-molded products.
9. These are catalysts in the formation of LDPE.
10. These newer catalysts are not as dangerous.
11. Examples are viscosity and toughness.
12. **Isotactic:** All groups on the same side of the polymer chain.
   **Syndiotactic:** Groups alternate on the chain.
   **Atactic:** Random arrangement of groups on the chain.
13. Isotactic chains are made by using the Ziegler-Natta catalyst.
14. The monomer **styrene**.
15. It is too brittle.
16. Butadiene is the main additive. Acrylonitrile is often used as well.
17. $(-HC=CCl-HC=CCl-)_n$.
18. Plasticizers make a substance more flexible.
19. Their primary function is to prevent thermal decomposition.
20. PVC does not contain lead in this application because of the potential for lead poisoning in water systems.
21. Antimony does not leach from pipes. However, it is expensive and makes processing more difficult.
22. Sodium antimonate is added for transparency.
23. Although plasticizers are useful, they increase the flammability of PVC.
24. Elastomers recover their shape after a deforming force.
25. World War II made natural rubber unavailable.
26. See the <u>section on elastomers in your textbook</u> for this structure.
27. As with nylon, the chains become more ordered.
28. There is an increase in entropy leading to a decrease in free energy. The result is a more relaxed system.
29. See the <u>Chemical Impact</u> in this section of <u>your textbook</u> for the complete reaction scheme.
30. **Solid-phase inhibition:** The polymer surface is changed to help prevent combustion.
   **Vapor-phase inhibition:** The chemistry of the combustion process is changed.

# CHAPTER 23

# Biochemistry

| Section | Contents | Page in this Study Guide |
|---|---|---|
| 23.1 | Proteins | 460 |
| 23.2 | Carbohydrates | 463 |
| 23.3 | Nucleic Acids | 466 |
| 23.4 | Lipids | 467 |

The introductory material in this chapter stresses the importance of biochemistry (the chemistry of living systems). It points out that at least 26 elements are essential to human life (see Table 23.1 in your textbook). The goal of this chapter is to investigate **the operation of cells on a molecular level**.

## 23.1 - Proteins

The following questions and exercises will help to review your understanding of the nature of proteins.

1. What are **proteins**?
2. What range of molar masses can proteins have?
3. Why are the acids that comprise proteins called **a-amino** acids? That is, what does "a" mean? What does "amino" mean?

---

## Example 23.1 A - The Common Protein Amino Acids

Match the name of the amino acid in the left-hand column with its "R" group in the right-hand column.

| Amino Acid | R group |
|---|---|
| 1. Glycine | |
| 2. Methionine | a. $-CH_2-CH_2-\overset{\displaystyle O}{\overset{\displaystyle \|}{C}}-NH_2$ |

3.  Glutamic Acid

4.  Glutamine                     b.   -CH$_2$-CH$_2$-C-OH
                                          (O double bond above C)

5.  Lysine                        c.   -CH$_2$-(C$_6$H$_5$)

6.  Phenylalanine                 d.   -(CH$_2$)$_4$-NH$_2$

                                  e.   -CH$_2$-CH$_2$-S-CH$_3$

                                  f.   -H

## Solutions

| Amino Acid | R group |
|---|---|
| 1.  Glycine | f.   -H |
| 2.  Methionine | e.   -CH$_2$-CH$_2$-S-CH$_3$ |
| 3.  Glutamic Acid | b.   -CH$_2$-CH$_2$-C-OH  (O double bond above C) |
| 4.  Glutamine | a.   -CH$_2$-CH$_2$-C-NH$_2$  (O double bond above C) |
| 5.  Lysine | d.   -(CH$_2$)$_4$-NH$_2$ |
| 6.  Phenylalanine | c.   -CH$_2$-(C$_6$H$_5$) |

---

4.  What is a peptide linkage?
5.  Define **dipeptide**.

---

# Example 23.1 B - Peptide Linkage

Draw the structure of the dipeptide formed from the condensation reaction between **leucine** and **phenylalanine**.

## Solution

As pointed out in your textbook, the peptide linkage occurs between the **carboxylic acid** end of one molecule and the **amine** end of the other. Also, as "standard procedure," the terminal amine is always put on the left and the terminal carboxylic acid is put on the right.

Leucine                 Phenylalanine
(Leu)                      (Phe)

(Leu-Phe)

---

## Example 23.1 C - Practice with Peptide Linkages

Draw the structure of the polypeptide with the sequence **Trp-Ser-Asp**.

## Solution

This is formed as a result of 2 peptide linkages, one between **Trp and Ser** and one between **Ser and Asp**.

Tryptophan
(Trp)

Serine
(Ser)

Aspartic Acid
(Asp)

(Trp-Ser-Asp)

---

6. How many sequences can possibly exist for a polypeptide chosen from 10 unique amino acids?  (Ans:  10! = $10 \times 9 \times 8 \times 7 \times 6 \times 5 \times 4 \times 3 \times 2 \times 1 = 87178291200 = \mathbf{8.71782912 \times 10^{10}}$ **sequences**)
7. What are the three levels of structure in proteins?
8. What kinds of bonding interactions are responsible for each level of structures?
9. What type of structure is an α-helix?
10. Give some practical examples of primary, secondary and tertiary structures.
11. Discuss the role of the **disulfide** linkage in getting "permanent waves" of hair.
12. Define **denaturation**.
13. List some causes of denaturation.

## Example 23.1 D - Protein Functions

Based on Table 23.1 in your textbook, fill in the function of each of the following proteins or class of proteins.

a.  Interferon  _____

b.  Ferritin  _____

c.  Transferrin  _____

d.  Insulin  _____

e.  Histones  _____

f.  Rhodopsin  _____

g.  Enzymes  _____

### Solution

a. protection   d. control      g. catalysis
b. storage      e. structure
c. transport    f. energy transformation

14. What differentiates enzymes from inorganic catalysts?
15. Describe the **lock-and-key model** of enzyme activity.
16. Define **inhibition**.
17. How does the nerve gas DIPF work?

Answers to review questions are at the end of this study chapter.

## 23.2 - Carbohydrates

The following questions will help you review the material in this section.

1. Why are carbohydrates so named?
2. What are monosaccharides?

3.  What is a **hexose**?
4.  What is necessary for **optical isomerism** to exist in a molecule?

---

## Example 23.2 - Chiral Carbons

How many chiral carbons are there in D-Glucose?

$$
\begin{array}{c}
_1CHO \\
H-_2C-OH \\
HO-_3C-H \\
H-_4C-OH \\
H-_5C-OH \\
_6CH_2OH
\end{array}
$$

D-Glucose

## Solution

A chiral carbon has **4 different substituents** attached to it.

Carbon **#1** has only 3 substituents. It is not chiral.

Carbon **#2 is chiral**.

Carbon **#3 is chiral**.

**Carbon #4 is chiral.**

```
              CHO
              |
        H —— C —— OH
              |
       HO —— C —— H
              |
       [H]—— C ——[OH]
              |
        H —— C —— OH
              |
             CH₂OH
```

**Carbon #5 is chiral.**

```
              CHO
              |
        H —— C —— OH
              |
       HO —— C —— H
              |
        H —— C —— OH
              |
       [H]—— C ——[OH]
              |
            [CH₂OH]
```

**Carbon #6 has two identical groups.  It is not chiral.**

```
              CHO
              |
        H —— C —— OH
              |
       HO —— C —— H
              |
        H —— C —— OH
              |
        H —— C —— OH
              |
       [H]—— C ——[OH]
              |
             [H]
```

In summary, **carbons 2, 3, 4 and 5 are chiral.**  Hexoses have $2^4 = 16$ optical isomers.

5. What are the bonds involved in cyclizing pentoses? hexoses?
6. Define **disaccharide**.
7. What is a **glycoside linkage**?
8. What is the function of α-amylase?
9. Describe the structure of **starch**.
10. Why is it advantageous for "fuel" storage to have starch as one long molecule instead of many small ones?
11. Why is cellulose **indigestible** by humans, but can be digested by certain animals?

Answers to review questions are at the end of this study chapter.

## 23.3 - Nucleic Acids

The following questions will help you review the material in this section.

1. What are the functions of DNA?
2. List the basic parts that make up nucleotides.
3. Why is a double helix structure important to the function of DNA?
4. Cytosine and guanine form hydrogen-bonding pairs. What in their structure makes this possible?
5. Outline the process for replication of DNA.

## Example 23.3 - Complimentary Sequences

A single strand of DNA contains the nucleotide sequence

$$A - A - G - T - T - G - C - C - A - T$$

List its complimentary strand.

### Solution

Adenine (A) and thymine (T) form complimentary pairs as do cytosine (C) and guanine (G). The complimentary strand would be

old   A - A - G - T - T - G - C - C - A - T

new   T - T - C - A - A - C - G - G - T - A

6. Define **gene**.
7. What is a **codon**? **anticodon**?
8. Describe the functions of mRNA and tRNA to protein construction.

Answers to review questions are at the end of this study chapter.

## 23.4 - Lipids

The following questions will help you review the material in this section.

1. What differentiates lipids from other biochemical substances?
2. List the **four types** of lipids.
3. What substances are combined to make fats?
4. What is a triglyceride?

## Example 23.4 A - Triglycerides

Write the chemical equation for the reaction between three lauric acid molecules (see Table 23.4 in your textbook) and a glycerol molecule.

### Solution

The reaction will form a triglyceride:

5. Outline the process of **saponification**.
6. How do soaps work (ie; outline the process).
7. Define **micelle**.
8. How is a detergent structurally different from a soap?
9. What is the advantage of using a detergent in areas with hard water?
10. What is the structure of phospholipids?
11. What are the biological functions of phospholipids?
12. Describe the **fluid mosaic model**.
13. Structurally, what do fats, phospholipids and waxes have in common?
14. How are waxes different from other lipids?
15. What are some properties of waxes?
16. What structures are common to steroids?

## Example 23.4 B - Steroids

Match the steroid with its function.

| Steroid | Function |
|---|---|
| 1. adrenocorticoid hormones | a. Starting material for the formation of bile acids, steroid hormones and vitamin D. |
| 2. sex hormones | b. Regulates water and electrolyte balance. |

3. cholesterol             c. Controls growth of reproductive organs.

4. bile acids              d. Aids the digestion of fats.

## Solution

1. b      2. c      3. a      4. d

---

**END OF CHAPTER TWENTY-THREE**

# ANSWERS TO REVIEW QUESTIONS

1. Proteins are "natural polymers" in our bodies that perform a variety of biological functions.
2. The molecular weights can range from 6000 to over 1,000,000.
3. They are called "$\alpha$-amino acids" because the amino group is attached to the $\alpha$-carbon.  Amino means an $NH_2$ group.
4. A peptide linkage occurs when the carboxyl hydroxide interacts with a hydrogen from an amine group.
5. A dipeptide involves 2 amino acids in a peptide linkage.
6. See our question #6?  The answer is there.
7. The levels are primary, secondary and tertiary.
8. Primary = peptide linkages, secondary = hydrogen bonding, tertiary = a variety of interactions including hydrogen bonding, ionic bonds and covalent bonds, among others.
9. An $\alpha$-helix represents a secondary structure.
10. See the discussion in the text for some examples.
11. The disulfide linkage in hair is broken and reformed to shape the hair.  (See the discussion adjacent to <u>Figure 23.8 in your textbook</u>.)
12. Denaturation involves breaking down the three-dimensional structure of a protein, thus rendering it inactive.
13. Heat and intense radiation are two causes.
14. Enzymes are faster and more selective than inorganic catalysts.
15. See <u>Figure 23.13</u> and accompanying discussion <u>in your textbook</u>.
16. Inhibition results when a substance other than the substrate is bound to the active site of the enzyme.  Its activity is thus quenched.
17. DIPF attaches to the active site of acetylcholinesterase thus inhibiting its ability to decompose toxic acetyl-choline.

## Section 23.2

1. Historically, they were thought to be hydrates of carbon.
2. Monosaccharides are simple sugars.
3. A hexose is a sugar with 6 carbon atoms.
4. Optical isomerism requires a chiral carbon.
5. The oxygen of the terminal OH group combines with the carbon of the ketone group.
6. A disaccharide is a combination of 2 simple sugars.
7. A C-O-C linkage between rings of glucose and fructose is a glycoside linkage.
8. The enzyme $\alpha$-amylase is found in saliva.  It catalyzes the decomposition of sucrose.
9. Starch is a polymer of $\alpha$-glucose.  (See <u>Figure 23.20 in your textbook</u>.)
10. There is less stress on the plant's internal structure.  (See the discussion on osmotic pressure in Chapter 11.)
11. We do not have the necessary enzymes, $\beta$-glycosidases, to break down cellulose.

## Section 23.3

1. DNA stores and transmits genetic information.
2. a.  A five-carbon sugar.          b.  A nitrogen-containing organic base.
   c.  A phosphoric acid molecule.
3. The double helix structure allows DNA to produce complementary strands.
4. Polar C=O and N-H bonds make these interactions possible.
5. See <u>Figures 23.24 - 23.26 in your textbook</u>.
6. A gene is a segment of DNA that contains the code for a specific protein.
7. A codon consists of 3 bases and codes for a specific amino acid.  An anticodon is a part of tRNA that decodes the "genetic message" form mRNA.
8. mRNA migrates from DNA to the cell cytoplasm where protein synthesis occurs.  tRNA decodes the genetic message from mRNA.

## Section 23.4

1. Lipids are not water soluble.
2. The four types are fats, phospholipids, waxes and steroids.
3. Polyhydroxy alcohol glycerol and long-chain carboxylic acids are combined to form fats.
4. Triglycerides are esters of glycerol.
5. Saponification is the decomposition of triglycerides by NaOH.
6. See Figures 23.28 and 23.29 in your textbook.
7. Micelles are aggregates of fatty acid anions that have nonpolar ends toward one another and polar ends toward the water solvent.
8. Detergents do not precipitate $Ca^{2+}$ and $Mg^{2+}$ from hard water. They typically contain a sulfonate group as their polar end.
9. See number 8.
10. See Figure 23.30 in your textbook.
11. Phospholipids form a significant portion of cell membranes.
12. The fluid mosaic model asserts that small uncharged particles (water, oxygen, etc.) can diffuse freely through the lipid bilayer while other substances interact with proteins in the membrane.
13. All three are esters.
14. Waxes are formed from monohydroxy alcohols rather than glycerol.
15. Waxes are low-melting solids.
16. Steroids have the fused carbon-ring structure shown in each molecule in Figure 23.33 in your textbook.